PHYSICS OF

NANOSTRUCTURES

PHYSICS OF NANOSTRUCTURES

Proceedings of the Thirty-Eighth Scottish Universities Summer School in Physics, St Andrews, July–August 1991.

A NATO Advanced Study Institute.

Edited by

J H Davies — Glasgow University
A R Long — Glasgow University

Series Editor

P Osborne — Edinburgh University

Copublished by
Scottish Universities Summer School in Physics
and
Institute of Physics Publishing, Bristol and Philadelphia

British Library Cataloguing-in-Publication Data. A catalogue record for this book is available from the British Library.

ISBN 0-7503-0170-8 (hbk)
0-7503-0169-4 (pbk)

Library of Congress Cataloging-in-Publication Data are available

Copublished by
SUSSP Publications
The Department of Physics, Edinburgh University
The King's Buildings, Mayfield Road, Edinburgh EH9 3JZ, Scotland
and
IOP Publishing Ltd, a company wholly owned by the
Institute of Physics, London.

IOP Publishing Ltd, Techno House, Redcliffe Way, Bristol BS1 6NX, UK
US Editorial Office: IOP Publishing Inc., The Public Ledger Building, Suite 1035, Independence Square, Philadelphia, PA 19106

Printed in Great Britain by Cromwell Press Ltd, Wiltshire

SUSSP Proceedings

1	1960	Dispersion Relations
2	1961	Fluctuation, Relaxation and Resonance in Magnetic Systems
3	1962	Polarons and Excitons
4	1963	Strong Interactions and High Energy Physics
5	1964	Nuclear Structure and Electromagnetic Interactions
6	1965	Phonons in Perfect and Imperfect Lattices
7	1966	Particle Interactions at High Energy
8	1967	Methods in Solid State and Superfluid Theory
9	1968	Physics of Hot Plasmas
10	1969	Quantum Optics
11	1970	Hadronic Interactions of Photons and Electrons
12	1971	Atoms and Molecules in Astrophysics
13	1972	Properties of Amorphous Semiconductors
14	1973	Phenomenonology of Particles at High Energy
15	1974	The Helium Liquids
16	1975	Non-linear Optics
17	1976	Fundamentals of Quark Models
18	1977	Nuclear Structure Physics
19	1978	Metal Non-metal Transitions in Disordered Solids
20	1979	Laser-Plasma Interactions: 1
21	1980	Gauge Theories and Experiments at High Energy
22	1981	Magnetism in Solids
23	1982	Laser-Plasma Interactions: 2
24	1982	Lasers: Physics, Systems and Techniques
25	1983	Quantitative Electon Microscopy
26	1983	Statistical and Particle Physics
27	1984	Fundamental Forces
28	1985	Superstrings and Supergravity
29	1985	Laser-Plasma Interactions: 3
30	1985	Synchrotron Radiation
31	1986	Localisation and Interaction
32	1987	Computational Physics
33	1987	Astrophysical Plasma Spectroscopy
34	1988	Optical Computing
35	1988	Laser-Plasma Interactions: 4
36	1989	Physics of the Early Universe
37	1990	Pattern Recognition and Image Processing
38	1991	Physics of Nanostructures
39	1991	High Temperature Superconductivity
40	1992	Quantitative Microbeam Analysis
41	1992	Spatial Complexity in Optical Systems
42	1992	High Energy Phenomenology

Key to photograph

Front row, from left to right:
John Beresford, David Finlayson, Alan Fowler, Frank Stern, Irene Robertson, Steve Beaumont, Andrew Long, Detlef Heitmann, Clivia Sotomayor Torres, Wolfgang Hansen, Michael Geller, Alexey Dmitriev, Youri Kumzerov, Sergey Romanov, Anders Gustafsson, Tony Herbert.

Second row, from left to right:
Xiaoguang Wu, Aiga Uhrig, Haiping Zhou, Eleftherios Skuras, Oguz Gulseren, Josep Pages Lozano, Claude Chapelier, Omar Abu-Zeid, Guido Goldoni, Oliver Kühn, Cecilia Garrido, Pei Dong Wang, Andreas Huber, Maria Gerling, Monica Pacheco, Michael Suhrke, Henrik Bruus.

Third row, from left to right:
Philip Calvert, Ruth Eyles, Martin Dawson, Steve Millidge, Richard Hornsey, David Cumming, Ian Osborne, Shing-Lin Wang, Richard Heemskirk, Maurizio Dabbicco.

Fourth row, from left to right:
Steven Richardson, Xiao Yuan Hou, Vincent Bayot, Wolfgang Demmerle, Sylvain Charbonneau, Thomas Swahn, Andrew Smart, Harold Grossner, Flavio Aristone, Gunther Berthold, Guiennadii Gusev, Franz Hirler, James Nicholls, Matthias Tewordt, Juan Jose Palacios.

Fifth row, from left to right:
Cris Eugster, Ron Marquardt, Mike Jackson, Ulrich Bockelmann, Mikhail Portnoi, Cristoph Lettau, Colombo Bolognese, Christian Peters, Michael Trott, Fintan Bolton, Marcel Schemmann, Alain Nogaret, Peter Selbmann, Andrew Peck, Alexei Orlov, Helmut Silberbauer, Konstantin Nikolic, Karl Brunner, Niaz Haider, Tomasz Skrabka, Marleen Van Hove, Ramon Cusco-Cornet, Luis Martin- Moreno, Doug Collins, Ferney Rodriguez.

Executive Committee

Dr A R Long	Glasgow University	Co-Director and Co-Editor
Prof. S P Beaumont	Glasgow University	Co-Director
Dr D M Finlayson	St Andrews University	Steward and Local Organiser
Dr C M Sotomayor Torres	Glasgow University	Treasurer
Mrs I Robertson	Glasgow University	Secretary
Dr J M R Weaver	Glasgow University	Social Secretary
Dr J H Davies	Glasgow University	Co-Editor

International Organising Committee

Dr A R Long	Glasgow University
Prof. S P Beaumont	Glasgow University
Dr G Bastard	Ecole Normale Supérieure
Prof. L Eaves	Nottingham University
Prof. J Kotthaus	Universität München
Dr F Stern	IBM T J Watson Research Center

Lecturers

Dr Gerald Bastard	Ecole Normale Supérieure
Prof. Harold G Craighead	Cornell University
Prof. Laurence Eaves	University of Nottingham
Dr Alan Fowler	IBM T J Watson Research Center
Dr Bart Geerligs	University of California at Berkeley
Dr Wolfgang Hansen	Universität München
Prof. Detlef Heitmann	Max-Planck Institut für Festkörperphysik
Dr Peter Main	University of Nottingham
Prof. Hiroyuki Sakaki	University of Tokyo
Dr Clivia Sotomayor Torres	University of Glasgow
Dr Frank Stern	IBM T J Watson Research Center
Prof. A Douglas Stone	Yale University
Dr Gregory Timp	AT&T Bell Laboratories

Preface

One of the most significant movements in semiconductor physics in the last twenty years or so has been the progressive miniaturisation of devices. On the one hand this has led to the explosive growth in computer technology; on the other, the opportunity to fabricate devices with characteristic dimensions significantly below 1 μm has led to the investigation of a whole range of novel physical phenomena. In turn, these new physical phenomena promise to give rise to new generations of devices working on completely different physical principles to their predecessors. It is the purpose of this volume, as it was of the 38th Scottish Universities Summer School in Physics on which it is based, to describe and analyse the fabrication and the electron transport and optical properties of such Nanostructures and to point the way towards future developments in the field.

The 38th SUSSP was held between 29th July and 9th August 1991, at the Department of Physics and Astronomy and John Burnet Hall in the University of St Andrews. The 78 participants from 18 countries were addressed by 13 lecturers, all internationally known experts in their particular fields. The lecturers then condensed their material into written form, and the result is these Proceedings.

As the study of nanostructures is led by technology, the early chapters by Sakaki and Craighead concentrate on the fabrication of samples: the preparation of atomically smooth layers of single crystals of nanometre thickness by molecular beam epitaxy and their subsequent patterning by high-resolution lithography. The electronic structure of nanostructures is covered from a number of different standpoints by Stern and Bastard, and Stone discusses theoretically the consequences for electron transport of making samples smaller than one or more of the characteristic scattering lengths in the nanostructure. Complementary experiments are described by Timp, together with a new method of maskless lithography using light pressure. Further experimental aspects of electron transport are covered by Main and Eaves including ballistic electron transport, motion in periodic potentials, non-local effects and new aspects of resonant electron tunnelling, among other topics. From ballistic electrons we move to highly localised electrons which conduct by hopping between states, and whose behaviour in small structures is reviewed by Fowler. The highly significant effects associated with the blockade induced by the electrostatic energy of a single electron in a nanostructure are described by Geerligs; these single-electron effects are prime candidates for exploitation in future devices. Finally Sotomayor Torres, Heitmann and Hansen cover a range of optical properties of arrays of nanostructures, including extensive discussion of the very powerful far infra-red techniques. The low dimensionality of many nanostructures, in which electrons are free to move in only two, one or even zero dimensions, has a profound effect on their electronic and optical properties. For example, the clear distinction between single-particle and collective effects, so valuable in three-dimensional systems, is lost and this is revealed clearly by optical experiments. The practical application of these effects is only just beginning, in such devices as lasers based on quantum wires or quantum dots.

As well as the lectures summarised in this volume, other important elements went to make a successful School. A number of discussion sessions on future developments were held, led by the lecturers, and by other experts invited to contribute to this aspect of the programme, notably Dr Haroun Ahmed of Cambridge University. The participants themselves contributed greatly to the programme by submitting some sixty short papers of which a third were given orally and the remainder in poster form. So high was the quality of these papers, and so important were the contribution they made to the success of the School, that the Organising Committee selected three representative contributions and these are published at the end of the volume, together with a list of the remaining submitted papers. The relationships formed at the School were cemented by an excellent Social programme, held in the benign weather conditions typical of St Andrews in August. Participants were introduced to those significant Scottish pastimes of hill-walking, dancing and golf, and to the delights of malt whisky.

Without the help of numerous organisations and individuals, there would be no possibility of holding a successful School. Foremost among these of course are our sponsers, NATO, the ESPRIT programme of the EEC, the Science and Engineering Research Council of the U.K. and SUSSP. Our International Committee offered wise advice when our programme was in course of preparation, and the programme was implemented by a local Committee who worked together very successfully. In particular it is appropriate to mention the never-failing helpfulness of Irene Robertson, Andrew Wines and Sandy Reeks in manning the Conference Office. Thanks are due to Professor Wilson Sibbett and the staff of the Department of Physics and Astronomy at St Andrews for placing their facilities at our disposal.

The Editors would like to thank the lecturers for their hard work in putting together an excellent set of contributions to this volume, in many cases under significant pressure of time, and particularly Peggy Owens who turned the majority of the text into LaTeX format for publication. Without her valiant efforts, the production schedule would have been much delayed.

John Davies
Andrew Long
Glasgow, January 1992

Contents

Selected contributed papers

Molecular Beam Epitaxy for the Formation of Nanostructures

H Sakaki

University of Tokyo
Tokyo, Japan

1 Introduction

Molecular beam epitaxy (MBE) is a method by which GaAs and other semiconductors are grown in an ultra-high vacuum environment by supplying molecular and/or atomic beams of Ga, As, and other constituents onto a cleaned crystalline substrate (Cho and Arthur 1975, Ueda 1982, Ploog and Graf 1984, Arthur 1985, Parker 1985, Foxon and Harris 1987, Marik 1989, Shiraki and Sakaki 1989, Tu and Harris 1991). Because one can prepare semiconductor films of high quality by MBE with a precision in thickness of one atomic layer or better, it is now widely used to form quantum wells (QWs), tunnelling barriers, superlattices and other quantum microstructures. In this chapter, we describe first the abruptness and flatness of interfaces in heterostructures grown by MBE and discuss those microscopic processes which determine the structure of the interface. We then examine the purity of heterostructures grown by MBE and discuss the mobility of electrons at low temperature in selectively doped n-$Al_xGa_{1-x}As$-GaAs heterostructures. Finally, we review some findings on the growth by MBE on patterned substrates and discuss attemps to achieve lateral definition of quantum structures.

2 Abruptness and flatness of interfaces

MBE is one of the best methods for the preparation of a very flat surface of one semiconductor on which another semiconductor of an arbitrary thickness can be deposited. This feature allows one to control the composition profile and thereby to implement an artificial potential $V(z)$ as a function of position z along the direction of stacking. This capability has been most generally used to form quantum wells with thickness L_w, where the kinetic energy E_z of electrons normal to the layer is quantised;

$E_z^{(1)} = (\hbar^2/2m)(\pi/L_w)^2$ for the lowest state in an infinitely deep well of width L_w.

A number of studies have shown, however, that interfaces of GaAs-AlAs (or AlGaAs) grown by MBE are generally abrupt but have a roughness Δ of one monolayer (ML), $a = 2.83$ Å. Since such an interface roughness of 1 ML leads to a variation in the thickness of the QW, a spatial fluctuation of $E_z^{(1)}$ or equivalently a random potential $V(x,y)$ in the x-y plane is inevitably produced and may affect various electronic processes in QW structures. Hence it is very important to clarify the origin and nature of such roughness, and to establish a method for controlling the atomic structure of interfaces. In addition, one should note that the atomic structure of an interface is usually a frozen image of the freshly grown surface, and therefore provides valuable information on the microscopic process of epitaxial growth. In the following we discuss major findings, particularly those obtained from reflection high-energy electron diffraction (RHEED), transmission electron microscopy (TEM), photoluminescence (PL) and mobility.

2.1 RHEED study of surfaces grown by MBE

In the early phase of research on MBE, it was assumed that the epitaxial growth of GaAs and related compounds proceeds in a layer-by-layer manner. In this most idealised model, a perfectly smooth surface is thought to be realised every time the total amount of deposited Ga (or Al) atoms is equal to an integral multiple of monolayer coverage. Indeed, the intensity of the specular beam in reflection high-energy electron diffraction (RHEED) is found to oscillate during growth as shown in Figure 1a (Harris *et al.* 1981, Wood 1981, Neave *et al.* 1983, 1985, Sakamoto *et al.* 1984, Sakaki *et al.* 1985, Madhukar *et al.* 1985, Tanaka *et al.* 1986, Tanaka and Sakaki 1987, 1988); this oscillation indicates that the surface morphology changes almost periodically between the high-intensity state (probably smooth) and the low-intensity state (probably rough) (Harris *et al.* 1981, Wood 1981, Neave *et al.* 1983, 1985, Sakamoto *et al.* 1984). It is apparent, however, that the measured RHEED intensity is not exactly periodic but shows a damped oscillation which tends asymptotically to a low-intensity level; this suggests that the surface morphology during the growth reaches a kind of steady-state condition (dynamic equilibrium) where the roughness of the surface or the average interval of steps is independent of time (Neave *et al.* 1983, 1985, Sakamoto *et al.* 1984, Sakaki *et al.* 1985, Tanaka and Sakaki 1988).

When the deposition of Ga is interrupted at a substrate temperature of typically 600°C, the RHEED intensity usually regains its strength with a typical time constant of tens of seconds and reaches another steady state condition (static equilibrium) as shown in Figures 1a and 1b. This suggests that a rearrangement (or relaxation) of atoms takes place in such a way that the surface morphology changes during the growth interruption, and the surface is likely to become smoother than that during growth.

Though the measurement of RHEED intensity provides a convenient way to monitor the surface morphology, it is still difficult to characterise it quantitatively and determine, for example, the height Δ and the lateral size L_i (or correlation length Λ) of island structures. This difficulty comes mainly from the intensity of the specular RHEED beam being dependent not only on the roughness but also on the angle of incidence, surface reconstruction, and other parameters (Neave *et al.* 1983).

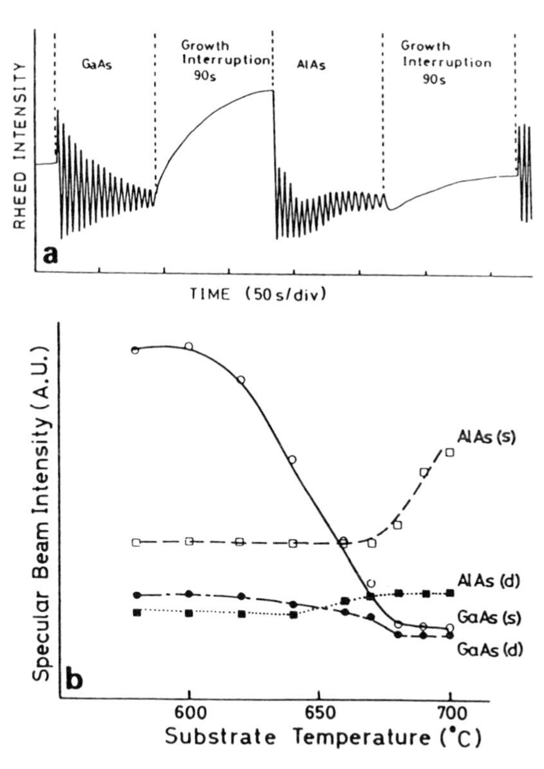

Figure 1: *(a) Typical oscillation of* RHEED *intensity during the growth of* GaAs *and* AlAs*, and its recovery during a growth interruption at* 600°C *with the ratio of fluxes* $J(As_4)/J(Ga) = 3.5$ *(Tanaka et al. 1986, Tanaka and Sakaki 1987). (b) Variation of* RHEED *intensity with temperature from dynamic GaAs surfaces during growth (solid circles and squares) and from static GaAs surfaces after a long growth interruption (open circles and squares) (Tanaka and Sakaki 1988).*

To overcome this problem, van Hove and Cohen (1987) have proposed a method to estimate the lateral size L_i of islands by measuring the widths of RHEED diffraction spots. From the Fourier transforms of measured widths of spots, they have found that L_i on GaAs surfaces is typically a few hundred Å under typical growth conditions (substrate temperature $T_{\text{sub}} \approx 600°\text{C}$, rate of growth approximately $0.5\,\mu\text{m hr}^{-1}$), and tends to increase when the growth rate and/or the incident As pressure are reduced. Similarly, Neave *et al.* (1985) have shown that the lateral size of islands, L_i, can be estimated by studying the RHEED oscillation on a vicinal surface (typically tilted 2° away from the (100) surface) where a quasi-periodic step structure is expected to be formed. They found that the RHEED oscillation disappears when the substrate temperature is raised above a certain value since there should be no change of surface roughness once the lateral size L_i of the island gets comparable with or greater than the average interval of quasi-periodic steps. By using this method, they estimated L_i to be around 100–200 Å under a typical growth condition, which is in accord with the results of van Hove and Cohen.

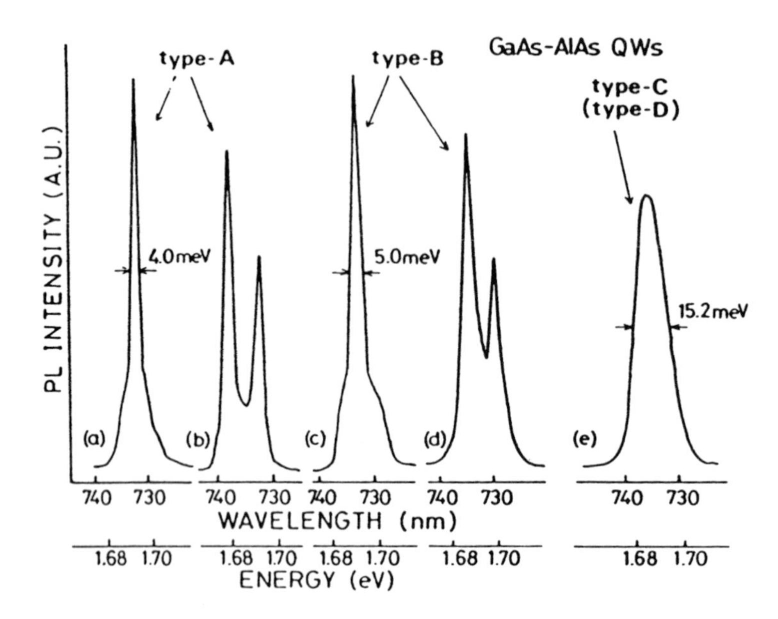

Figure 2: *Photoluminescence spectra taken at* 77 K *from* 48 Å GaAs – 51 Å AlAs *quantum wells prepared by four different techniques at* 600°C. *Growth interruption is done at both interfaces for type A* QWs, *only at the top (*AlAs-on-GaAs*) interface for type B* QWs, *only at the bottom (*GaAs-on-AlAs*) interface for type C* QWs, *and at neither interfaces for type D* QWs *(Tanaka et al. 1986, Tanaka and Sakaki 1987).*

2.2 Study of QW interfaces by photoluminescence

The presence of interface roughness in QW structures can be detected most conveniently by the study of photoluminescence (PL) (Sakaki *et al.* 1985, Madhukar *et al.* 1985, Tanaka *et al.* 1986, Tanaki and Sakaki 1987, 1988, Fukunaga *et al.* 1985, Hayakawa *et al.* 1985). Spatial fluctuations of the quantised energy levels leads to broadening of the linewidth $\Delta(h\nu)$ of luminescence spectra at low temperature, particularly when the width of the well L_w is reduced. See, for example, the spectra of Figure 2 and the solid circles of Figure 3, which are the PL linewidths measured on GaAs-AlAs QWs. Since the PL reflects the quantised energy probed by the excitons whose Bohr diameter D_{ex} is of the order of 100–200 Å, the broadening of PL is caused mainly by roughness whose lateral scale is comparable with D_{ex}. Hence the broadening of PL indicates that at least one of the two interfaces has the roughness whose lateral scale $L_i \approx 100$–200 Å. In the following, we denote the interface to be 'smooth' when the lateral scale L_i of roughness is far greater than D_{ex}, whereas we call it 'pseudo-smooth' when L_i is far smaller than D_{ex} (Figure 4).

It has been found that a growth interruption (GI) of tens of seconds before the formation of the interface of QW structures leads to a modification of the structure of the interface, resulting in a drastic reduction of the PL linewidth (Sakaki *et al.* 1985). As shown in Figures 2a and 2b, very sharp PL spectra are seen in type A QWs where GIs of 90 s were done prior to the formation of both top (AlAs-on-GaAs) and bottom (GaAs-

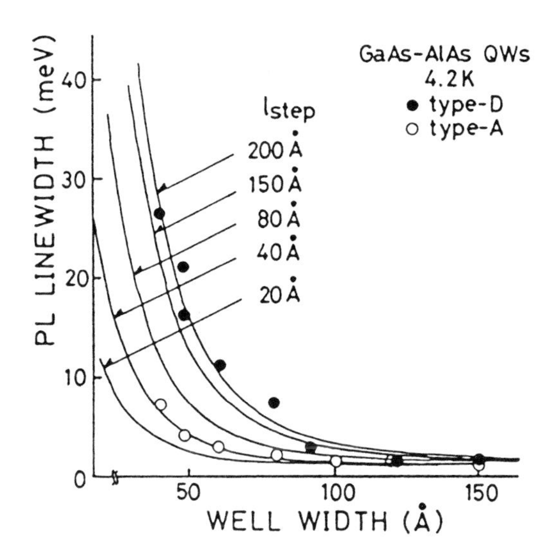

Figure 3: *Predicted photoluminescence (PL) linewidths of* GaAs-AlAs *quantum wells (solid lines) as a function of well width, calculated according to the model of Singh et al. (1984) for a fluctuation of the interface with amplitude* $\Delta L_w \approx 2.8$ Å *and various lateral sizes,* l_{step}. *Also shown are the measured* PL *linewidths at* 4.2 K *from quantum wells prepared by conventional* MBE *without growth interruption (solid circles) and from those prepared by refined* MBE *growth with interruption (open circles) (Tanaka et al. 1986, Tanaka and Sakaki 1987).*

on-AlAs) interfaces, whereas PL spectra of type D QWs prepared with no GI is broad as shown in Figure 2e. These GaAs-AlAs QWs have a barrier width of $L_b = 50.9$ Å and a well width of $L_w = 48.1$ Å, and were grown at 600°C with a growth rate of $0.3\,\mu\mathrm{m\,h^{-1}}$ for GaAs and AlAs with a V/III flux ratio of 2.5–3.5. The measured PL linewidths (open circles in Figure 3) show that the effective interface roughness L_w^* of QWs [defined by $\Delta h\nu/(\partial E_z^{(1)}/\partial L_w)$] is reduced to as little as 0.2–0.25 ML by a modified growth method — Interruption of DEposition for Atomic Layer Smoothing (IDEALS) (Sakaki *et al.* 1985, Madhukar *et al.* 1985, Tanaka *et al.* 1986, Tanaka and Sakaki 1987, 1988, Fukunaga *et al.* 1985, Hayakawa *et al.* 1985).

From the following study (Sakaki *et al.* 1985, Tanaka *et al.* 1986, Tanaka and Sakaki 1987), it has been established that the reduction of PL linewidth is mainly caused by the smoothing of the top interface during GI, whereas the bottom interface remains pseudo-smooth irrespective of GI, as will be explained below. To establish this, we have studied two other types of QW: type B QWs with GI only at the top interfaces and type C QWs prepared with the GI only at the bottom interfaces. As shown in Figures 2c and 2d, the PL spectra of type B QWs are quite similar to those of type A, while the PL of type C QWs are identical with those of type D as shown in Figure 2e. Since the PL linewidth reduction results only from the GI at the top interfaces, it is clear that the GI enhances the rearrangement of Ga atoms on the GaAs surface in such a way that the lateral interval L_i of the atomic steps gets much larger than the exciton diameter D_{ex}.

In contrast, the fact that the bottom interface does not broaden the PL linewidths

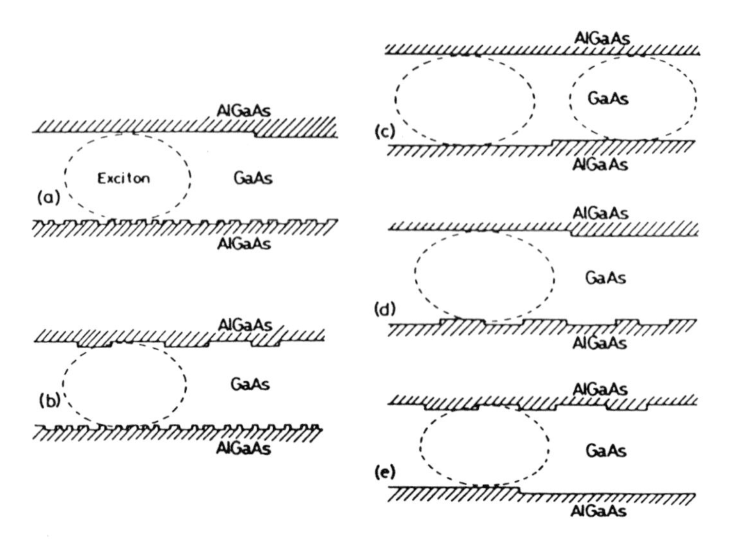

Figure 4: *Atomic models of interfaces of various quantum well structures. When the lateral size* Λ *of roughness is close to the exciton diameter* D_{ex}*, the luminescence broadens most severely. Interfaes with* $\Lambda \gg D_{ex}$ *are referred to as truly smooth, whereas those with* $\Lambda \ll D_{ex}$ *are called pseudo-smooth interfaces; the* PL *spectra become narrow in either case (Tanaka et al. 1986, Tanaka and Sakaki 1987).*

indicates that the AlAs surfaces or the bottom interfaces are either smooth or pseudo-smooth irrespective of the growth interruption. The pseudo-smooth structure is far more likely, since Al atoms are known to be more adhesive to the surface, as suggested in the damped RHEED oscillation data of Figure 1 as well as its higher melting temperature. Hence, l_{step} at the bottom interface is concluded to be far smaller than D_{ex}, and the bottom interfaces behave as pseudo-smooth in the sense that excitons feel little statistical fluctuation. Models of the GaAs-AlAs interface structures are illustrated in Figures 4a and 4b. By comparing the measured PL linewidth with theoretical calculations, l_{step} at the bottom interface is estimated to be 40 Å and l_{step} at the top interface to be 100–200 Å (when continuously grown), as shown in Figure 3.

It has been pointed out independently by Briones *et al.* (1987) and Horikoshi *et al.* (1986) that a periodic shut-off of As_4 beams for one second or so in synchronisation with the mono-layer deposition cycle of GaAs is effective in prolonging the oscillatory behavior of RHEED and also in improving the quality of the crystal when it is prepared at low substrate temperature. This modified growth procedure [often referred to as migration enhanced epitaxy when the Ga beam is alternately modulated] is found to be effective in reducing the PL linewidths of QWs to a level far narrower than those prepared with continuous growth, but considerably wider than those prepared with interruption. This suggests that the smoothing can indeed be achieved by creating periodically a deficit of As on the surface.

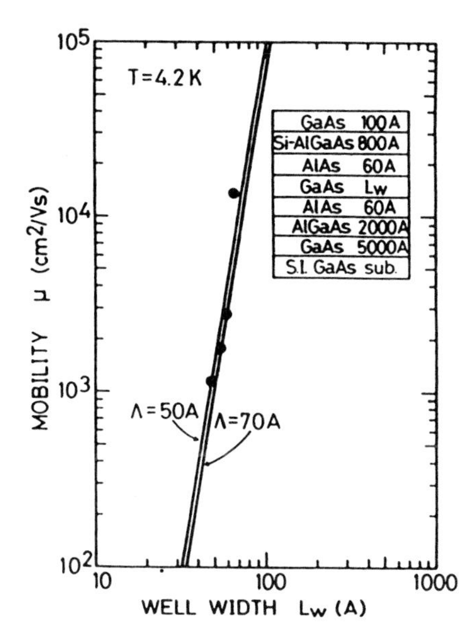

Figure 5: *Mobility of electrons μ in selectively doped n-AlGaAs-AlAs-GaAs-AlAs quantum wells at low temperatures as a function of the width L_w of the well. Solid lines show the calculated mobility, assuming that it is dominated by interface roughness with a height $\Delta = 2.8$ Å and lateral correlation length $\Lambda = 50$–70 Å. The inset shows the composition of the layers (Sakaki et al. 1987).*

2.3 Study of QW interfaces by mobility

The amplitude and correlation length of the interface roughness can also be characterised also by studying the mobility of electrons in very thin QW structures, since the interface roughness causes a large fluctuation of the quantised energy, leading to a strong momentum scattering (Sakaki *et al.* 1987, Noda *et al.* 1990). This is quite different from selectively doped n-$Al_xGa_{1-x}As$-GaAs single heterojunctions, where the interface roughness contributes very little to the mobility because of the looser confinement of electrons (Hirakawa and Sakaki 1986).

To investigate the roughness, we prepared a series of selectively doped QW samples whose structure is shown in the inset to Figure 5. The thickness L_w of the well is varied from 40 to 70 Å. The MBE growth was done at 590°C with an interruption of 60 s at the top interface, while the bottom interface was continuously grown. Hence the top interface is smoothed so that the lateral size of the atomic steps is far longer than 200 Å. In contrast, the bottom interface has some roughness as shown in Figure 4a and should contribute to the scattering. In Figure 5 mobilities μ measured at 4.2 K are plotted as a function of L_w. Note that μ falls very rapidly as L_w decreases, suggesting that the roughness dominates the mobility.

Interface roughness scattering can easily be evaluated theoretically, when the roughness $\Delta(\mathbf{r})$ is characterised by a height Δ and a lateral scale or correlation length Λ with Gaussian fluctuations (Sakaki *et al.* 1987). When the barriers are sufficiently high, the

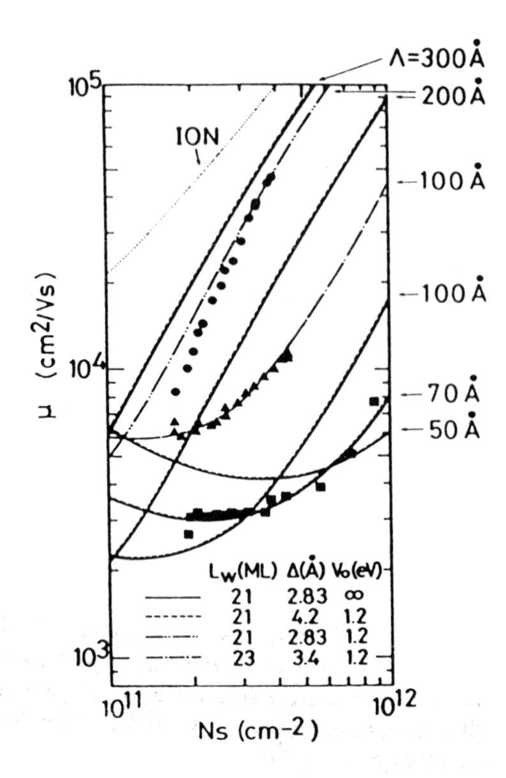

Figure 6: *Dependence of mobility on (areal) concentration N_s of electrons. Measured data are plotted as filled circles for sample (a) with L_w = 21 ML, filled squares for sample (b) with L_w = 21 ML, and filled triangles for sample (c) with L_w = 23 ML. Samples (b) and (c) were prepared with usual MBE, whereas sample (a) was prepared with modulated beam (migration enhanced) MBE. Solid, broken, and chained lines are the mobilities calculated at zero temperature for different correlation lengths Λ, assuming that scattering from interface roughness dominates. The dotted line is the mobility calculated at zero temperature for a QW with 21 ML assuming that impurity scattering dominates (Sakaki et al. 1987).*

potential fluctuation ΔV caused by the roughness is given by $\Delta V = (\partial E_z^{(1)}/\partial L_w)\Delta$ which is proportional to Δ/L_w^3. Since the probability of scattering is proportional to $(\Delta V)^2$, the mobility $\mu_{\mathrm{IR}} \propto L_w^6/\Delta^2$ when it is dominated by interface roughness scattering. This prediction is in good agreement with experiment. If one assumes Δ to be one monolayer (3 Å), one can determine the correlation length from the measured data and find Λ of the bottom interface to be around 50–100 Å. In fact one can determine Λ without making any assumptions, by measuring and analysing the mobility as a function of electron concentration N_s, since the dependence of mobility on N_s reflects the dependence of the scattering rate on wave-number (Noda *et al.* 1990). Figure 6 shows an example of such measurements and, by comparing them with theoretical lines, one finds Λ to be 50–70 Å at the bottom interface, which agrees fairly well with the result of our PL study.

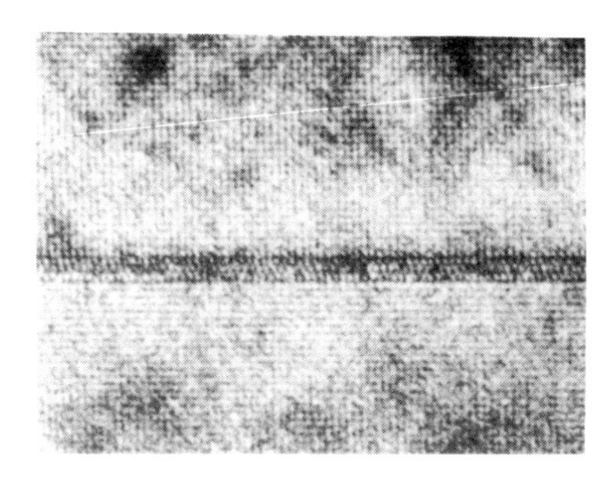

Figure 7: *Lattice image of a* AlAs *film 3 monolayers thick (the stripe in the lower half) embedded in* GaAs, *constructed from* (200) *diffraction beams. The image is seen from (or with the incident electron beam along) the* [100] *direction (Ichinose et al. 1987).*

2.4 Study of QW interfaces by electron microscopy

From the studies of RHEED, PL, and mobility described up to now, it has been shown that the structure of top (AlAs-on-GaAs) interfaces (or GaAs surfaces) is characterised by a roughness of about one monolayer in height Δ; its lateral scale when continuously grown is typically 100–300 Å and can be extended far beyond the diameter of the exciton when grown with interruption (much greater than 200 Å). In contrast, the structure of bottom (GaAs-on-AlAs) interfaces (or AlAs surfaces) is pseudo-smooth irrespective of interruption with the lateral scale of roughness being 40–100 Å. When a substrate temperature far higher than 580°C is used for the growth, the size of the islands of AlAs tends to increase, particularly when prepared with interruption as suggested in Figure 1b. However the GaAs surface starts to roughen, possibly by the frequent detachment of atoms from the islands.

A number of attempts have been made to characterise the structure of the interface directly by high-resolution transmission electron microscopy (TEM). This effort, however, has achieved only a limited success, primarily for two reasons. Firstly, TEM provides only a cross-sectional image of a sliced sample, which is averaged over a few hundred Å along the direction of propagation of the electron beam, and therefore cannot represent faithfully structures with lateral size smaller than a few hundred Å. Secondly, it is not easy to achieve both high resolution and high material contrast (between GaAs and AlAs) simultaneously unless the thickness and orientation of TEM specimen, the defocusing condition and other factors are all optimised (Ichinose *et al.* 1987, Ourmazd *et al.* 1989, Ikarashi *et al.* 1990). Figure 7 shows a TEM image of a heterostructure, where three monolayers of AlAs are embedded inside a GaAs crystal which was prepared with growth interruption (Ichinose *et al.* 1987). Note that the AlAs-on-GaAs interface is quite smooth, in agreement with our model. In contrast, the GaAs-on-AlAs interface is somewhat fuzzy, indicative of the pseudo-smooth nature of the interface. Some workers, however, claim that interfaces prepared with growth interruption contain very small island structures when studied using a chemical imaging technique (Ourmazd *et al.* 1989).

In summary, our understanding of atomic structures of interfaces has deepened considerably as a consequence of data accumulated from by a variety of methods, including RHEED, photoluminescence, mobility and TEM. Further work, including scanning tunnelling microscopy, micro-RHEED and other methods, is likely to be necessary to establish a full understanding of the interface.

3 Impurities and mobility in doped heterostructures

The high mobility of electrons confined in two-dimensional (2D) states in selectively doped (SD) n-$Al_xGa_{1-x}As$-GaAs heterojunctions and quantum wells has been widely exploited. Applications include both practical devices, such as high-speed field-effect transistors, and basic transport studies, two examples being the quantum Hall effect and the quantised conductance of ballistic point contacts. In this section, we describe first the effect of residual impurities (mostly carbon) on the limiting mobility at low temperature of electrons in SD heterojunctions. We then examine the effect of segregation of intentionally introduced impurities (mostly silicon) and its influence on the mobility of 2D electrons in SD quantum wells and in inverted heterojunctions, where the dopants are incorporated below the 2D channel.

3.1 Selectively doped heterojunctions and residual impurities

To study systematically the mobility of 2D electrons in SD n-AlGaAs-undoped AlGaAs-undoped GaAs heterojunctions, we prepared five different samples by MBE at a substrate temperature of 580°C (Hirakawa and Sakaki 1986). The thickness W_{sp} of the undoped AlGaAs spacer layer between n-AlGaAs (doped with Si to $3.4\times10^{17}\,cm^{-3}$) and undoped GaAs is varied as a parameter from 0 to 300 Å. Figure 8 shows the dependence of the measured mobility on temperature. At high temperatures, $T > 100\,K$, the dependence of the mobility on temperature is quite similar to that of high purity n-GaAs, which indicates that the mobility is dominated mainly by scattering by polar optical phonons. At intermediate temperatures, $50\,K < T < 100\,K$, the mobility is influenced by two more scattering mechanisms: ionised impurity scattering and deformation potential scattering. In the low temperature region, $T < 10\,K$, mobility is found to increase monotonically with W_{sp}, ranging from $6\times10^4\,cm^2\,V^{-1}\,s^{-1}$ to $1.5\times10^6\,cm^2\,V^{-1}\,s^{-1}$ for W_{sp} between 0 and 300 Å. This indicates that the scattering by remote Si donors in AlGaAs plays the decisive role and that the modulation (or selective) doping scheme is effective in reducing this scattering.

To evaluate the influence of both intentionally introduced Si donors and residual impurities in SD heterojunctions, we have calculated the mobility at $T = 0$ when it is limited by ionised impurity scattering (see also the chapter by Stern). The results are shown by solid lines in Figure 9 as a function of W_{sp} with the residual (acceptor) impurity concentration N_{res} as a parameter. Although the details of the results may depend on various parameters such as the concentration of Si donors in n-AlGaAs, which is set to be constant ($3.4\times10^{17}\,cm^{-3}$), the following results are mostly valid semi-quantitatively. When $N_{res} = 0$, the mobility μ is dominated by the remote donors in n-AlGaAs, and shows a monotonic increase with W_{sp}, reaching $10^7\,cm^2\,V^{-1}\,s^{-1}$ for

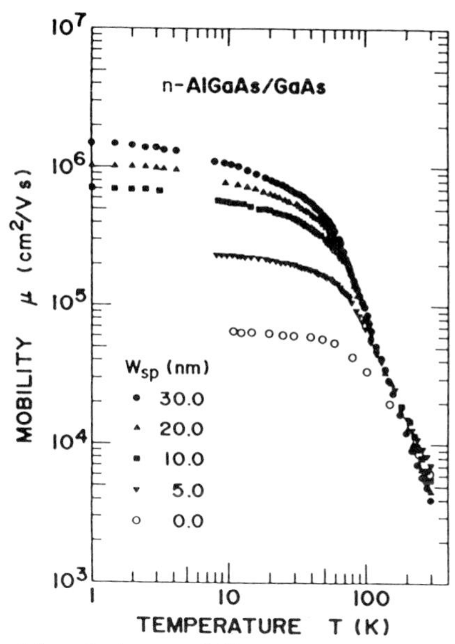

Figure 8: *Dependence of the electron mobility on temperature in selectively doped single heterojunctions for various thicknesses W_{sp} of the spacer layer (Hirakawa and Sakaki 1986).*

$W_{sp} > 800$ Å. The presence of residual acceptors lowers the mobility, keeping it below $5 \times 10^5\,cm^2\,V^{-1}\,s^{-1}$ when $N_{res} = 1 \times 10^{15}\,cm^{-3}$. Our data points, shown by four circles in Figure 9 , are well explained by assuming $N_{res} = 1.5 \times 10^{14}\,cm^{-3}$.

Several recent works have reported a higher maximum mobility; one example is $5 \times 10^6\,cm^2\,V^{-1}\,s^{-1}$ at 1.4 K in samples with W_{sp} = 700–800 Å (English *et al.* 1987, Pfeiffer *et al.* 1989). These data, when compared with the theoretical curve, suggest that the concentration of residual impurities is about $5 \times 10^{13}\,cm^{-3}$ or less for the former sample. Figure 9 shows also that an increase of N_{res} leads to a dramatic decrease of μ, particularly in samples with thick spacer layers. Hence one can evaluate the residual impurity concentration of GaAs grown by MBE by growing a series of samples of heterojunctions with different spacer layers and comparing their mobilities with theoretical curves. For example, the early work reporting a maximum mobility around $2 \times 10^4\,cm^2\,V^{-1}\,s^{-1}$ suggests that the residual impurity concentration was $10^{16}\,cm^{-3}$ or higher at that time (Dingle *et al.* 1978).

When interpreting the mobility, careful attention must be paid to the (areal) concentration N_s (in cm^{-2}) of 2D electrons in the channel of specific samples, since the mobility depends sensitively on N_s at low temperature. An increase of N_s at low temperatures usually results in an increase of the Fermi energy E_F of the electrons, which leads to a reduction of the impurity scattering rate. Hence, if N_s is increased by applying a positive bias to a gate electrode without changing the impurity concentration, this results in the increase of μ. The broken lines in Figure 10 show the impurity-dominated mobility calculated for such a case as a function of N_s for various values of W_{sp} and $N_{res} = 0$. For comparison, experimental data are shown in Figure 10 which are obtained by changing N_s with gate voltage. The measured data are in fairly good

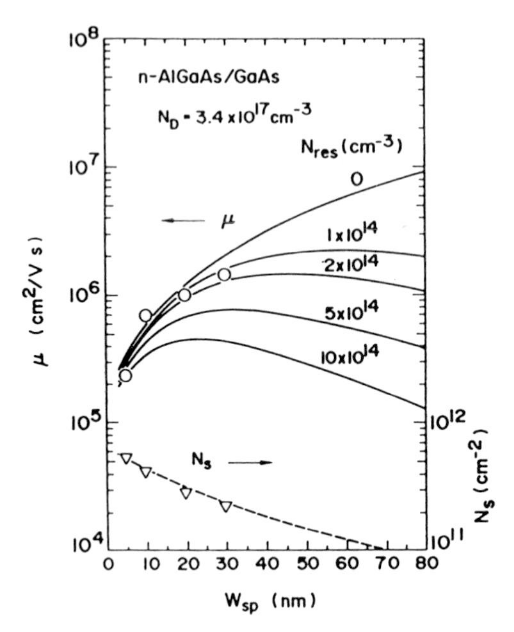

Figure 9: *Mobility of electrons in selectively doped heterojunctions limited by ionised impurity scattering, as a function of the thickness of the spacer layer. Circles represent experimental data and the solid lines are theoretical curves for different concentrations N_{res} of residual impurities and a fixed concentration of donors (3.4×10^{17} cm^{-3}) in the n-AlGaAs. The concentration N_s of electrons is also shown. (Hirakawa and Sakaki 1986).*

agreement with theory. The increase of μ that is often observed under (or after) the illumination of heterojunctions at low temperatures is primarily caused by the increase of N_s by persistent photoconductivity, which leads to a reduction of the scattering rate.

Note that in Figure 9 that the maximum electron concentration N_s in the channel (without the gate) decreases when W_{sp} is increased. This decrease of N_s results in a decrease of E_F, and leads to an increase in the scattering rate by impurities. This is particularly true when the mobility is limited by residual impurities in GaAs, and explains why the mobility of impure samples drops as larger spacer layers are used as shown in Figure 9. In samples of high purity, however, where μ is dominated by the remote impurities in AlGaAs, an increase of W_{sp} reduces the scattering potential more effectively than it enhances the scattering rate and therefore results in a higher mobility.

In some of the early literature, it was pointed out that a high electron mobility in excess of 10^6 cm^2 V^{-1} s^{-1} was achieved only when a thick (more than 3 μm) GaAs buffer layer was grown, or a superlattice buffer layer was inserted before forming the heterojunction. It is found that neither is necessary provided that the concentration of residual impurities is reduced to a low level, below 4×10^{14} cm^{-3}. Indeed, the data shown in Figures 8–10 were all obtained in heterojunction samples with 1 μm thick GaAs grown without resorting to a superlattice buffer layer.

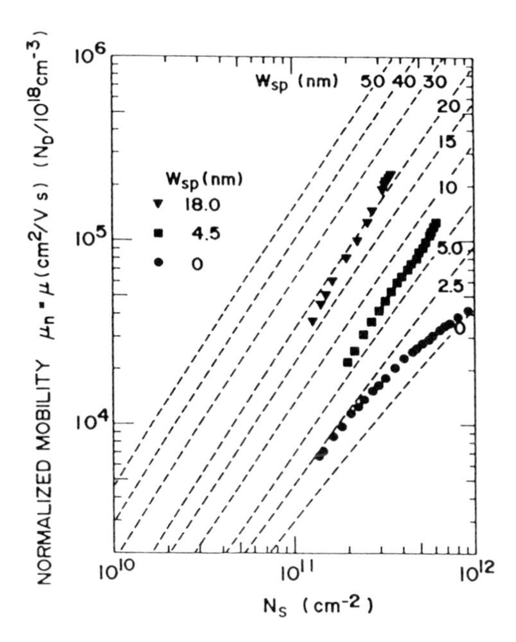

Figure 10: *Mobility at low temperature of two-dimensional electrons in an n-*AlGaAs-GaAs *selectively doped heterojunction plotted as functions of electron concentration N_s for various thicknesses W_{sp} of the spacer layer. Dashed lines are calculated for the case where μ is dominated by intentional donors in the n-*AlGaAs *(Tanaka et al. 1986, Tanaka and Sakaki 1987).*

3.2 Inverted heterojunctions and impurity segregation

The effectiveness of modulation doping for the reduction of impurity scattering was first demonstrated in multi-quantum well (QW) structures which consisted of 200 Å GaAs and 200 Å n-AlGaAs (Pfeiffer *et al.* 1989). The highest mobility achieved in the initial experiment was rather low, around 20 000 cm^2 V^{-1} s^{-1}. A much higher mobility was later obtained in selectively doped (normal) heterojunctions (SD-HJ), consisting of an n-AlGaAs layer grown on top of undoped GaAs (Hirakawa and Sakaki 1986, English *et al.* 1987, Pfeiffer *et al.* 1989). The inferiority of MD-QW to SD-HJ structures was ascribed to the poor quality of the GaAs-on-n-AlGaAs (inverted) heterojunction. Indeed, it was demonstrated that the mobility of electrons measured in an inverted HJ was found to be extremely low, a few to several times 10^3 cm^2 V^{-1} s^{-1}, if the standard growth conditions were adopted.

Possible causes of the problems with inverted HJs have been examined. Some workers have speculated that the extra scattering is due to the intrinsic roughness of the interface at the inverted HJ, which results from the roughening of the growth front of AlGaAs. Others have attributed it to ionised impurity scattering at the heterojunction, which is caused either by the segregation of intentional donor impurities from n-AlGaAs to undoped GaAs or by the enhanced accumulation of residual impurities at the heterojunction. A hybrid of these two or the additional scattering by impurity-induced

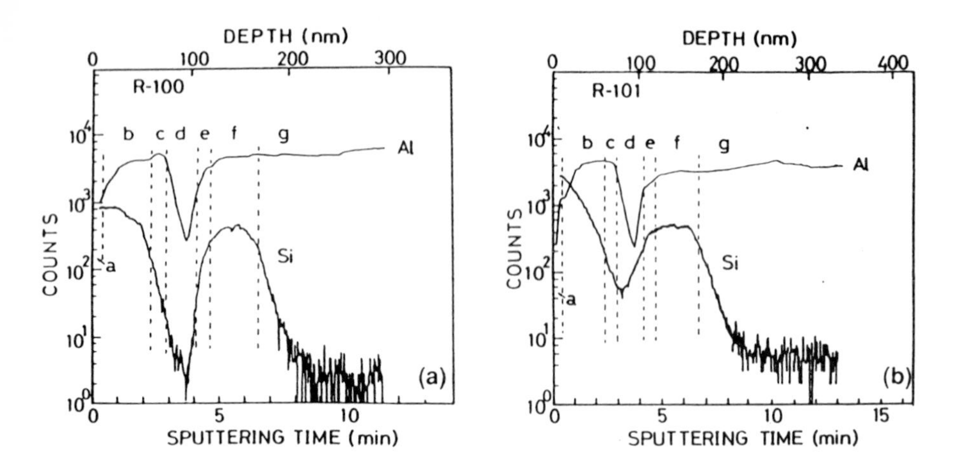

Figure 11: *Depth* (SIMS) *profile of* Si *donors and* Al *in n*-AlGaAs-GaAs-*n*-AlGaAs *quantum wells doped on both sides, grown (a) at* $T_s = 550°\text{C}$ *and (b) at* $T_s = 630°\text{C}$*, showing that Si donors segregate towards the surface at higher* T_s *(Inoue et al. 1985).*

roughening and by the impurities themselves was also considered as a possible reason.

It has been found experimentally that the mobility in single QWs with $L_w > 100$ Å is rather high, above $10^5\,\text{cm}^2\,\text{V}^{-1}\,\text{s}^{-1}$, when only the top AlGaAs layer is doped (see, for example, the study of mobility in Section 2.3), and the mobility gets low only when Si donors are introduced in the bottom AlGaAs layer (Sasa *et al.* 1984). This observation rules out the possibility that the fluctuations at the inverted heterojunction, typically an atomic monolayer, are responsible for the observed reduction of mobility, as long as the QW thickness L_w is greater than 80 Å.

It was also found experimentally that the use of a low substrate temperature during growth, $T_s < 550°\text{C}$, together with a low growth rate (about 0.3–0.5 μm h^{-1}) is effective in improving the mobility in MD-QWs where both the top and bottom sides are doped (Inoue and Sakaki 1984, Inoue *et al.* 1984, 1985). Indeed SIMS measurements have shown that the segregation of Si impurities can be suppressed at low substrate temperatures, as shown in Figure 11. Hence it is now possible to prepare an inverted GaAs-*n*-AlGaAs heterojunction with a quality comparable to that of a normal junction.

These observations lead to the conclusion that the degradation of mobility in inverted heterojunctions is mainly caused by donor impurities. Whether this degradation is caused directly by the enhancement of Coulomb scattering by ionised impurities, or results indirectly from the interface roughness associated with impurity-induced disorder, may depend on the specific growth conditions adopted.

4 Growth of laterally defined structures

There exists a growing interest in extending the study of quantum effects from ultrathin layered structures to laterally defined microstructures such as quantum wires, quantum boxes, and in-plane superlattices. Although the majority of these investigations have been done by patterning otherwise uniform two-dimensional carrier systems by lithography (see the chapter by Craighead), there have recently been some attempts to exploit various epitaxial growth approaches to form lateral structures with a feature size on the nanometre scale. In this section, we describe briefly the principle and current state of three approaches: (a) selective growth on facets on patterned substrates, (b) formation of quantum wires on an edge of the exposed surface of a layered structure, and (c) selective growth on vicinal crystallographic planes.

4.1 Growth on patterned substrates

Atomic and/or molecular species arriving at the foremost surface of the substrate with a net normal flux J_n migrate along the surface for some time τ_{inc} before being incorporated into the crystal. During this initial period, these unbound species with areal density $N = J_n \tau_{\mathrm{inc}}$ move randomly by diffusion through the (diffusion) length $L_D \approx (2D\tau_{\mathrm{inc}})^{1/2}$, where D is the diffusion constant. If the surface of the substrate and the incoming net flux are both uniform over a distance of several times L_D, the diffusion process creates no net flow $\mathbf{J}_{\mathrm{ip}}$ of the material species along the surface, since the diffusion flux $\mathbf{J}_+$ along one direction is exactly cancelled by the opposite flux $\mathbf{J}_-$.

If either the incoming normal flux J_n or any properties of the surface of the substrate depend on position over the distance L_D, material species begin to flow along the surface with flux $\mathbf{J}_{\mathrm{ip}}$ (Nilsson *et al.* 1989, Nakamura *et al.* 1991). Then the local growth rate becomes position-dependent since it is given by the normal flux J_n minus the net flow $\operatorname{div} \mathbf{J}_{\mathrm{ip}}$ going out along the surface. Hence the growth takes place selectively, which provides the possibility of preparing a variety of laterally defined structures.

In Figures 12a–d we show one example of such selective growth, where the cross-sectional SEM images are presented for a GaAs-AlAs heterostructure grown by MBE on a patterned (100) GaAs substrate with stripes of width 6 μm running along the [110] direction (Nakamura *et al.* 1991). Note that the growth on the top of each hill (reverse-mesa) structure proceeds in such a way that a facet structure is formed with two (111)B side surfaces and a (100) top surface. This result indicates that the rate of growth R of GaAs on (111)B surfaces is far smaller than that on (100) surfaces with the ratio $R(111)/R(100)$ being in the range $\frac{1}{30}$ – $\frac{1}{3}$, even when the ratio of incoming fluxes of Ga is set at 0.6. This substantial reduction of the rate of growth on (111)B has been studied in detail and shown to be caused mainly by the migration of deposited Ga atoms from the (111)B surface to the top (100) surface, which is particularly enhanced when the incoming As flux is reduced.

Note also in Figure 12b that the rate of growth of GaAs on the (100) surface, $R(100)$, reaches a maximum at the edge of the hill ($x = 0$) and decreases gradually as the distance x from the edge increases. This dependence can be fitted by the form $R_0 + \Delta R \exp(-x/L_D)$ and shows clearly that Ga atoms flow from the (111)B surface into the (100) surface and are incorporated on the (100) surface within a diffusion length L_D of typically 1 μm.

When AlGaAs is grown on a similar patterned substrate, in-plane diffusion is usually less efficient with a typical diffusion length of a few hundred Å or less. Hence the

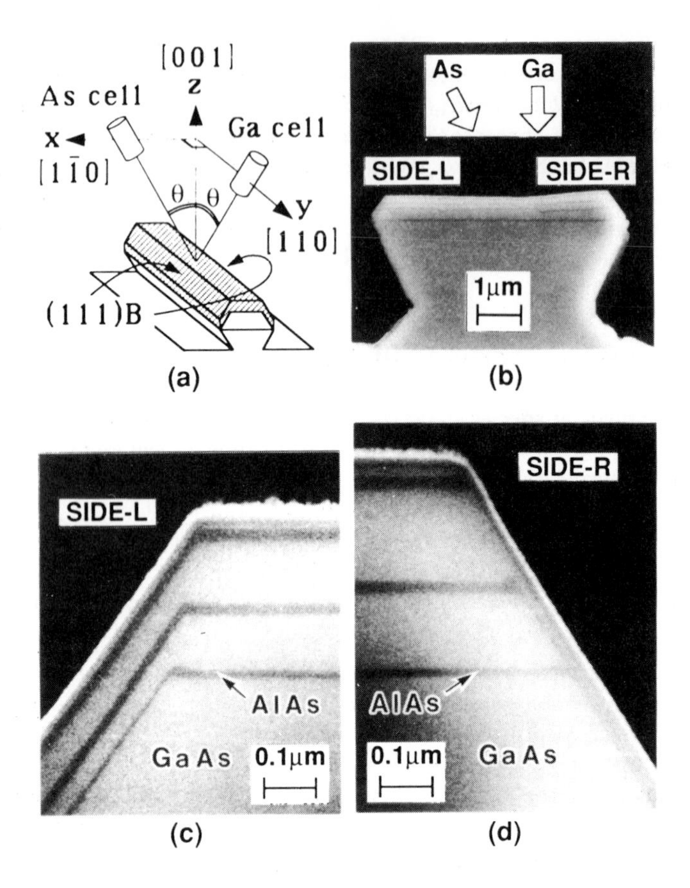

Figure 12: *Growth of facets on patterned heterostructures. (a) Schematic of the growth conditions, showing the geometry of the* Ga *and* As_4 *beams with respect to the reverse mesa stripe on a (100) substrate. Both beams are at an angle* $\theta = 32 \pm 4°$ *from* [001]. *The projections of the beams onto* (001) *are normal to each other. The unhatched region on the reverse mesa stripe was grown with rotation of the* Mo *sample holder, while the hatched region was deposited with the geometry shown here. (b) Cross-sectional SEM image of the grown layer; the shoulder regions are shown in (c) and (d) at higher magnification. Here, brighter regions are* GaAs *and darker regions are* AlAs. *The flux* $F_{\mathrm{As(111)B}}$ *of* As *on side* R *was* $\frac{1}{8}$ *of that on side* L, *whereas the flux* $F_{\mathrm{Ga(111)B}}$ *of* Ga *was almost same on both sides;* $d_{\mathrm{(111)B}}$ *on side* R *is about* $\frac{1}{10}$ *of that on side* L *(Nakamura et al. 1991).*

crystallographic selectivity of growth rate is substantially suppressed, as demonstrated in Figure 12b–d.

There are several different ways to use facet structures for the formation of laterally defined quantum structures. The most straightforward way is to continue the growth of GaAs structures such as that shown in Figures 12b and 12c until the width of the top (100) surface becomes as small as a few tens of nanometres. Then one can, in principle, grow first a thin uniform AlAs barrier layer followed by triangular GaAs, which works as a quantum wire or a quantum box. In practice, however, it is not easy to control the key width of these structures accurately, since a slight change in the local growth rate leads to a substantial fluctuation of wire and/or box widths. A similar facet structure can be prepared also with MOCVD or hydrogen-assisted MBE by performing selective epitaxial growth on a GaAs substrate which is with a line-and-space pattern of Si_3N_4 or SiO_2 (Fukui and Ando 1989, Kawabe 1991). This is because the growth of GaAs on the surface of Si_3N_4 or SiO_2 is inhibited, mainly due to the enhanced sublimation of source materials.

An alternative way of using facet growth to form a quantum wire is to make a V-groove structure, either by etching or growth, and to deposit GaAs selectively in the bottom of the groove to form the wire. The effectiveness of this approach was demonstrated first by Kapon *et al.* using MOCVD and also MBE (Kapon *et al.* 1989a, b). Recently a refined technique has been reported by Tsukamoto *et al.* (1992) using MOCVD.

A third way to achieve lateral confinement by the use of facet growth is to prepare undoped quantum well structures having facet surfaces and form 'edge quantum wire' structures by overgrowing the appropriate layer. This approach will be described in more detail in the following section.

4.2 Edge quantum wires and related structures

It was first pointed out by the author (Sakaki 1980) that a quantum wire with cross-sectional dimensions of $100\,\text{Å} \times 100\,\text{Å}$ could in principle be formed if one used the 'edge wire structure' of Figure 13a, where electrons are confined two-dimensionally by a quantum well (QW) along the z-direction and by an electrostatic potential along y. It is simplest to achieve confinement in metal-oxide-semiconductor structures, but this is not applicable to most compound semiconductors because their surfaces are very delicate and pin the Fermi level readily if they are contaminated or disordered. Hence one must first prepare a clean surface of undoped GaAs (or other) QW structures with minimal damage, and then overgrow n-AlGaAs (or its equivalent) to form the edge wire structure.

Although chemical etching and subsequent cleaning was first proposed as a method to expose the edge of QW structures, the resulting surface was usually found to show pinning of the Fermi level and this method could not readily be applied. Recent progress towards a clean and damage-free etching technology in ultra-high vacuum has, however, solved several key problems and may make this etching approach a viable one in the future.

Two alternative methods have successfully been tried to expose the edge of quantum well structures with sufficiently low contamination and disorder. One is the facet growth scheme and the other is the cleavage scheme. As described in Section 4.1, the growth of facets of GaAs and AlGaAs under appropriate conditions allows novel QW structures to be prepared with their (111)B surface exposed, as shown in Figure 12. Hence one can form a wire structure by overgrowing an n-AlGaAs layer on its top. The effectiveness of this approach was demonstrated first by Fukui and Ando (1989) with MOCVD and later by Nakamura *et al.* (1991) with MBE. There are some indications, however, that

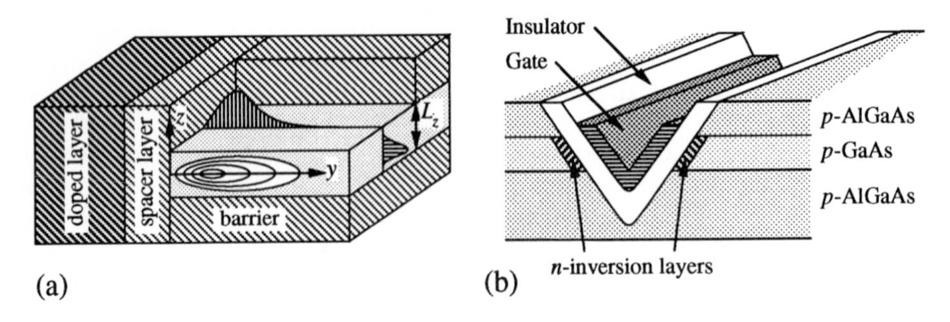

Figure 13: *(a) The concept of edge quantum wire structures, where electrons are confined by the double-heterojunction quantum well in one direction and by the electrostatic quasi-triangular potential in the other direction. (b) The original proposal to form an edge quantum wire by etching a V-groove and subsequently forming a channel on the surface.*

the edge surface is influenced by contaminating impurities to some extent.

As a second alternative, Stormer *et al.* (1991) have adopted cleavage and successfully prepared a very clean and damage-free (110) edge surface. An n-AlGaAs layer has been grown on its top to form an in-plane superlattice.

4.3 Selective growth on vicinal surfaces

When a single crystal of GaAs is polished with the surface slightly tilted by an angle α from the (100) orientation, atomic step structures will appear with an average period $p = (2.8/\tan\alpha)$ Å, which is about 80 Å when $\alpha = 2°$. Petroff *et al.* (1984) proposed the deposition of half-monolayers of GaAs and AlAs alternately on such a surface so that monolayer stripes of GaAs and AlAs with width $p/2$ are formed selectively along each step (Figure 14). By repeating the same process over and over, one can in principle prepare a periodic structure with compositional modulation along the crystal surface, which can be used to form quantum wire structures.

Fukui *et al.* (1988) demonstrated the feasibility of this approach by MOCVD. Subsequently, Petroff *et al.* (1989) and Tanaka and Sakaki (1989) showed the success of MBE in achieving a lateral modulation of the potential. The quality of the resulting structures is still far from perfect, however, since the growth can be perfect only when the following requirements are all satisfied: (i) the atomic steps must be straight and equally spaced; (ii) all the deposited Ga (or Al) must be incorporated only along the step, and straight monolayer stripes of GaAs (or AlAs) must be formed with equal widths; and (iii) the total sum of deposited GaAs and AlAs must be precisely one monolayer per cycle, to keep the stacking direction normal to the surface. Any deviation from these conditions could lead to the degradation of structural quality.

Although there exist several schemes to fulfill the third condition automatically by adopting such structures as serpentine superlattices and quantum well structures with inserted grids (Tanaka and Sakaki 1989), the first and second conditions cannot com-

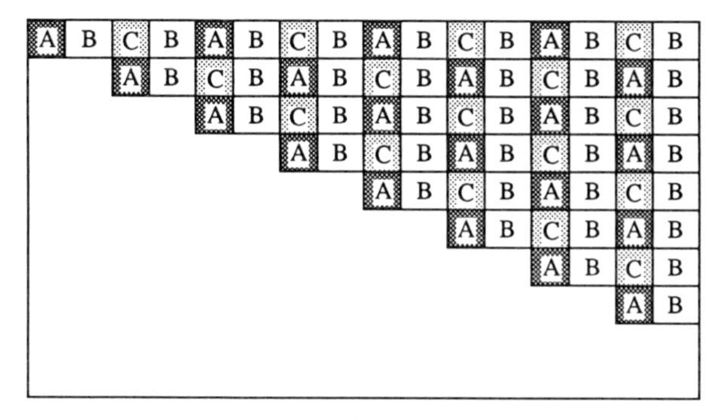

Figure 14: *Conceptual illustration of a tilted superlattice structure and its fabrication by the alternate deposition of fractional monolayer of* GaAs *(A),* AlAs *(B), and* AlGaAs *(C) on a vicinal* (100) *plane of* GaAs *(Petroff et al. 1984, 1989).*

pletely be met even with the most advanced technology now available. Hence, although cross-sectional transmissional electron micrographs have disclosed clearly the presence of compositional modulation, they have shown also the fuzziness and irregularity of the boundaries. Further work is certainly necessary to clarify the mechanisms of growth on stepped surfaces and to improve their controllability.

References

Arthur J (ed), 1985. Proc. 3rd Int. Conf. on MBE, San Francisco, 1985, *J. Vac. Sci. Technol. B*, **3**, 509–809.

Briones F, Golmayo D, Gonzalez L, and Ruiz A, 1987. *J. Cryst. Growth*, **81**, 19.

Cho A Y, and Arthur J R, 1975. *Prog. Solid State Chem.*, **10**, 157.

Dingle R, Stormer H L, Gossard A C, and Wiegmann W, 1978. *Appl. Phys. Lett.*, **33**, 665.

English J H, Gossard A C, Stormer H L, and Baldwin K W, 1987. *Appl. Phys. Lett.*, **50**, 1826.

Foxon C T, and Harris J J (eds), 1987. Proc. 4th Int. Conf. on MBE, York, UK, 1986, *J. Cryst. Growth*, **81**, 1–580.

Fukui T, and Ando S, 1989. *Electron. Lett.*, **25**, 410.

Fukui T, Saito H, and Tokura Y, 1988. *Jpn. J. Appl. Phys.*, **27**, L1320.

Fukunaga T, Kobayashi K L T, and Nakashima H, 1985. *Jpn. J. Appl. Phys.*, **24**, L510.

Harris J J, Joyce B A, and Dobson P J, 1981. *Surf. Sci.*, **103**, L90.

Hayakawa T, Suyama T, Takahashi K, Kondo M, Yamamoto S, Yano S, and Hijikita T, 1985. *Appl. Phys. Lett.*, **47**, 952.

Hirakawa K, and Sakaki H, 1986. *Phys. Rev. B*, **33**, 8291.

Horikoshi Y, Kawashima M, and Yamaguchi H, 1986. *Jpn. J. Appl. Phys.*, **25**, L868.

Ichinose H, Furuta T, Sakaki H, and Ishida Y, 1987. *J. Electron Microsc.*, **36**, 82.

Ikarashi N, Sakai A, Baba T, Ishida K, Motohisa N, and Sakaki H, 1990. *Appl. Phys. Lett.*, **57**, 1983.

Inoue K, and Sakaki H, 1984. *Jpn. J. Appl. Phys.*, **23**, L61.

Inoue K, Sakaki H, and Yoshino J, 1984. *Jpn. J. Appl. Phys.*, **23**, L767.

Inoue K, Sakaki H, Yoshino J, and Yoshioka Y, 1985. *Appl. Phys. Lett.*, **46**, 973.

Kapon E, Hwang D M, and Bhat R, 1989a. *Phys. Rev. Lett.*, **63**, 2715.

Kapon E, Simhony S, Bhat R, and Hwang D M, 1989b. *Appl. Phys. Lett.*, **55**, 2715.
Kawabe M, 1991. Private communication (presented at Conf. on Solid State Devices and Materials, Tokyo, August 1991).
Madhukar A, Lee T C, Yen M Y, Chen P, Kim J Y, Ghaisas S V, and Newman P G, 1985. *Appl. Phys. Lett.*, **46**, 1148.
Marik R J (ed), 1989. *Molecular Beam Epitaxy in III-V Semiconductor Materials and Devices*, Elsevier, pp. 218–330.
Nakamura Y, Koshiba S, Tsuchiya M, Kano H, and Sakaki H, 1991. *Appl. Phys. Lett.*, **59**, 200.
Neave J H, Dobson P J, Joyce B A, and Zhang J, 1985. *Appl. Phys. Lett.*, **47**, 100.
Neave J H, Joyce B A, Dobson P J, and Norton N, 1983. *Appl. Phys. A*, **31**, 1.
Nilsson S, van Gieson E, Arent D J, Meier H P, Walter W, and Forster T, 1989. *Appl. Phys. Lett.*, **55**, 972.
Noda T, Tanaka M, and Sakaki H, 1990. *Appl. Phys. Lett.*, **56**, 51.
Ourmazd A, Taylor D W, Cunningham J, and Tu C W, 1989. *Phys. Rev. Lett.*, **62**, 933.
Parker E H C (ed), 1985. *Technology and Physics of* MBE, New York, Plenum Press.
Petroff P M, Gaines J, Tsuchiya M, Simes R, Coldren L, Kroemer H, English J H, and Gossard A C, 1989. *J. Cryst. Growth*, **95**, 260.
Petroff P M, Gossard A C, and Weigmann W, 1984. *Appl. Phys. Lett.*, **45**, 620.
Pfeiffer L, West K W, Stormer H L, and Baldwin K W, 1989. *Appl. Phys. Lett.*, **55**, 1888.
Ploog K, and Graf K, 1984. *Molecular Beam Epitaxy of III-V Compounds — A Comprehensive Bibliography*, Berlin, Springer.
Sakaki H, 1980. *Jpn. J. Appl. Phys.*, **19**, L735.
Sakaki H, Noda T, Hirakawa K, Tanaka M, and Matsusue T, 1987. *Appl. Phys. Lett.*, **51**, 23, 1934.
Sakaki H, Tanaka M, and Yoshino J, 19085. *Jpn. J. Appl. Phys.*, **24**, L417.
Sakamoto T, Funabashi H, Ohta K, Nakagawa T, Kawai N S, and Kojima T, 1984. *Jpn. J. Appl. Phys.*, **23**, L657.
Sasa S, Saito J, Nambu K, Ishikawa T, and Hiyamizu S, 1984. *Jpn. J. Appl. Phys.*, **23**, L573.
Shiraki Y, and Sakaki H (eds), 1989. Proc. 5th Int. Conf. on MBE, Sapporo, 1988, *J. Cryst. Growth*, **95**, 1–637.
Singh J, Bajaj K K, and Chandhui S, 1984. *Appl Phys. Lett.*, **44**, 805.
Stormer H L, Pfeiffer L N, Baldwin K W, West K W, and Spector J, 1991. *Appl. Phys. Lett.*, **58**, 726.
Tanaka M, and Sakaki H, 1987. *J. Cryst. Growth*, **81**, 153.
Tanaka M, and Sakaki H, 1988. *Superlatt. Microstruct.*, **4**, 237.
Tanaka M, and Sakaki H, 1989. *Appl. Phys. Lett.*, **54**, 1326.
Tanaka M, Sakaki H, and Yoshino J, 1986. *Jpn. J. Appl. Phys.*, **25**, L155.
Tsukamoto S, Nagamune Y, Nishioka M, and Arakawa Y, 1992. *J. Appl. Phys.*, **71**, 1.
Tu C W, and Harris J S (eds), 1991. Proc. 6th Conf. on MBE, La Jolla, USA, 1990, *J. Cryst. Growth*, **111**.
Ueda R (ed), 1982. Collected Papers of the 2nd Int. Symp. on Molecular Beam Epitaxy and Clean Surface Techniques, Tokyo, 1982, Japanese Society of Applied Physics.
van Hove J M, and Cohen P I, 1987. *J. Cryst. Growth*, **81**, 13.
Wood C E C, 1981. *Surf. Sci.*, **108**, L441.

Lateral Patterning of Nanostructures

H G Craighead

Cornell University
Ithaca, New York, U.S.A.

1 Introduction

There is growing interest, as represented by this volume, in the physical properties of extremely small structures. The experimental realisation of new effects relies on the ability to create high quality structures and devices. Much attention has been directed towards semiconductors, particularly GaAs, in which a direct gap and long carrier scattering lengths enable a wide range of experiments to be performed. Metallic and superconductor studies that involve reduced dimensionality may require even smaller dimensions (Reed and Kirk 1989).

The growth of thin films and epitaxial layers can be controlled with extreme precision, approaching single atomic layer dimensions. Control of lateral dimensions has not reached this level of precision. The purpose of this paper is to analyse the practical limits of lateral patterning techniques and to suggest an approach that may be viable for the extension of essentially conventional processes for patterning below 10 nm.

Planar processes, originally developed for the semiconductor electronics industry, can be adapted to dimensions of a few nanometres in a limited class of material systems. However, compound semiconductors and other materials that make up low dimensional structures have a high surface-to-volume ratio and are susceptible to damage and surface effects. Great care must therefore be taken to create devices that are not destroyed by surface and process induced damage.

Our understanding of material processing in the pursuit of ultra-small structures is continually advancing. For example, the applicability of electron beam lithographic instruments continues to increase and the processes needed to exploit the capabilities of electron beams are developing. Recent advances in focused ion beam technology have been significant and, while not yet resulting in the smallest structure generation, this

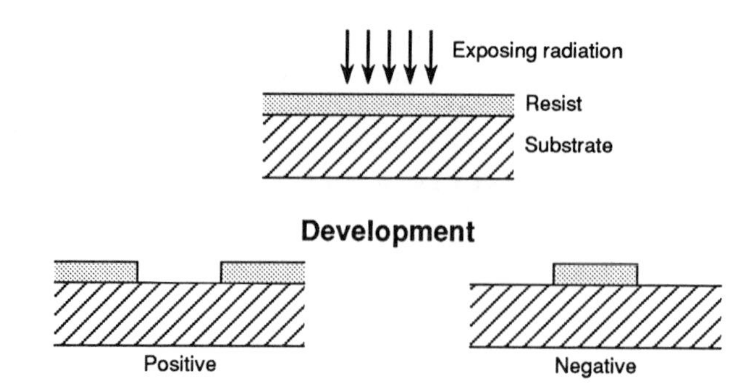

Figure 1: *Schematic diagram of the basic lithographic steps of resist exposure and development.*

area is moving rapidly. The invention of the scanning tunnelling electron microscope has recently provided a new method of directly imaging structures at the atomic level, and it is being explored as a fabrication tool involving the movement of one atom at a time. Most of the small structures examined have been defined by electron beam lithography, and the smallest semiconductor structures have been created by transferring e-beam patterns into III–V semiconductors by ion etching or growth on etched substrates.

It is instructive to consider the conventional but advanced forms of lithography as practised today. As it has developed for the semiconductor industry, lithography is the creation of a pattern in a resist layer, usually an organic polymer film, on a substrate material. A latent image, consisting of a chemical change in the resist, is created by exposing the desired area with some form of radiation. The pattern is developed by selectively removing either the exposed areas of resist (for a positive resist) or the unexposed areas (for a negative resist). These processes are shown schematically in Figure 1. The development of a polymeric resist is usually done with a solvent that brings out the pattern based on the solubility difference created by the exposure.

2 Overview of lithography

Electrons, ions, and photons can all be used for exposure in the lithographic process. Because of intrinsic scattering mechanisms, in no case can the energy provided by the exposing radiation be confined to arbitrarily small volumes. Firstly, the radiation or particle beam can not be confined arbitrarily because of diffraction, Coulomb repulsion among particles, optical limitations or other effects. Secondly, the energy incident on a resist layer spreads in the film, because of intrinsic scattering mechanisms in the solid.

The figures of merit for a resist/developer system include: the sensitivity to the exposing radiation, that dictates the speed of exposure; the contrast, that describes the exposure dose dependence of the resist response; the resolution, dictating the minimum feature size; and the suitability of the resist for pattern transfer. While the sensitivity and contrast of a system are vital in determining the speed and process latitude, it is the resolution and pattern transfer utility that are most important to research level nanostructure fabrication.

Inorganic resist materials exist and have been applied in the high resolution field. Related processes include vapour or plasma development that operate on difference in gas phase or plasma reactivity rather than liquid development based on solubility differences. Self-developing resists are of some importance in the high resolution area. For these materials the exposing radiation volatises the resist, creating a pattern in the resist without a development step. By this method, structures on the order of unit cell dimensions were made some years ago (Muray *et al.* 1984).

3 Photon lithography

The use of photons for high resolution lithographic exposure is a broad field that includes most of the lithography in use today. This is the technique widely used by the semiconductor industry and illustrates the general technique of lithography. It can be taken to include X-ray lithography and deep and near-UV photolithography. Only the shortest wavelengths are directly relevant to high resolution fabrication. The projection and proximity printing processes to be described are analogous to processes that could be used at nanometre dimensions for parallel printing with electrons or ions.

In projection lithography a mask containing the desired pattern is demagnified by an optical system and projected on the resist layer. This is desirable since there is no damaging contact between the mask and wafer, and the fabrication of the mask is not so critical since it is demagnified. The resolution is limited however by diffraction and the quality of the optics. The diffraction limit for the minimum resolvable grating period is usually taken as the Rayleigh criterion for the overlap of the diffraction peaks, which is approximately

$$\text{Minimum line resolvable spacing} \approx \frac{\lambda}{NA} \tag{1}$$

where λ is the wavelength and NA is the numerical aperture of the optical system, which is of the order of one. The shortest wavelength to be used with conventional optics is about 193 nm from an excimer laser source. The use of wavelengths much shorter than this is limited by absorption in optical materials. While the technology constantly improves, with better optics, improved contrast enhanced resists and sources of shorter wavelength, the limit of far-field diffraction is a fundamental one that prevents the use for feature sizes smaller than the wavelength of the light. Photolithography is essentially ruled out for fabrication at dimensions below 100 nm.

Projection systems employing X-rays would reduce diffraction effects, but the problem of fabricating optics for use in the X-ray region remains. Work in this area is advancing, with the development of multilayer mirrors and X-ray mask technology.

Proximity printing allows the 1 : 1 replication of a mask pattern. The effects of near field diffraction can be arbitrarily reduced by decreasing the distance between mask and substrate. Qualitatively, the minimum linewidth is

$$d \approx \sqrt{\lambda s} \tag{2}$$

where s is the separation between mask and substrate.

Proximity printing can be done effectively by X-ray exposure where the shorter wavelengths, of the order of 1 nm, allow much greater mask-substrate separation. With short wavelength X-rays, diffraction can effectively be ignored for attainable gaps of the order of a few micrometres. However, the exposure of polymeric resists by X-rays takes place primarily through the generation of photoelectrons which then expose the resist by the same mechanisms as an electron beam. The range of the photoelectrons in the resist will limit the resolution. Since the photoelectrons can have energies up to that of the X-rays, their range can be greater than 100 nm for the shorter wavelength X-rays. We can see qualitatively that there is a trade-off between diffraction and the range of photoelectrons in limiting the resolution. Perhaps the most significant problem with X-ray printing is the creation of a durable and stable mask of high contrast. All materials absorb X-rays, so the transparent area of the mask must be very thin or nonexistent. An absorbing metal such as gold supported on a thin film such as SiN_4 is one type of mask. Some of the smallest features replicated by X-rays were 17.5 nm in extent, fabricated by using edge evaporated metal as a mask. Some other high resolution process is necessary to generate the mask features; in most cases this means electron beam lithography.

4 Electron beam lithography

Electron beam lithography has now developed substantially and been used extensively for the generation of photolithographic masks. Schemes for electron beam 1 : 1 printing similar to those using X-rays or ions have been explored but generally not for the creation of nanostructures. Because of the ready availability of small electron sources and the high quality of electron optics, electron beams can be focussed to dimensions less than 1 nm and scanned with great accuracy.

In scanned electron beam lithography, a small electron source is imaged on the substrate through a series of electromagnetic lenses in an electron-optical column. The beam can be rapidly scanned over the sample to trace out any desired pattern. In the normal exposure of organic polymer resists, the electron beam causes the breaking of bonds or induces additional chemical bonds to be formed. This results in different molecular sizes in the exposed versus the unexposed areas, and this can be translated into differences in solubility in an appropriate developer.

The problem with electron beams is to confine the deposition of energy to a small area in the resist. There are several scattering processes in the material that lead to the deposition of energy at distances remote from the initial impact point of the electron. This leads to the well-known proximity and resolution limiting effects.

The scattering effect of longest range is the backscattering of electrons, resulting in large-angle scattering with energies near that of the incident electron. The range of these backscattered electrons can be many micrometres for typical electron beam energies. This range has been determined to have a power law dependence, proportional to $(\text{Energy})^{1.7}$, over a wide electron energy range. This presents the greatest problems for energies where the electron range is comparable to the feature spacing (Jackel *et al.* 1984).

Because of small-angle scattering events, forward scattering tends to spread the incident beam passing through the resist. The scattering angles decrease with electron energy, and the spread of the beam is obviously more important for thicker resist layers.

The generation of low energy secondaries has an effect analogous to that of photoelectrons in the X-ray case. This results in a cylindrical exposed volume around a δ-function beam. Keyser (1981) and others have performed Monte Carlo calculations to model this effect. This is a fundamental problem with electron beam exposure that can be reduced only by using resists with shorter ranges for these low energy (up to 50 eV) electrons, or which are exposed only by higher energy primary electrons. There have been suggestions that this is the case for the fluoride and oxide self-developing films, but these processes are not yet fully understood. The narrowest lines that have been formed in a polymeric resist (PMMA) were about 10 nm wide, which is consistent with the estimates of the range of lateral secondary electrons in these systems.

Thin resists are desirable, as mentioned before, for limiting the effects of forward scattering. This has been exploited in multi-level resists where the thin active layer is deposited on top of a thicker spacer or planarising layer. This spacer layer also reduces the effect of the backscattered electrons and adds a thicker usable resist thickness when the exposed pattern is transferred. Lines of 50 nm and smaller widths have been created with a three-level resist.

A method of eliminating the difficulties of electrons backscattered from the substrate is to remove the underlying substrate layer and work with a thin supporting film. This is how all the patterning of highest resolution has been done. In addition to eliminating the backscattered electrons and improving the process latitude, as described above, working on a membrane transparent to electrons allows the use of high resolution transmission electron microscopes for both exposure and imaging. It is with this type of instrument that one obtains the smallest electron beam sizes, down to 0.2 nm. The high resolution imaging capability of the transmission electron microscope can be used to image the ultra-small objects with this type of thin film sample. This is significant because the smallest structures may be difficult to resolve or have little contrast when viewed by conventional scanning electron microscopy.

A variety of resist systems have been studied on thin films. These include PMMA, Langmuir-Blodgett films, contamination films, fluoride films and alumina. Some remarkable work has been done creating functioning field effect transistors on thin films of III-V semiconductors. After much work on thin films, it became apparent from studies on bulk materials that the fundamental resolution limit of electron beam lithography was not affected by the presence of the substrate. This can readily be seen, since 10 nm wide patterns can be developed in a single layer of PMMA on solid semiconductor substrates, as small as those on thin support films. It is, however, more difficult to work on bulk materials, because of the unwanted backscattering of electrons. Also the imaging is more difficult with only the electron scattering contrast mechanisms of the SEM. There are, however, overwhelming benefits in working on bulk materials in terms of the types and durability of structures that can be created and studied. Many new types of devices and structures have been created on bulk substrates.

With electron beams it is possible to expose such small areas that the molecular size of the resist can become significant. This can be of the order of 10 nm for large organic molecules. The grain size of a polycrystalline layer to be patterned can also become larger than the size of the desired features. Clearly there are many new materials considerations to be addressed at nanometre dimensions. Another consideration important for the ultra-small size range is the time it takes to write a pattern by a scanning serial process. This time is inversely related to the size of the pixel. The smaller the electron beam size, the lower the current available for exposure and the longer the exposure time. The resists of highest resolution also require higher exposures. Fortunately the areas involved in exposures for devices are small. The time for writing is insignificant for experimental devices where only a few objects are being written, and the benefits and success of scanned electron beam lithography have been great.

A recently studied alternative approach to scanned electron beam lithography is to use technology related to the scanning tunnelling microscope to scan a low energy electron beam in close proximity to the surface. The use of low energy beams allows only thin conducting layers to be modified by the electron beam. There are exciting possibilities for manipulating individual atoms, as demonstrated by IBM researchers. For dimensions larger than this, the flexibility and capabilities of scanned beams as opposed to scanned sources are, as yet, more significant for fabricating devices for research.

As a variation on the scanned beam technique, another group at IBM is developing a microfabricated electron source and column to provide electron beams of high brightness, low energy and less than 10 nm diameter (Muray *et al.* 1992). A high resolution reduction projection electron beam lithography system is under study by AT&T Bell Laboratories, specifically for applications in nanofabrication (Koops 1989, Berger and Gibson 1990). The above techniques are candidates for exposure tools at below 10 nm. Inorganic resist or patterning layers have been demonstrated using a few monolayers of native oxide on GaAs (Clausen *et al.* 1990). This oxide layer is altered by exposure to an energetic electron beam. This type of process using grown or deposited thin layers of inorganic material as the layer sensitive to the beam is a possible component of a patterning scheme for general use at the nanometre scale.

5 Ion Beams

The use of ion beams is diverse, with a great range of ion species and energies. Both collimated beams of large area and scanned focused beams have been applied. The interaction of the ion beam with a solid can be to expose resist, which is most appropriate for high energy low mass ions. It is also possible to etch away material, to change the chemistry by implantation, to deposit films selectively, or to induce structural changes within a selected area (Kubena *et al.* 1989).

For ion lithographic printing, the effects of diffraction are negligible, because of the extremely small de Broglie wavelengths. Masks of high contrast are a challenge to fabricate, since the ions are absorbed in all solids. The use of ion channelling masks consisting of absorbers on oriented single crystals and of shadow masks supported by grids has been demonstrated. These masks are usually defined by electron beam lithography. Considerations of ion scattering in solids are similar to those for electrons, but are less severe.

Scanned focussed beams can be used to draw patterns of any desired shape. The limits of the size of the ion source and the quality of focussing optics have been the factors limiting focussed beam resolution. The quality of ion optics is poor compared to light optics and aberrations have limited the size of the spot. Advances with liquid metal field emitters have been significant in producing bright ion sources. Scanned focussed systems with liquid metal sources have been demonstrated with beam diameters as small as about 20 nm.

In the exposure of resist by scanned ion beams there is no proximity effect of the type seen for electron beams, and energy is transferred very efficiently to the resist. By appropriate choices of resist processes, it is possible to expose thin imaging layers very efficiently without ion damage to the substrate. Resist features as small as 20 nm have been demonstrated by exposure using focussed systems with liquid metal ion sources. Direct ion milling with scanned focussed beam or chemically assisted ion milling processes can be used to etch materials directly with impressive resolution approaching the size of the beam. However the rates achieved limit this approach to removal of small volumes of material.

By implantation-enhanced disordering within selected areas, lateral quantum structures have been created in GaAs-AlGaAs heterostructures. These techniques allow for lateral modulation of the composition and the band-gap, without exposing surfaces.

6 Pattern transfer

The definition of high resolution patterns in a resist layer is only part of the fabrication process. In fact, transfer of the pattern is frequently the step limiting resolution in fabrication. While holes in thin films as small as 2 nm have been generated by high voltage electron beam lithography, they have not been transferred into a solid material for study. Also the effects of damage induced by the process on ultra-small structures can be devastating, and this must be taken into consideration.

The reactive techniques are generally superior, since they allow selective etching of materials by proper choice of the chemistry. The general technique of transfer requires an additive or subtractive process, typically a step of etching or deposition and lift-off. The practical limitation in the lift-off procedure is on the aspect ratio and thickness of the film to be patterned in the lift-off step. For a successful removal of the deposited film, it should be discontinuous over the resist edge. This will not in general be the case if the deposited film is thick compared to the resist, and lift-off will be impossible. For best results, a resist profile with an under-cut shape is desirable, but even so the resist thickness must be greater than that of the deposited film. As noted above, the best resolution is obtained with thin resist. The multi-layer resists have been used to improve the resist performance by using a thin imaging layer on a thicker underlayer. The pattern exposed and developed in the thin imaging layer can be transferred into the underlying layer by a high resolution process such as reactive ion etching. Appropriately choosing the etching chemistry can greatly increase the utility of the resist. As an alternative additive transfer process, electro-plating has been used to create 0.8 nm diameter gold wires in narrow cavities.

Directed ion etching methods of ion milling, reactive ion etching, ion beam etching and ion beam assisted etching have all been used as high resolution etching processes. These are highly anisotropic because they use a directed beam of ions. Ion milling does

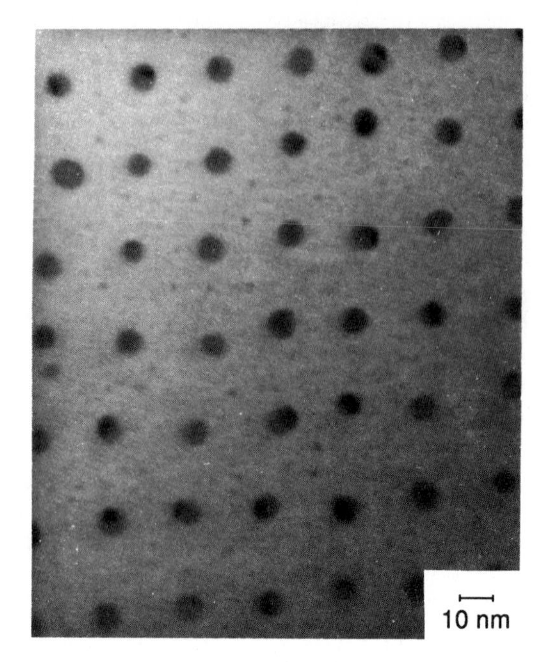

Figure 2: *Transmission electron micrograph of Ag particles written using an electron beam.*

not have the selectivity of etch rates, but has been used (Craighead and Mankiewich 1982) with contamination lithography to create very narrow, 8 nm wide Nb wires and 7 nm diameter Ag disks (Figure 2).

In ion milling, the atoms removed will deposit on surfaces in the line of sight of the sputtered surface. In the reactive process, the reaction products are volatile and do not deposit on exposed surfaces. In etching GaAs, for example, a relatively non-reactive metal can be used as a mask to prevent the underlying material from being removed in a plasma containing Cl (see for example the etched structures of Figure 3, Scherer and Craighead 1986). Etching narrow grooves can still be a problem because of the transport into long narrow cavities required. The structures of highest aspect ratio in semiconductors have been made by ion beam assisted etching.

7 Conclusion and Discussion

Current technologies present a diverse choice of methods for lateral patterning at high resolution. For creating experimental structures in GaAs-AlGaAs and other semiconductors, a more restricted set of techniques consisting largely of direct-write electron beam lithography combined with liftoff and etching are used, with some application of scanned ion beams and forms of photolithography. Fairly conventional approaches allow reliable fabrication of metal conductors and etched structures down to widths of 20 nm. Smaller structures can be created if care is taken to minimise the difficulties of surface

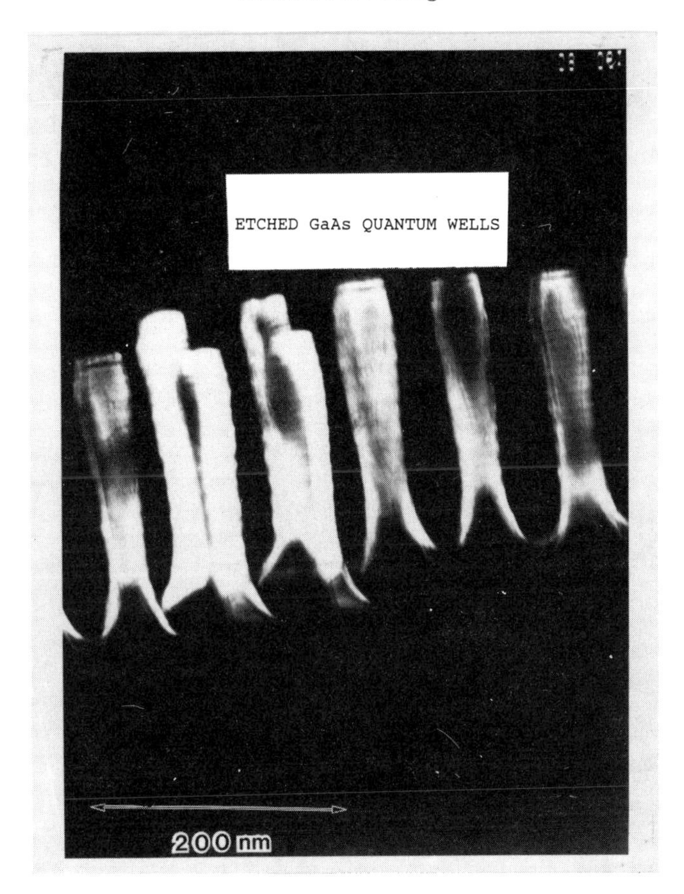

Figure 3: *Transmission electron micrograph of etched* GaAs-AlAs *columns defined by electron beam lithography and reactive ion etching.*

damage, stability and other problems associated with the small dimensions (Craighead 1984). A careful interplay between materials and fabrication issues are required for design and interpretation of physical measurements at reduced dimensions.

References

Berger S, and Gibson J M, 1990. *Appl. Phys. Lett.*, **57**, 153.
Clausen E M, *et al.*, 1990. *Appl. Phys. Lett.*, **57**, 1043.
Craighead H G, 1984. *J. Appl. Phys.*, **55**, 4430.
Craighead H G, and Mankiewich P M, 1982. *J. Appl. Phys.*, **53**, 7186.
Jackel L D, *et al.*, 1984. *Appl. Phys. Lett.*, **45**, 698.
Keyser D F, 1981. *J. Vac. Sci. Tech.* , **19**, 1.
Koops H W P, 1989. *Microcircuit Engineering 88*, North-Holland.
Kubena R L, *et al.*, 1989. *J. Vac. Sci. Tech. B*, **7**, 1798.
Muray A, Isaacson M, and Adesida I, 1984. *Appl. Phys. Lett.*, **45**, 1289.

Muray L P, *et al.*, 1992. *Proc. SPIE 1991*, to be published in *J. Vac. Sci. Tech.*, 1992.
Reed M, and Kirk W P, 1989. *Proc. Int. Symposium on Nanostructure Physics and Fabrication*, Boston, Academic Press.
Scherer A, and Craighead H G, 1986. *Appl. Phys. Lett.*, **49**, 1384.

Theory of Electrons in Low-Dimensional Systems

F Stern

IBM Thomas J. Watson Research Center
Yorktown Heights, New York, U.S.A.

1 Introduction

This chapter introduces some of the basic concepts associated with electrons in low-dimensional systems, with a bias toward simple examples. The rapid strides in fabrication technology have made it possible to realise structures of increasing sophistication and decreasing size. This has made it possible to turn textbook examples into real devices and phenomena.

After a brief discussion of several contexts in which effects of reduced dimensionality arise, the remainder of the chapter deals with three main topics. Sections 2–4 give basic concepts of electron states in low-dimensional systems and describe some of the consequences of reduced dimensionality for screening and for bound states. Section 5 deals with charge transfer in heterostructures, and Section 6 discusses electrical transport, with the primary focus on low-temperature mobility of electrons in GaAs-$Al_xGa_{1-x}As$ heterostructures. Then a number of approaches to modeling the charge densities, potentials, and energy levels of confined electrons are described. Finally, special properties of parabolic potentials are briefly noted.

Some of the material in this chapter is taken from lectures presented at the 1990 Spring College in Condensed Matter, in Trieste, on 'Physics of low-dimensional semiconductor structures' (Stern 1992).

1.1 Reduced dimensionality

Reduced dimensionality arises when at least one dimension of a physically active region is small compared to a relevant scale length. We shall consider two examples, one dealing with transport properties and one dealing with spatial quantisation effects.

For transport processes, two physical scale lengths are particularly important. One is the mean free path, the average distance between collisions that change the direction of current flow. In high-mobility samples this length can be of order μm or larger. Samples smaller than this length are in the ballistic regime, discussed elsewhere in this volume.

A second scale length important for transport is the phase-breaking length, the distance in which wave function coherence is lost. At high temperatures, where phonon scattering dominates, coherence is lost in a distance given by the mean free path between phonon scattering events. At lower temperatures the inelastic scattering rate—given by phonon scattering or by carrier-carrier scattering—becomes small compared to the elastic scattering rate, and only a small fraction of scattering events will change the phase. The phase-breaking length is then often determined by the inelastic scattering length, $L_{\text{in}} = (2dD\tau_{\text{in}})^{1/2}$, which is an average distance carriers diffuse between inelastic collisions in d dimensions. The diffusion constant D is usually controlled by elastic scattering processes like impurity scattering, and is approximately independent of temperature, while the inelastic scattering time is temperature dependent. At sufficiently low temperatures the inelastic scattering length can be of order 1 to 10 μm, placing many samples in the low-dimensionality regime for transport. Corrections to transport properties at low temperatures occur for all dimensionalities but are more pronounced in lower-dimensional systems. This subject, vastly oversimplified in this brief discussion, is associated with weak localisation phenomena. It had a rapid development beginning about twelve years ago and has led to a large literature dealing with low-dimensional aspects.

In this chapter we are mainly concerned with a third physical scale length, namely the Fermi wavelength, $2\pi/k_F$, where k_F is the carrier wave vector at the Fermi surface. When a sample dimension becomes comparable to this distance (or to the corresponding distance without the factor 2π), the motion in that direction becomes quantised, resulting in changes in the energy spectrum and in the dynamical properties of the system. When only one dimension of the sample is small compared to this distance we speak of a system that is (dynamically) two-dimensional because carrier motion is unconstrained in two space dimensions. Many properties of such systems are reviewed in the article by Ando *et al.* (1982), particularly in relation to silicon inversion layers. Some basic ideas will be presented here for two-dimensional electron systems as well as for systems that are constrained in two or three dimensions and are therefore dynamically one- or zero-dimensional. A dynamically one-dimensional system is often called a quantum wire, and a dynamically zero-dimensional system is often called a quantum dot or quantum box.

There are many ways to realise carrier confinement, as will be clear from examples in this chapter and other chapters. In an earlier discussion (Stern 1992) I have categorised some of the most important to include geometrical confinement, for example through mesa etching, compositional confinement, as in heterostructures and semiconductor-insulator structures, electrostatic confinement, as in various gated structures, and selective growth on tilted surfaces. Combinations of these methods, as well as less common methods such as particle beam patterning and strain confinement, continue to lead to novel structures.

2 Electron states in low-dimensional systems

One of the simplest illustrations of low-dimensional behaviour is the particle in a box, here taken as a rectangular structure with a flat potential inside (taken to be zero) and infinitely high, mathematically sharp walls. Here, and throughout most of this chapter, we use the effective mass approximation and assume that the respective effective masses for motion in the k_x, k_y, and k_z directions are m_x, m_y, and m_z. The wave functions are approximated as products of Bloch functions, taken near the relevant band edge, and envelope functions which are presumed to be slowly varying on an atomic scale although that approximation is severely tested at times. For most purposes the underlying Bloch function is omitted but it must be included for some purposes, such as the calculation of optical transitions; this is discussed in the chapter by Bastard. In the effective mass approximation the envelope wave function of a state with quantum numbers i and j and wave vector k_x in a rectangular wire of dimensions a and b is taken to be

$$\zeta_{i,j,k_x}(x,y,z) = \sqrt{\frac{4}{ab}} \sin\frac{\pi i y}{a} \sin\frac{\pi j z}{b} \exp(ik_x x), \quad i,j = 1,2,3,\ldots. \tag{1}$$

Motion along the wire is taken to be free-electron-like, leading to continua of states—called subbands—with energies

$$E_{i,j}(k_x) = \frac{i^2\pi^2\hbar^2}{2m_y a^2} + \frac{j^2\pi^2\hbar^2}{2m_z b^2} + \frac{\hbar^2 k_x^2}{2m_x}. \tag{2}$$

The energies of the subbands for $k_x = 0$ are nondegenerate (not including spin) in general, but for a square wire with $a = b$ the states i,j and j,i will be degenerate when $i \neq j$. Other degeneracies can occur, depending on the ratio of a to b. The density of states per unit energy for the one-dimensional subband with quantum numbers i and j in a wire of length L is

$$\rho_{i,j}(E) = 2 \times 2\frac{L}{2\pi}\frac{dk_x}{dE} = \frac{L}{\pi}\left(\frac{2m}{\hbar^2}\right)^{1/2}(E - E_{i,j})^{-1/2}, \tag{3}$$

where the first factor 2 arises because there are states with both positive and negative values of k_x and the second factor 2 is the spin degeneracy. We assume that there is only a single conduction band valley, as in GaAs, here. The inverse square root behaviour in Equation (3) is to be contrasted with the two-dimensional result

$$\rho_i(E) = \frac{m}{\pi\hbar^2}, \quad E > E_i, \tag{4}$$

that the density of states is a constant above the subband edge, and with the three-dimensional result that the density of states varies as the square root of the energy measured from the band edge.

Somewhat surprisingly, no simple closed-form solution for the envelope wave functions and energy levels of electrons in a rectangular wire embedded in a material with a finite barrier is available. We consider a treatment for such a case below.

If the simple rectangular quantum wire is truncated to become a rectangular parallelepiped, or quantum box, then the envelope wave function will have sinusoidal factors in all three space dimensions, by obvious extension of Equation (1), and the energy levels will be discrete.

Another simple case is a wire with a circular cross-section. The Schrödinger equation in polar coordinates takes the form

$$-\frac{\hbar^2}{2}\left[\frac{1}{R}\frac{\partial}{\partial R}\frac{R}{m}\frac{\partial}{\partial R}+\frac{1}{R^2}\frac{\partial^2}{\partial\phi^2}\right]\zeta(R,\phi)=\left[E-V(R,\phi)\right]\zeta(R,\phi), \tag{5}$$

where we have assumed free-carrier motion in the x direction, taken the effective mass m to be isotropic but not necessarily constant in the y-z plane, and used $R=(y^2+z^2)^{1/2}$. If the potential inside the wire is zero and the barrier at radial distance R_0 from the centre is infinite, and if the effective mass m is constant, then the envelope functions are

$$\zeta_{l,n,k_x}(R,\phi,x)=\frac{1}{R_0\sqrt{\pi L}}\left[\frac{1}{J_{l+1}(j_{l,n})}J_l\left(\frac{j_{l,n}}{R_0}R\right)\right]\mathrm{e}^{\mathrm{i}l\phi}\,\mathrm{e}^{\mathrm{i}k_x x},\quad l=0,1,2,\ldots, \tag{6}$$

where J_l is the Bessel function of order l and $j_{l,n}$ is its nth zero. The energy levels are

$$E_{l,n}(k_x)=\frac{\hbar^2 j_{l,n}^2}{2mR_0^2}+\frac{\hbar^2k_x^2}{2m}. \tag{7}$$

The subbands with angular momentum $l\hbar$ are nondegenerate if $l=0$ and are doubly degenerate for nonzero values of l.

It is instructive to compare the integrated density of states for a square wire and a circular wire of the same cross-sectional area $a^2=\pi R_0^2$. The asymptotic density of states at large quantum numbers depends mainly on the area and only weakly on the geometry. However, there is a significant effect of the boundary conditions. The number of states below a given energy found with the boundary condition that the envelope function vanishes on the boundary is smaller than the two-dimensional value by an amount proportional to the ratio of the wire perimeter to its area, which vanishes as the wire gets larger. If, instead, the boundary condition that the normal derivative of the envelope function vanish at the boundary is used, then the number of states is correspondingly larger than the two-dimensional value. While the relative differences vanish for very large quantum numbers, they are noticeable even when one hundred or more quantum states lie below the given energy.

Level broadening and thermal occupation effects must usually be small compared to the energy differences between subbands to make it possible to observe the energy level structure associated with the quantum confinement. Thus small dimensions, which enlarge the energy scale as indicated in the particle-in-a-box expressions in Equations (2) and (7), and low temperatures, which tend to minimise level broadening, are normally required.

In most realisations of wire and dot structures the barrier at the boundaries should be treated as finite, and the approximation that the envelope wave function vanishes at the boundary is likely to fail. That is the case, for example, when the confining potential is provided by a band offset as in a GaAs-$Al_xGa_{1-x}As$ heterostructure. In such cases a more general Schrödinger equation must be solved. The general question of boundary conditions at a heterointerface is beyond the scope of this chapter. The discussion here deals only with the simplest case, as in GaAs-$Al_xGa_{1-x}As$ heterojunctions with $x \leq 0.4$, for which both materials have their conduction band minimum at the centre of the Brillouin zone. In that case the boundary condition to be used, chosen to preserve probability current as calculated from the envelope function, is

$$\frac{1}{m_1}\zeta_1{}' = \frac{1}{m_2}\zeta_2{}' \tag{8}$$

(Ben Daniel and Duke 1966), where m_1 and m_2 are the respective effective masses and the prime denotes the derivative in the direction normal to the interface. The envelope function and its derivatives in the plane of the interface are continuous at the interface.

For the circular wire, explicit solutions involving Bessel and Neumann functions can be written down for the finite barrier case. The eigenvalues are found from the solutions of a transcendental equation for the matching at the boundary, using Equation (8). In addition to the discrete solutions, which vanish at infinity, there is also a continuum of solutions with energies greater than the barrier height. The density of states in the continuum has broad maxima corresponding to resonances or virtual states.

The two examples described here are typical of a particle-in-a-box. There is also considerable interest in parabolic potentials, which arise naturally in electrostatic confinement and can be fabricated via compositional variations in quantum wells. Some properties of systems with parabolic potentials will be discussed in Section 8.

Once we leave simple models like those just described, calculation of the energy level structure becomes more complicated. The problem can be divided into two parts: a description of the potential, and the calculation of the energy levels and other properties of the electronic system. If the system has carriers present, these two problems are coupled and must be solved self-consistently. Examples of such solutions are given below.

One may well question the use of the effective mass approximation for structures that can be only a few atomic layers in extent in one or more dimensions. As is often the case, the approximation turns out to be rather robust. Nevertheless, it will lose validity for monolayer superlattices or for crystallites with a small number of atoms. Approximations that invoke the discrete atomic structure are then required, although in such cases it is more difficult to include effects associated with external fields or other long-range phenomena. Gell *et al.* (1987) discuss the connection between a microscopic and a bulk picture for a short-period superlattice.

3 Screening

The response of an electronic system to a weak perturbing potential can be represented by the dielectric function or permittivity, which depends in general on the variation of the perturbation in space and in time. In this chapter we focus on the static response; the frequency dependence of the dielectric response, as reflected for example in plasmon dispersion, is discussed in the chapters by Hansen and Heitmann. We also simplify the problem by omitting most of the nonlocal aspects of dielectric response, which have been discussed, for example, by Dahl and Sham (1977) for two-dimensional systems.

3.1 Two-dimensional systems

The polarisability associated with a static perturbation of wave vector q is given for a two-dimensional electron system at low temperatures with only one subband occupied by (Stern, 1967)

$$\chi(q) = \frac{me^2}{\pi\hbar^2 q^2}\left\{1 - \left[1 - \left(\frac{2k_F}{q}\right)^2\right]^{1/2}\right\}, \quad q > 2k_F. \tag{9}$$

The term with a square-root is absent for $q < 2k_F$. The real physical system is three-dimensional, so we must introduce some spatial dependence for the third dimension. Suppose that we have an inversion layer, or heterojunction channel, or quantum well charge layer, which is uniform in the x-y plane and has a charge density proportional to $g(z)$ determined by doping, band offsets, and external potentials. $g(z)$ is normalised: $\int g(z)dz = 1$. Now suppose that there is a weak potential variation $\delta\phi(q,z)$ imposed, taken to be sinusoidal with wave vector q in the x-y plane. We first convert this perturbation to a two-dimensional one by averaging over the charge density, to get an average perturbation

$$\delta\overline{\phi}(q) = \int \delta\phi(q,z)\, g(z)\, dz. \tag{10}$$

The induced charge associated with this perturbation follows from Maxwell's equations together with the polarisability. In the spirit of the simple approximations used here, that assume only a single subband to be occupied and avoid the complications of a full nonlocal treatment, the induced charge is

$$\delta\rho(q,z) = -\tfrac{1}{2}\epsilon q_s\, \delta\overline{\phi}(q)\, g(z), \tag{11}$$

where at low temperatures $q_s = (q^2/2\epsilon)\chi(q)$, and ϵ is the permittivity of the medium in which the electron plane is embedded. If we further simplify the result by replacing the polarisability by its long-wavelength limit, and now consider a perturbing potential produced by an external fixed charge Ze at a point $(0,0,z_0)$, we find that the Poisson equation for the perturbing potential is

$$\nabla\cdot[\epsilon(z)\,\nabla\delta\phi(R,z)] - 2\epsilon q_s \delta\overline{\phi}(R)\, g(z) = -Ze\delta(x)\,\delta(y)\,\delta(z-z_0). \tag{12}$$

The explicit solution of Equation (12) for silicon inversion layers was given by Stern and Howard (1967). They approximated the charge distribution by $g(z) = \zeta_{\mathrm{FH}}^2(z)$, where ζ_{FH} is the Fang-Howard (1966) variational envelope function

$$\zeta_{\mathrm{FH}}(z) = \left(\frac{b^3}{2}\right)^{1/2} z \exp\left(-\frac{bz}{2}\right), \tag{13}$$

which goes to zero at the Si-SiO_2 interface. Although the interface barrier (*i.e.* band offset) in most heterostructures is considerably smaller than the 3 eV barrier between Si and SiO_2, the Fang-Howard function is sometimes used there also for simplicity.

In the long-wavelength limit, the screening parameter q_s has the more general form

$$q_s = \frac{e^2}{2\epsilon}\frac{dN_s}{dE_F}, \tag{14}$$

which reduces to the earlier result at low temperatures, and gives $q_s = (e^2/2\epsilon)(N_s/k_BT)$ at high temperatures. Note, however, that at high temperatures more than one subband will usually be occupied, and each subband will contribute to the screening. A two-dimensional approach gradually loses validity as the temperature increases.

The screening parameter has rather large values for typical two-dimensional systems. It is about 2×10^7 cm^{-1} for a silicon inversion layer (if the average dielectric constant of Si and SiO_2 is used and a factor 2 is added for valley degeneracy), and about 2×10^6 cm^{-1} for a GaAs heterolayer. However, screening in a two-dimensional system is not as strong as screening in three dimensions, because the screening charge cannot fully surround the perturbing charge. The explicit expressions are too cumbersome to give here (see Appendix B of Stern and Howard 1967), but the perturbation falls off at large distances as the third power of distance, rather than exponentially. Thus the intuition developed for three-dimensional systems cannot be transferred to two-dimensional systems. In particular, q_s^{-1} does not set a length scale.

For a degenerate two-dimensional system at low temperature, the abrupt change in screening at a wave vector of $2k_F$ leads to oscillations in the potential, whose leading term is proportional to $R^{-2}\sin 2k_FR$ (Stern 1967). The corresponding three-dimensional Friedel oscillations are proportional to $r^{-3}\cos 2k_Fr$ (Fetter and Walecka 1971). Some other aspects of the dielectric response of a two-dimensional electron system in the hydrodynamic approximation have been presented by Fetter (1973).

3.2 One-dimensional systems

Screening in one-dimensional electron systems has been treated by many authors (see, for example, Das 1974, Friesen and Bergersen 1980, Lee and Spector 1985). The static polarisability is given by

$$\chi(q) = \frac{2m}{\pi\hbar^2 q} \ln\left|\frac{q-2k_F}{q+2k_F}\right| . \tag{15}$$

The changes in charge distribution and potential induced by an external perturbation depend on the initial charge distribution in the wire. Unlike the two-dimensional case, where the assumption of a layer of zero thickness simplifies the treatment, the one-dimensional case leads to a divergence if the wire radius goes to zero. Explicit results are too cumbersome to present here, except for the one-dimensional analog of Friedel oscillations, with a potential that varies as $x^{-1}\cos 2k_Fx$ (Cheng *et al.* 1991).

4 Bound states

As the dimensionality of a system decreases, the binding energy of electrons to impurities and the analogous exciton binding energy increase. For short-range potentials, it is known that in three dimensions there is a minimum strength required before an attractive potential has a bound state. In two dimensions an arbitrarily weak potential has a bound state, although the binding energy is extremely small for a weak potential. Bound states in wires and dots have been examined by many authors, including Bryant (1984), Osório *et al.* (1988), Zhu *et al.* (1990), and Gold and Ghazali (1990).

One of the best-known results is that the binding energy of an electron of effective mass m to a Coulomb centre of charge e in two dimensions is $me^4/8\pi^2\epsilon^2\hbar^2$, four times as large as the corresponding three-dimensional result. Here ϵ is the permittivity of the medium in which the electron plane is embedded, and we assume that the attractive centre is also located in the plane. For electrons confined to a wire or dot of small radius, the binding energy diverges logarithmically or as the inverse of the radius, respectively.

In silicon inversion layers and in other gated heterostructures, the carrier density can be varied and can be very different from the density of impurities, either residual or purposely introduced. Vinter (1980) showed that when the density of electrons in a (001) silicon inversion layer greatly exceeds the density of impurities, the impurity level is four-fold degenerate and very shallow.

The same considerations that enhance binding to a fixed attractive centre also enhance the attraction between an electron and a hole, as studied for example by Keldysh (1979) in a quasi-two-dimensional context. Excitonic effects are central to the understanding of optical properties of wires and dots, discussed in the chapters by Hansen and Heitmann.

5 Charge transfer

Most experiments involving nanostructures focus on a particular portion of a device in which the electrical or optical phenomena of interest take place. The actual device generally has a more complex structure which serves to define and support this 'active' region. Carriers can be introduced into the active region optically, by direct local doping, by remote doping (also called modulation doping), electrostatically by capacitive induction across insulating regions, by tunneling through barriers, or in other ways. A good understanding of a given device or phenomenon often requires a detailed understanding of the process by which charge is transferred. This section deals with a particularly simple case, charge transfer in remotely doped heterojunctions. More complicated cases are considered in Section 7.

A remotely doped heterojunction is illustrated in Figure 1. The simplest example has a semi-infinite GaAs region in the half-space $z > 0$, with $Al_xGa_{1-x}As$ occupying the half-space $z < 0$. Electrons are supplied to the GaAs by donors located in the region $z < -d_{sp}$ in the $Al_xGa_{1-x}As$. The undoped spacer layer of thickness d_{sp} serves to increase the distance between the impurities and the electrons, enhancing the mobility, as discussed in the following section. There will be some residual impurities in the GaAs, which we here assume to be acceptors with density N_A. The donors are taken here to be simple donors with density N_{Db} and binding energy E_{Db}. Three other energies enter in the consideration of charge transfer. One is the conduction band offset ΔE_c at the GaAs-$Al_xGa_{1-x}As$ interface; the second is the energy E_0 of the lowest subband in the GaAs, measured from the conduction band edge at the interface; and the third is the Fermi energy relative to the bottom of the lowest subband, $E_F - E_0$. At low temperatures, the case considered here, the Fermi level in the $Al_xGa_{1-x}As$ can be considered to lie at the position of the deep donor (this assumes that there are some compensating acceptors in the $Al_xGa_{1-x}As$). We then have the situation shown in Figure 1, where the charge transfer into the GaAs is determined by the condition that

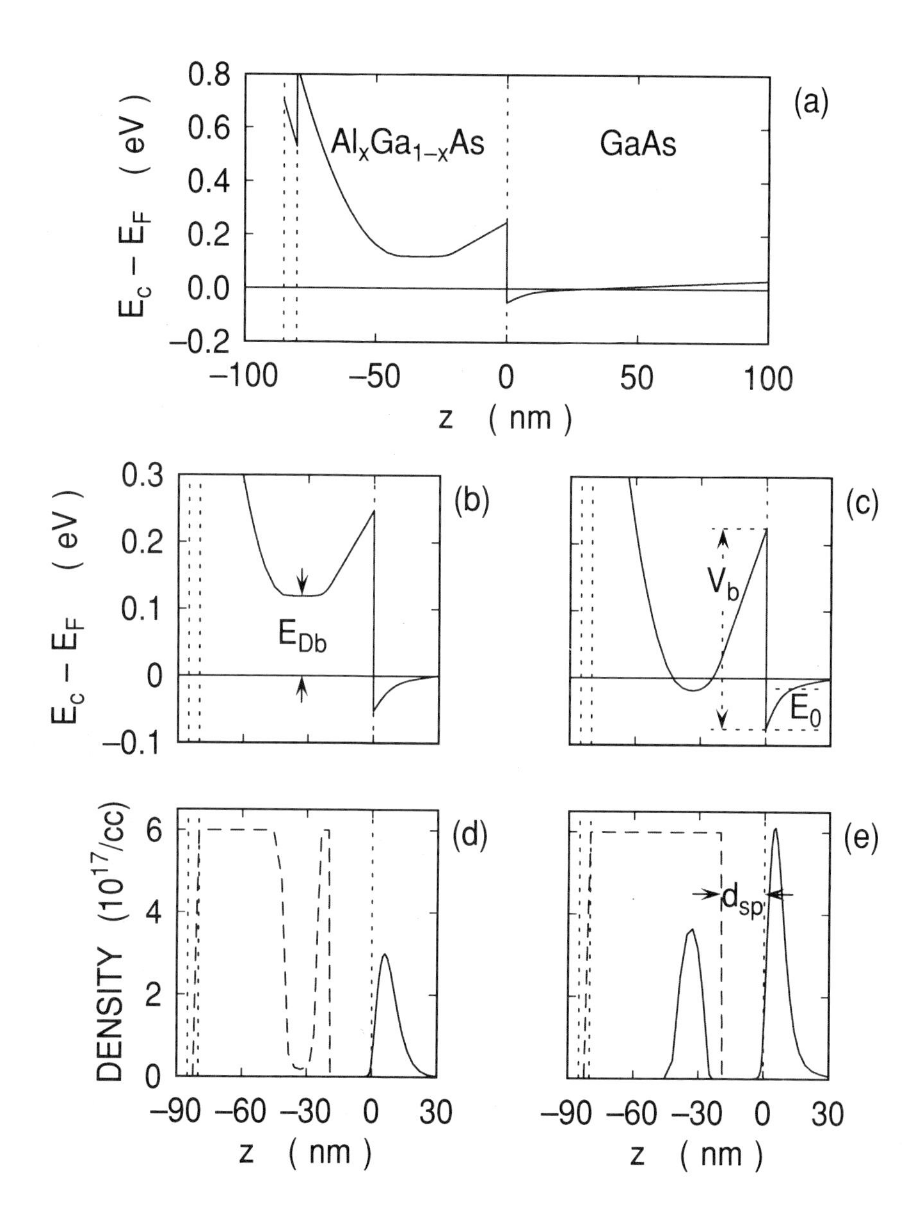

Figure 1: *Band bending at a* GaAs-$Al_xGa_{1-x}As$ *heterojunction. (a) Band bending in the dark; a* GaAs *cap,* $Al_xGa_{1-x}As$ *barrier, and the* GaAs *layer in which the two-dimensional channel forms are shown. (b,c) Band bending in the dark and after illumination, respectively. (d,e) Electron density (full lines) and ionised donor density (dashed lines) in the dark and after illumination, respectively.*

the Fermi level on both sides of the heterojunction is the same. This leads to the condition

$$E_{Db} + \frac{e^2 N_t^2}{2\epsilon_b N_{Db}} + \frac{e^2 d_{\rm sp} N_t}{\epsilon_b} = \Delta E_c - E_0 - (E_F - E_0), \qquad (16)$$

where ϵ_b is the permittivity of the barrier material, $N_t = N_s + N_d$, N_s is the areal density of electrons in the GaAs channel and

$$N_d = \frac{1}{e}\sqrt{2\epsilon_c N_{Ac}(E_c - E_{Fc} - e\phi_s)} \qquad (17)$$

is the areal density of charges in the GaAs depletion layer. Here N_{Ac} is the net acceptor density in the GaAs, ϵ_c is the permittivity of the GaAs, $E_c - E_{Fc}$ is the energy difference between the conduction band edge and the Fermi level in the GaAs bulk, and ϕ_s is the substrate bias if any, *i.e.* the voltage at the substrate contact relative to the channel. Some correction terms to Equation (17) are given, for example, in Equation (3.13) of Ando *et al.* (1982).

To illustrate these ideas, consider a GaAs-$Al_xGa_{1-x}As$ heterojunction with a *p*-type GaAs substrate, and with an $Al_xGa_{1-x}As$ barrier with AlAs fraction $x = 0.35$, for which we assume a conduction band offset of $\Delta E_c = 0.3\,\mathrm{eV}$ and a donor binding energy of $E_{Db} = 0.12\,\mathrm{eV}$. Assume that the net acceptor density in the GaAs is $N_{Ac} = 10^{14}\,\mathrm{cm}^{-3}$, leading to $N_d = 4.6\times10^{10}\,\mathrm{cm}^{-2}$ charges in the depletion layer, and that the donor doping density in the $Al_xGa_{1-x}As$ is $N_{Db} = 1\times10^{18}\,\mathrm{cm}^{-3}$. There is no substrate bias in this example. We want to calculate the thickness of the undoped spacer layer that will give two different electron densities N_s in the GaAs channel of 1 and $5\times10^{11}\,\mathrm{cm}^{-2}$. The subband energies E_0, 24 and 46 meV respectively, have been calculated by Stern and Das Sarma (1984). The Fermi energy relative to the bottom of the lowest subband is obtained from the density of states, from Equation (4), to be 3.4 and 17 meV respectively, for the two values of N_s. With these assumptions, all the terms in Equation (16) are known except the term proportional to the spacer thickness $d_{\rm sp}$, and we find the calculated spacer thicknesses to be 74 and 13 nm, respectively, for channel densities of 1 and $5\times10^{11}\,\mathrm{cm}^{-2}$. These number are given as examples of the method, not as a claim of great precision. For a gated structure these arguments must be extended to include the band bending associated with the Schottky barrier and the applied gate voltage. Quantisation effects in the potential well in the $Al_xGa_{1-x}As$, shown in Figure 1c, were studied by Vinter (1983).

Donors in $Al_xGa_{1-x}As$, particularly for AlAs fractions $x > 0.2$, are deep levels whose wave functions are derived not only from the Γ conduction band valley but also from the L and X valleys. In addition, they have the remarkable property that excited electrons must overcome a substantial energy barrier to recombine, leading to persistent photoconductance at low temperatures. The special properties of the deep donor, originally called a *DX* centre because it was thought to be a complex of a donor with a defect (Lang *et al.* 1979), have attracted considerable interest and have been reviewed, for example, by Mooney (1990) and by Theis *et al.* (1991). The theoretical finding by Chadi and Chang (1989) that the deep donor is a so-called negative-U centre now appears to be supported by most of the available experiments. This model leads to a lower energy when half the donors are positively charged and half are negatively charged than in the conventional picture in which all the donors are neutral, for an uncompensated layer in equilibrium at low temperatures.

The persistent photoconductance has been used by many authors, for example Störmer *et al.* (1981) and Weimann and Schlapp (1985), to vary (generally to increase) the carrier concentration at low temperatures by exposing the sample briefly to illumination and then turning the light off. The increase in carrier concentration can be understood by a simple modification of the arguments that led to Equation (16). After illumination, which is assumed to be strong enough to excite all the deep donors to their positively charged state, the Fermi level rises from a position near the deep donor level to a position close to the conduction band edge. This increases the 'driving force' that generates charge in the GaAs channel from a value approximately equal to the difference between the band offset and the donor binding energy to a value close to the band offset, and leads to values in at least qualitative agreement with experiment. A discussion based in a conventional deep donor picture was given by Stern (1986), but needs to be modified for the statistics associated with a negative-U centre (see, for example, Yoffa and Adler 1975, Theis *et al.* 1991). This model, as illustrated in Figure 1e, can lead to carriers in the $Al_xGa_{1-x}As$ in a sample that has been exposed to light at low temperatures, although the density of such carriers depends on the number of doping atoms in the $Al_xGa_{1-x}As$ in excess of the number needed to compensate the surface Schottky barrier and to supply electrons to the GaAs channel. Some samples show evidence for the parallel conduction suggested by such a model, but many do not. Quantisation effects in the $Al_xGa_{1-x}As$, alluded to earlier, or localisation effects as studied by Katsumoto *et al.* (1987), may help to limit the amount of parallel conduction that would otherwise be expected.

6 Transport in heterolayers

The physics of carrier transport in confined structures has a substantial literature, both theoretical and experimental. It can be thought to have its origin in the physics of surface scattering and surface transport, well described in the book by Many *et al.* (1965), and in the physics of thin films. Some aspects of transport in two-dimensional electron systems have been reviewed by Ando *et al.* (1982), and other aspects of transport are described elsewhere in this volume. Here we focus on the low-temperature mobility in two-dimensional electron systems, and touch very briefly on some aspects of carrier heating when large fields drive carriers out of the ohmic regime.

6.1 Coulomb scattering

In the regime where a treatment of transport based on the Boltzmann equation is applicable, the conductance is given by the usual expression, $\sigma = ne\mu$, where n is the carrier density per unit volume, area, or length, depending on the dimensionality, and $\mu = e\tau_t/m$, where m is the effective mass, which is assumed to be isotropic. The physics of the conductance properties is mainly in the scattering time τ_t, given for two-dimensional systems by

$$\tau_t = \int_0^\infty \tau(E)\, E \left(-\frac{\partial f_0}{\partial E}\right) dE \bigg/ \int_0^\infty E \left(-\frac{\partial f_0}{\partial E}\right) dE \tag{18}$$

where $1/\tau(E)$ is the effective rate of scattering, assumed isotropic, of carriers of energy E and f_0 is the equilibrium (Fermi-Dirac) distribution function at temperature T. This expression differs by a factor of $E^{1/2}$ from the corresponding result in three dimensions (see, for example, Smith 1959) because of the difference in density of states.

Note that scattering rates add, but that the thermal averaging in Equation (18) is over the scattering time, the reciprocal of the scattering rate. This leads to the awkward result that Matthiessen's rule, which states that the mobility is the reciprocal of the sum of reciprocal mobilities of individual scattering mechanisms, is generally invalid. For accurate results, particularly when scattering processes with different energy dependences are present, the integration indicated in Equation (18) is required. If the scattering rates have no explicit or implicit temperature dependence, *i.e.* if phonon scattering is negligible and the effects of temperature-dependent screening (Stern 1980) can be ignored, then expanding the integrand of Equation (18) near the Fermi level leads to the result that the temperature dependence of mobility at low temperatures is given by

$$\mu(T) = \mu(0)\left[1 + \frac{\pi^2}{6}p\left(p + \frac{3}{2}\right)\left(\frac{k_B T}{E_F}\right)^2\right], \tag{19}$$

where the scattering time varies with energy as E^p (Stern and Das Sarma 1985). An increasing mobility with increasing temperature was indeed found by Lin *et al.* (1985, see also Lin and Tsui 1986) under conditions in which impurity scattering dominates. At higher temperatures, or in purer samples, the more usual decrease of mobility with increasing temperature is seen.

There are many corrections to the simple Boltzmann approach used here, arising from quantum corrections associated with localisation effects and with carrier-carrier interactions. They lead to rather small logarithmic corrections to conductance at very low temperatures in two-dimensional electron systems (see, for example, Kawaji 1986) but are more important in one-dimensional transport.

In addition to working in the Boltzmann transport regime, we also make some further approximations in describing Coulomb scattering of carriers by ionised impurities. In particular, we use the Born approximation, which assumes that the carrier wave function is perturbed only weakly during the scattering, and we assume that the scattering centres are uncorrelated, so that the phase relations between different scattering centres cancel, leading to a scattering rate given simply by a sum of scattering rates due to individual scatterers, with no interference effects. These approximations may not be valid, and several alternative approaches are discussed below.

The scattering rate $\tau^{-1}(E)$ is given by

$$\tau^{-1}(E) = N_C v \int_{-\pi}^{\pi} \sigma(\theta)\,(1 - \cos\theta)\,d\theta, \tag{20}$$

where N_C is the density of scatterers, v is the velocity of the carriers, and $\sigma(\theta)$ is the cross-section for scattering through an angle θ. The factor $(1 - \cos\theta)$ reflects the fact that small-angle scattering has a reduced effect on the current flow. The scattering cross-section in the Born approximation is proportional to the square of the Fourier component of the scattering potential with wave-vector $S = 2k\sin\theta/2$, where the wave-vector of the carriers is $k = (2mE/\hbar^2)^{1/2}$. See, for example, Stern and Howard (1967) for details and for a discussion of the validity of the Born approximation for silicon inversion layers.

In considering the broadening of the energy in the single-particle Green function, even small-angle events are believed to contribute, and one can define a single-particle relaxation time τ_s using the same expressions as above, but without the factor of $(1 - \cos\theta)$. There is experimental evidence, given for example by Harrang *et al.* (1985) and by Fang *et al.* (1988), that τ_s enters in the damping of Shubnikov–de Haas conductance oscillations. An extensive analysis and interpretation of such experiments has recently been given by Coleridge (1991).

Explicit expressions for Coulomb scattering rates in GaAs-$Al_xGa_{1-x}As$ heterojunctions have been given by Ando (1982b). Stern (1983) combined these results with the constraints imposed by charge-transfer considerations discussed above. He found that if the doping in the $Al_xGa_{1-x}As$ is fixed, and only the thickness of the undoped spacer layer is varied, the low-temperature mobility is expected to reach a maximum as the thickness of the spacer layer increases and then to decline. The initial increase is due to the weakening of the Coulomb scattering as the distance between the electrons and the scatterers increases. Eventually, though, when the spacer layer thickness is sufficiently large, this scattering becomes very weak and the dominant scattering is due to residual ionised impurities in the GaAs. Then the main effect is the increase in scattering rate as the density, and with it the Fermi velocity, decrease with further increase in spacer layer thickness. This is a consequence of the increase in scattering cross-section with decrease in Fermi velocity. While the general trend predicted by this simple approach is consistent with the trend of experiments (Störmer *et al.* 1981, Weimann and Schlapp 1985, Saku *et al.* 1991), a number of extensions to the theory are needed.

At sufficiently high carrier concentrations, carriers will occupy several subbands and the simple treatment outlined here must be amended to include the effect of screening by carriers in all the occupied subbands, as well as the scattering between subbands. The screening effect is now nontrivial even at low temperatures because the Fermi wave vector in each subband has a different value. Mori and Ando (1979) showed that the scattering effect can dominate, leading to a reduction in the mobility of electrons in a silicon inversion layer when the second subband is occupied. Corresponding effects in a GaAs-$Al_xGa_{1-x}As$ heterojunction have been considered, both experimentally and theoretically, in a detailed paper by Fletcher *et al.* (1990).

Coulomb scattering in wires can be discussed using the formalism described here, with suitable modification for the change in geometry and for the change in screening. Sakaki's conjecture (1980) that the restriction of momentum-changing processes in wires to only one angle, *i.e.* backscattering, would lead to an increase in mobility in wires, aroused considerable interest. Recently Nixon *et al.* (1991) have shown that a more detailed treatment of the scattering process can lead to results different from those of a conventional Boltzmann-like approach when applied to constrictions of varying lengths in the presence of potential fluctuations associated with random impurities. Indeed, a consequence of the geometry of a quantum wire, in which there is no way to bypass a large fluctuation, is that the resistance of a sufficiently long wire is expected to diverge (see, for example, Erdös and Herndon, 1982). Some aspects of conductance in quantum wires and quantum constrictions are discussed elsewhere in this volume.

Three important physical processes complicate the simple considerations we have described here. One is the persistent photoconductance associated with deep donors in $Al_xGa_{1-x}As$. This effect changes the occupation of states in the $Al_xGa_{1-x}As$ and can give rise to additional screening from free or lightly bound carriers there, supplementing the screening by carriers in the GaAs channel that is included in the theory described here.

An even more serious complication follows if the deep donor in $Al_xGa_{1-x}As$ is a negative-U centre. The most important consequence for mobility in a GaAs-$Al_xGa_{1-x}As$ heterojunction is that doped regions which are neutral will be expected to have equal numbers of positive and negative sites. This will lead to a considerable increase in the scattering, and a decrease in the mobility, compared to the result obtained with a conventional donor model unless the doping is carefully adjusted to have only the value needed to compensate the surface Schottky barrier and to provide carriers to the the GaAs, with very few extra donors. However, Saku *et al.* (1991) do not find it necessary to limit the donor concentration in this way to achieve high mobilities.

The third important correction to the simple treatment of Coulomb scattering we have outlined comes from correlation effects, in which the occupation of charged sites is not random. Some correlations can come from deviations of the locations of donor impurities from random sites, but we ignore such effects here. The dominant effect is the possibility that when there are choices available for the charging of sites, the charges will prefer arrangements with the lowest energy. Such arrangements generally have weaker potential fluctuations, and will lead to a higher mobility than would be expected from a random arrangement. These ideas are well known in bulk systems (see, for example, Mycielski 1986). They have been invoked in a phenomenological way by a number of authors (Lassnig 1988, van Hall 1989). More physical treatments have been given by Pikus and Efros (1989) and Efros *et al.* (1990), and show that very significant increases in mobility can be obtained. Here too there is need for a systematic comparison with experiment.

An intuitive treatment of the departure from a simple Born-approximation treatment of scattering in two-dimensional electron systems that assumes that only one scatterer is active in a given scattering event has been given by Ridley (1988).

At very low electron densities, quantum corrections to transport lead to strong localisation and hopping conduction at low carrier densities. Results for silicon inversion layers are given by Ando *et al.* (1982). More recently, such effects have been reported in GaAs-$Al_xGa_{1-x}As$ heterojunctions (Jiang *et al.* 1988), and theoretical treatments have been given by Efros (1989) and by Gold (1989b).

6.2 Interface roughness scattering

Carriers in most nanostructures are confined near one or more interfaces, which can lead to additional scattering if the interfaces are not ideal. The model of interface roughness scattering originally developed in the context of magnetically induced states at metal surfaces (Prange and Nee 1968) and extended to silicon inversion layers (Matsumoto and Uemura 1974) assumes that the interface is a sharp boundary whose deviations from planarity are characterised by an autocorrelation function with a presumed form

$$\langle \delta z(\mathbf{r}')\, \delta z(\mathbf{r}+\mathbf{r}') \rangle = \Delta^2 \exp\left[-\left(\frac{r}{\Lambda}\right)^2\right], \tag{21}$$

where Δ and Λ characterise the r.m.s. height of the roughness and its lateral correlation length, respectively.

The scattering rate due to roughness scattering in the Born approximation is proportional to the Fourier component of the roughness autocorrelation function corresponding to the change of wave-vector in the scattering event. Explicit expressions have been given in the two papers just cited and in Ando *et al.* (1982).

A number of authors (Sugano *et al.* 1980, Hahn and Henzler 1984, Goodnick *et al.* 1985) have related experimental measurements of interface roughness to the mobility in silicon inversion layers. Some measurements of roughness favor an exponential correlation, rather than the Gaussian form given in Equation (21). A detailed comparison of measured low-temperature mobility of silicon inversion layers with theory has recently been given by Kruithof *et al.* (1991).

Heterojunctions between GaAs and $Al_xGa_{1-x}As$ usually have rather smooth interfaces, although interfaces in which GaAs is grown on $Al_xGa_{1-x}As$, called inverted interfaces, can be rougher than interfaces in which $Al_xGa_{1-x}As$ is grown on GaAs. The effect of interface roughness in GaAs-$Al_xGa_{1-x}As$ heterojunctions was discussed by Ando (1982b). A GaAs quantum well is bounded by both a normal and an inverted interface, and roughness scattering becomes important for wells less than about 10 nm thick (Sakaki *et al.* 1987, Gold 1989a). Similar effects in HgTe-CdTe superlattices have been studied by Meyer *et al.* (1991). Interface roughness also influences the linewidth of exciton luminescence in quantum wells, but this aspect will not be discussed here.

Roughness effects in quantum wires in a magnetic field have been considered by Akera and Ando (1991). A treatment in the context of weak localisation was given by van Houten *et al.* (1988).

6.3 Phonon scattering

Phonon scattering is usually the dominant scattering process at room temperature, and is an important scattering mechanism at all but the lowest temperatures. Phonon scattering is one of the inelastic processes that destroy the coherence of carriers, and therefore may determine the phase-breaking length discussed in Section 1. Phonons also enter in energy loss processes of hot electrons, discussed very briefly in the following subsection.

The temperature dependence of phonon scattering derives mainly from the occupation of phonon modes, determined by the lattice temperature if the phonons are in equilibrium, or by the dynamics of phonon generation and decay processes in some systems far from equilibrium. In the Bloch-Grüneisen regime, usually at very low temperatures, energy and wave-vector conservation conditions lead to additional constraints that change the temperature dependence (see, for example, Price 1984, and Störmer *et al.* 1990).

For details regarding phonon scattering processes in heterolayers, see, for example, Price (1981). A detailed experimental study of electron mobility in GaAs-$Al_xGa_{1-x}As$ heterojunctions, including phonon scattering effects, has been given by Hirakawa and Sakaki (1986a). The theory of phonon scattering in quantum wires has been treated by a number of authors, including Riddoch and Ridley (1984) and Fishman (1987).

6.4 Hot-carrier effects

Most experiments and calculations on low-dimensional systems involve low-temperature ambients, so 'hot' must be considered a relative term. Hot-carrier effects can arise in many ways, but the most common ones are optical excitation at photon energies well above the bandgap and transport in devices under non-ohmic conditions. We consider the latter case here, but only to give a few entry points to the literature.

A detailed experimental study of electron heating in GaAs heterojunctions was carried out by Hirakawa and Sakaki (1988), and a theoretical investigation that included quantisation effects was carried out by Yokoyama and Hess (1986). An early experimental study of hot carrier effects in silicon inversion layers is the work of Fang and Fowler (1970).

Hot-carrier effects control the performance of field-effect transistors and other semiconductor devices, and have been extensively studied (see, for example, Jacoboni and Lugli 1989). If carriers gain enough energy to escape over barriers, such as the 3 eV barrier between the conduction bands of Si and SiO_2 or the conduction band offset in a GaAs-$Al_xGa_{1-x}As$ quantum well, they can affect device characteristics significantly, by changing the terminal characteristics as in real space transfer (see, for example, Hess 1988), or by introducing traps or other defects in the oxide, which leads to long-term degradation.

Hot-carrier effects can be used as a means to determine the strength of inelastic processes. For example, Hirakawa and Sakaki (1986b) used measurements of carrier heating to infer the strength of the deformation potential for scattering of electrons in GaAs by acoustic phonons.

7 Modelling of confined electrons

The validity of the effective mass approximation has been discussed briefly above. Here we proceed, within the framework of this approximation, to discuss calculations for envelope wave functions and energy levels. We begin with a dynamically two-dimensional system, for which the equations to be solved involve only one space dimension. The potential that enters in the Schrödinger equation for the electron envelope function can be written as a sum of several terms:

$$V(z) = -e\phi(z) + V_{\rm cbo}(z) + V_{\rm im}(z) + V_{\rm xc}(z), \tag{22}$$

where $\phi(z)$ is the electrostatic potential obtained from the solution of the Poisson equation, $V_{\rm cbo}(z)$ is the position-dependent effective conduction band edge that contains any conduction band offsets, $V_{\rm im}(z)$ is the image potential, and $V_{\rm xc}(z)$ is the exchange-correlation potential. The image potential represents the effect of the dielectric discontinuity on a point charge and is given for a carrier of charge Ze located a distance d from an abrupt interface by

$$V_{\rm im}(z) = \frac{\epsilon_1 - \epsilon_2}{\epsilon_1 + \epsilon_2} \frac{Z^2 e^2}{16\pi\epsilon_1 d}, \tag{23}$$

where ϵ_1 and ϵ_2 are the permittivities of the material in which the charge is located, and of the material on the far side of the interface, respectively. Note that the image potential is always attractive if the carrier is on the low-permittivity side of the interface, and repulsive if it is on the high-permittivity side. The effect of the image potential is small but not negligible in silicon inversion layers but is less important—and can usually be neglected—in GaAs-$\rm Al_xGa_{1-x}As$ heterostructures. Many-body effects cannot be rigorously represented as a contribution to the effective potential in a one-electron Schrödinger equation, but such an approximation, for example an effective potential related to the local density, is often made. Additional details about image potentials and many-body effects for two-dimensional electron systems are given in Ando *et al.* (1982). Applications to GaAs-$\rm Al_xGa_{1-x}As$ heterojunctions are given for example, in Ando (1982a) and in Stern and Das Sarma (1984), where self-consistent solutions of the Poisson and Schrödinger equations are described.

Variational approximations are often the simplest way to obtain approximate solutions, especially for the ground state, and their use in two-dimensional systems goes back to the variational function, Equation (13), used by Fang and Howard (1966). The Fang-Howard trial function is not a very accurate representation of the self-consistent solution, but alternative functions are somewhat less tractable analytically. As full numerical solutions become more and more widely accessible, especially for two-dimensional electron systems where the equations only involve one space dimension, it is often less necessary to rely on variational functions. Additional details and examples can be found, for example, in Stern (1972), Ando *et al.* (1982), and Grinberg (1985).

Many workers have obtained self-consistent solutions of the Poisson and Schrödinger equations for two-dimensional electron systems. These can generally can be managed with relatively modest computational resources, but the situation becomes more difficult when carriers are confined in more than one spatial dimension. In some cases the potential is known, as for example in a structure that uses geometrical or compositional confinement. In other cases, particularly those involving electrostatic confinement, determination of the potential may require analytical or numerical methods. An example of a numerical calculation is given below. Analytical methods for describing a quantum wire constrained by lateral diffusions were presented by Shik (1985). A detailed analysis of quantum wires near threshold has been given by Davies (1988) and an approximate treatment above threshold has been given by Shikin (1989).

One complication in models for confined structures is the treatment of Fermi level pinning at free surfaces, like those that occur when a heterostructure is etched to form a mesa-like structure with exposed sides. This can lead to a depletion of charge near the edge of the structure, as discussed for example by Davies (1988) and Luscombe and Luban (1990).

The Schrödinger equation for carriers in a wire must be solved in two space dimensions, with free-carrier motion assumed in the third direction. Exact solutions can be obtained for only a few cases, some of which have already been mentioned. To make further progress, it can be useful to formulate the problem as a set of coupled one-dimensional equations, as done for example by Brum and Bastard (1988) in treating a rectangular GaAs wire buried in $Al_xGa_{1-x}As$. If the normalised z-dependent solutions $\chi_n(z)$ for an infinitely wide strip and the corresponding eigenvalues E_n are known, then the wave functions in the wire can be expanded in terms of a complete set of solutions:

$$\zeta^{(p)} = L^{1/2}\, e^{ik_x x} \sum_n \alpha_n^{(p)}(y)\, \chi_n(z). \tag{24}$$

After the free-electron behaviour in the x direction is separated out, the Schrödinger equation becomes a set of coupled equations over the index m (which could, in principle, include a continuum),

$$-\frac{\hbar^2}{2m}\frac{\partial^2}{\partial y^2}\alpha_m^{(p)}(y) + \sum_n \Delta V_{nm}(y)\, \alpha_m^{(p)}(y) = \left[E^{(p)} - E_m\right] \alpha_m^{(p)}(y), \tag{25}$$

where the spatial variation of the effective mass has been dropped for simplicity. The matrix element which enters in Equation (25) is

$$\Delta V_{nm}(y) = \int \chi_m(z)\, \Delta V(y,z)\, \chi_n(z)\, dz, \tag{26}$$

where $\Delta V(y,z)$ is the perturbing potential, *i.e.* the difference between the actual potential and the potential that appears in the Schrödinger equation for the $\chi_n(z)$. If the lateral dimension of the wire is larger than the thickness of the layer whose eigenstates are used for the expansion above, only a few coupled one-dimensional equations are needed to get accurate solutions. The case considered by Brum and Bastard (1988), in which carriers were supplied from a plane of donors outside the quantum wire, also required a self-consistent treatment to determine the charge in the wire and the potential from the doping density and the geometry of the structure. A multiband envelope function expansion method based on a Fourier-series expansion, applicable to one-, two-, or three-dimensional confinement, has recently been described by Baraff and Gershoni (1991).

Numerical self-consistent calculations for electronic states in quantum wires have been carried out by Laux and Stern (1986) for electrons in silicon and by Laux *et al.* (1988) and Kojima *et al.* (1989) for electrons in GaAs. Numerical aspects of such calculations are discussed, for example, by Laux (1987) and by Kerkhoven *et al.* (1990). Numerical simulations for quantum wires in GaAs show structure in qualitative agreement with capacitance studies by Smith *et al.* (1987).

Most calculations for electron states in wires have used the effective mass approximation and the Hartree approximation, in which each electron moves in the average potential of all electrons. That approximation ignores exchange and correlation effects, which lead to substantial changes in the energy levels in two-dimensional electron systems. Corresponding, and perhaps even larger, changes are expected in wires and dots, although Laux *et al.* (1988) found that a local density treatment of exchange and correlation effects led to relatively small changes in the energy spacings connected with lateral quantisation for quantum wires in GaAs with effective width of order 100 nm.

The effects of electron-electron interaction in a quantum box with two electrons have been investigated by Bryant (1987). For a box 100 nm on a side, but with vanishing thickness, he found that the interaction effects are comparable to one-electron energies. The relative importance of interactions decreases as the size of the box decreases. Maksym and Chakraborty (1990) investigated many-electron interactions for quantum dots with three and four electrons.

There have been a number of calculations for electronic states in quantum dots, generally using model potentials, some of which are cited elsewhere in this chapter. Here we want to mention the self-consistent Hartree calculations by Kumar *et al.* (1990), which were motivated by the capacitance experiments of Smith *et al.* (1988) and Hansen *et al.* (1989). They used GaAs-$Al_xGa_{1-x}As$ heterostructure samples, patterned the GaAs cap layer using electron beam lithography to form a square array of mesas, and deposited a metallic gate over the entire array. A negative voltage on the gate depletes the charges from donors in the $Al_xGa_{1-x}As$ except under the middle of the mesas, where there are isolated groups of electrons which constitute the quantum dots. The capacitance and more particularly its derivative with respect to gate voltage show oscillatory structure associated with the small size of the dots—the period of the structure increases as the dot size decreases. A similar structure has recently been studied by Ashoori *et al.* (1991). Kumar *et al.* (1990) have carried out numerical self-consistent solutions of the Poisson and Schrödinger equations (in the Hartree approximation) for this structure both with and without a magnetic field. The lateral potential variation is approximately parabolic when there are no electrons in the dot and flattens out as electrons are introduced, but not as much as for the wires mentioned above.

Most of the discussion in the present chapter deals with conduction band states. For states in the valence band a more elaborate treatment is usually required because of the degeneracy at the top of the valence band in most of the semiconductors of interest, with its accompanying nonparabolicity and anisotropy. See the chapter by Bastard in this volume.

8 Parabolic potentials

The special properties of parabolic potentials, which result in optical inter-subband transition energies that are unaffected by the presence of carriers, are described elsewhere in this volume. It is interesting to note that parabolic potentials, while they can be approximated by molecular engineering as done, for example, by Miller *et al.* (1984), arise naturally through the workings of the Poisson or Laplace equations, as described in the previous section. Among the first people to find the special properties of parabolic potentials were Ruden and Döhler (1983), in the context of doping superlattices.

The energy levels and optical properties of quantum wires and quantum dots in a magnetic field are discussed elsewhere and will not be repeated here. A graphic description of the current flow associated with the states of a quantum dot in a magnetic field has been given by Lent (1991). The paper by Darwin (1931), which gives a detailed discussion of the states of a quantum dot in a magnetic field (the basic results had been found by Fock, 1928), was concerned with the diamagnetism of three-dimensional electrons. It is tempting to think, though, that he had some premonition of the nanostructure physics to come.

Acknowledgments

I am indebted to the organisers of the 38th Scottish Universities Summer School in Physics for the opportunity to participate. Steven Laux and Arvind Kumar have created the programs for the two- and three-dimensional modeling work we have done together, and I have benefited greatly from our collaboration. Thomas Theis, Patricia Mooney, and Thomas Morgan have helped to educate me about the properties of deep donors in $Al_xGa_{1-x}As$. I am indebted to many colleagues, including Ray Ashoori, Gerald Bastard, Jose Brum, Peter Coleridge, John Davies, Alan Fowler, Craig Lent, Don Monroe, Wolfgang Porod, and Uri Sivan for preprints and for valuable discussions.

References

Akera H, and Ando T, 1991. *Phys. Rev. B*, **43**, 11 676.
Ando T, 1982a. *J. Phys. Soc. Japan*, **51**, 3893.
Ando T, 1982b. *J. Phys. Soc. Japan*, **51**, 3900.
Ando T, Fowler A B and Stern F, 1982. *Rev. Mod. Phys.*, **54**, 437.
Ashoori R C, Silsbee R H, Pfeiffer L N, and West K W, 1991. In *Proc. Int. Conf. on Nanostructures and Mesoscopic Systems*, Eds. Reed M A, and Kirk W P, Boston, Academic Press, pp. 323–334.
Baraff G A, and Gershoni D, 1991. *Phys. Rev. B*, **43**, 4011.
BenDaniel D J, and Duke C B, 1966. *Phys. Rev. B*, **152**, 683.
Brum J A, and Bastard G, 1988. *Superlatt. and Microstruct.*, **4**, 443.
Bryant G W, 1984. *Phys. Rev. B*, **29**, 6632.
Bryant G W, 1987. *Phys. Rev. Lett.*, **59**, 1140.
Chadi D J, and Chang K J, 1989. *Phys. Rev. B*, **39**, 10063.
Cheng D, Ulloa S E, and Mochan W L, 1991. *Bull. Am. Phys. Soc.*, **36**, 824.
Coleridge P T, 1991. *Phys. Rev. B*, **44**, 3793.
Dahl D A, and Sham L J, 1977. *Phys. Rev. B*, **16**, 651.
Darwin C G, 1931. *Proc. Cambridge Phil. Soc.*, **27**, 86.
Das A K, 1974. *Solid State Commun.*, **15**, 475.
Davies J H, 1988. *Semicond. Sci. Technol.*, **3**, 995.
Efros A L, 1989. *Solid State Commun.*, **70**, 253.
Efros A L, Pikus F G, and Samsonidze G G, 1990. *Phys. Rev. B*, **41**, 8295.
Erdös P, and Herndon R C, 1982. *Adv. Phys.*, **31**, 65.
Fang F F, and Fowler A B, 1970. *J. Appl. Phys.*, **41**, 1825.
Fang F F, and Howard W E, 1966. *Phys. Rev. Lett.*, **16**, 797.
Fang F F, Smith T P, and Wright S L, 1988. *Surf. Sci.*, **196**, 310.
Fetter A L, 1973. *Ann. Phys. (NY)*, **81**, 367.
Fetter A L, and Walecka J D, 1971. *Quantum Theory of Many-Particle Systems*, McGraw Hill.
Fishman G, 1987. *Phys. Rev. B*, **36**, 7448.
Fletcher R, Zaremba E, D'Iorio M, Foxon C T, and Harris J J, 1990. *Phys. Rev. B*, **41**, 10649.
Fock V, 1928. *Z Phys.*, **47**, 446.

Friesen W I, and Bergersen B, 1980. *J. Phys. C: Solid State Phys.*, **13**, 6627.
Gell M A, Ninno D, Jaros M, Wolford D J, Kuech T F, and Bradley J A, 1987. *Phys. Rev. B*, **35**, 1196.
Gold A, 1989a. *Z Phys. B*, **74**, 53.
Gold A, 1989b. *Appl. Phys. Lett.*, **54**, 2100.
Gold A, and Ghazali A, 1990. *Phys. Rev. B*, **41**, 7626.
Goodnick S M, Ferry D K, Wilmsen C W, Liliental Z, Fathy D, and Krivanek O L, 1985. *Phys. Rev. B*, **32**, 8171.
Grinberg A A, 1985. *Phys. Rev. B*, **32**, 4028.
Hahn P O, and Henzler M, 1984. *J. Vac. Sci. Technol. A*, **2**, 574.
Hansen W, Smith T P, Lee K Y, Brum J A, Knoedler C M, Hong J M, and Kern D P, 1989. *Phys. Rev. Lett.*, **62**, 2168.
Harrang J P, Higgins R J, Goodall R K, Jay P R, Laviron M, and Delescluse P, 1985. *Phys. Rev. B*, **32**, 8126.
Hess K, 1988. *Solid-State Electron.*, **31**, 319.
Hirakawa K, and Sakaki H, 1986a. *Phys. Rev. B*, **33**, 8291.
Hirakawa K, and Sakaki H, 1986b. *Appl. Phys. Lett.*, **49**, 889.
Hirakawa K, and Sakaki H, 1988. *J. Appl. Phys.*, **63**, 803.
Jacoboni C, and Lugli P, 1989. *The Monte-Carlo Method for Semiconductor Device Simulation*, Springer.
Jiang C, Tsui D C, and Weimann G, 1988. *Appl. Phys. Lett.*, **54**, 2100.
Katsumoto S, Komori F, Naokatsu S, and Kobayashi S, (1987). *J. Phys. Soc. Japan*, **56**, 2259.
Kawaji S, 1986. *Surf. Sci.*, **170**, 682.
Keldysh L V, 1979. *Zh. Eksp. Teor. Fiz. Pis'ma Red.*, **29**, 716 [*JETP Lett.*, **29**, 658].
Kerkhoven T, Galick A T, Ravaioli U, Arends J H, and Saad Y, 1990. *J. Appl. Phys.*, **68**, 3461.
Kojima K, Mitsunaga K, and Kyuma K, 1989. *Appl. Phys. Lett.*, **55**, 862.
Kruithof G H, Klapwijk T M, and Bakker S, 1991. *Phys. Rev. B*, **43**, 6642.
Kumar A, Laux S E, and Stern F, 1990. *Phys. Rev. B*, **42**, 5166.
Lang D V, Logan R A, and Jaros M, 1979. *Phys. Rev. B*, **19**, 1015.
Lassnig R, 1988. *Solid State Commun.*, **65**, 765.
Laux S E, 1987. In *Proc. Fifth Int. Conf. on Numerical Analysis of Semiconductor Devices and Integrated Circuits (NASECODE V)*, Ed. Miller J J H, Boole Press, Dublin, pp. 270–275.
Laux S E, and Stern F, 1986. *Appl. Phys. Lett.*, **49**, 91.
Laux S E, Frank D J, and Stern F, 1988. *Surf. Sci.*, **196**, 101.
Lee J, and Spector H N, 1985. *J. Appl. Phys.*, **57**, 366.
Lent C S, 1991. *Phys. Rev. B*, **43**, 4179.
Lin B J F, and Tsui D C, 1986. *Surf. Sci.*, **174**, 397.
Lin B J F, Tsui D C, and Weimann G, 1986. *Solid State Commun.*, **56**, 287.
Luscombe J H, and Luban M, 1990. *Appl. Phys. Lett.*, **57**, 61.
Maksym P A, and Chakraborty T, 1990. *Phys. Rev. Lett.*, **65**, 108.
Many A, Goldstein Y, and Grover N B, 1965. *Semiconductor Surfaces*, North-Holland.
Matsumoto Y, and Uemura Y, 1974. *Jpn. J. Appl. Phys.*, Suppl. 2, Pt. 2, 367.
Meyer J R, Arnold D J, Hoffman C A, and Bartoli F J, 1991. *Appl. Phys. Lett.*, **58**, 2523.
Miller R C, Kleinman D A, and Gossard A C, 1984. *Phys. Rev. B*, **29**, 7085.
Mooney P M, 1990. *J. Appl. Phys.*, **67**, R1.
Mori S, and Ando T, 1979. *Phys. Rev. B*, **19**, 6433.
Mycielski J, 1986. *Solid State Commun.*, **60**, 165.

Nixon J A, Davies J H, and Baranger H U, 1991. *Phys. Rev. B*, **43**, 12638.
Osório F A P, Degani M H, and Hipólito O, 1988. *Phys. Rev. B*, **37**, 1402.
Pikus F G, and Efros A L, 1989. *Zh. Eksp. Teor. Fiz.*, **96**, 985 [*Sov. Phys. JETP*, **69**, 558].
Prange R E, and Nee T-W, 1968. *Phys. Rev.*, **168**, 779.
Price P J, 1981. *Ann. Phys. (NY)*, **133**, 217.
Price P J, 1984. *Solid State Commun.*, **51**, 607.
Riddoch F A, and Ridley B K, 1984. *Surf. Sci.*, **142**, 260.
Ridley B K, 1988. *Semicond. Sci. Technol.*, **3**, 111.
Rimberg A J, and Westervelt R M, 1989. *Phys. Rev. B*, **40**, 3970.
Ruden P, and Döhler G H, 1983. *Phys. Rev. B*, **27**, 3538 and 3547.
Sakaki H, 1980. *Jpn. J. Appl. Phys.*, **19**, L735.
Sakaki H, Noda T, Hirakawa K, Tanaka M, and Matsusue T, 1987. *Appl. Phys. Lett.*, **51**, 1934.
Saku T, Hirayama Y, and Horikoshi Y, 1991. *Jpn. J. Appl. Phys.*, **30**, 902.
Shik A Ya., 1985. *Fiz. Tekh. Poluprovodn.*, **19**, 1488 [*Sov. Phys. Semicond.*, **19**, 915].
Shikin V B, 1989. *Pis'ma Zh. Eksp. Teor. Fiz.*, **50**, 150 [*JETP Lett.*, **50**, 167].
Smith R A, 1959. *Semiconductors*, Cambridge University Press.
Smith T P, Arnot H, Hong J M, Knoedler C M, Laux S E, and Schmid H, 1987. *Phys. Rev. Lett.*, **59**, 2802.
Smith T P, Lee K Y, Knoedler C M, Hong J M, and Kern D P, 1988. *Phys. Rev. B*, **38**, 2172.
Stern F, 1967. *Phys. Rev. Lett.*, **18**, 546.
Stern F, 1972. *Phys. Rev. B*, **5**, 4891.
Stern F, 1980. *Phys. Rev. Lett.*, **44**, 1469.
Stern F, 1983. *Appl. Phys. Lett.*, **43**, 974.
Stern F, 1986. *Surf. Sci.*, **174**, 425.
Stern F, 1992. In *Physics of low-dimensional semiconductor structures*, Eds. Butcher P, March N, and Tosi M, Plenum Press.
Stern F, and Das Sarma S, 1984. *Phys. Rev. B*, **30**, 840.
Stern F, and Das Sarma S, 1985. *Solid-State Electron.*, **28**, 211.
Stern F, and Howard W E, 1967. *Phys. Rev.*, **163**, 816.
Störmer H L, Gossard A C, Wiegmann W, and Baldwin K, 1981. *Appl. Phys. Lett.*, **39**, 912.
Störmer H, Pfeiffer L N, Baldwin K W, and West K W, 1990. *Phys. Rev. B*, **41**, 1278.
Sugano T, Chen J J, and Hamano T, 1980. *Surf. Sci.*, **98**, 154.
Theis T N, Mooney P M, and Parker B D, 1991. *J. Electron. Mater.*, **20**, 35.
van Hall P J, 1989. *Superlatt. and Microstruct.*, **6**, 213.
van Houten H, Beenakker C W J, van Wees B J, and Mooij J E, 1988. *Surf. Sci.*, **196**, 144.
Vinter B, 1980. *Surf. Sci.*, **98**, 197.
Vinter B, 1983. *Solid State Commun.*, **48**, 151.
Weimann G, and Schlapp W, 1985. *Appl. Phys. Lett.*, **46**, 411.
Yoffa E J, and Adler D, 1975. *Phys. Rev. B*, **12**, 2260.
Yokoyama K, and Hess K, 1986. *Phys. Rev. B*, **33**, 5595.
Zhu J-L, Xiong J-J, and Gu B-L, 1990. *Phys. Rev. B*, **41**, 6001.

Hole Eigenstates in Semiconductor Heterostructures

G Bastard

Ecole Normale Supérieure
Paris, France

1 Introduction

We present a brief discussion of the eigenstates of holes in semiconductor heterostructures and focus our attention on their influence on the optical polarisation effects in quantum wells and wires.

The electronic and optical properties of semiconductor heterostructures have been extensively discussed in several reviews (Weisbuch and Vinter 1991, Allan *et al.* 1986, Beenakker and van Houten 1991, Bastard 1988, Bastard *et al.* 1991). Most of the material presented during the lectures can be found in Bastard (1988) and Bastard *et al.* (1991). In these notes we shall primarily deal with the peculiarities of the hole eigenstates which have a significant influence on the optical properties and which were not covered in these two works: the polarisation properties of the optical transitions in quantum wells and wires. In all what follows we adopt the envelope description of the heterostructure eigenstates; electronic energy levels are discussed by Stern elsewhere in this volume. For our purpose we shall assume that they can written as

$$F_{c\mathbf{k}\sigma} = |iS\sigma\rangle\,\Psi_{n\mathbf{k}}, \tag{1}$$

where $\sigma = \uparrow$ or $\downarrow$, S is the Γ_6 periodic part of the bulk Bloch function and $\Psi_{n\mathbf{k}}$ is an envelope function. For a single quantum well this can be written as

$$\Psi_{n\mathbf{k}}(\mathbf{r}) = A^{-1/2}\exp\left(\mathrm{i}\mathbf{k}\cdot\boldsymbol{\rho}\right)\chi_n(z), \tag{2}$$

where $\boldsymbol{\rho} = (x, y)$, $\mathbf{k}$ is the in-plane wave-vector of the electron and χ_n is the envelope probability amplitude along the growth axis. For a bound state χ_n decays exponentially in the barriers. For a symmetrical quantum well χ_n has a definite parity: even for the ground state, odd for the first excited state, and so on. The electron energy associated with $F_{c\mathbf{k}\sigma}$ is independent of σ and will be written as

$$\varepsilon_{n\mathbf{k}} = E_n + \frac{\hbar^2\mathbf{k}^2}{2m^*} + \varepsilon_g . \tag{3}$$

Explicit calculations of E_n and χ_n are possible for rectangular quantum wells (see, for example, Bastard 1988 and Bastard *et al.* 1991).

2 Valence bands in low-dimensional systems

The four-fold degeneracy of the extremum of the valence band in the bulk (Γ_8 point) is lifted by one-dimensional confinement in quantum wells and bi dimensional quantum wires. In the envelope function framework, the valence eigenstates are the solutions of

$$\left[H_v(\mathbf{k} = -i\boldsymbol{\nabla}, p_z) + V_{\rm conf}(\mathbf{r})\right]\Psi = \varepsilon\Psi, \tag{4}$$

where the envelope function Ψ is a 4×1 spinor and H_v is the 4×4 Luttinger matrix. The potential energy term $V_{\rm conf}(\mathbf{r})$ is proportional to the identity matrix in the Γ_8 basis. The complete $\mathbf{k}$-dependent valence wavefunction is

$$F_v = \sum_{m_J} \left|\tfrac{3}{2}, m_J\right\rangle \Psi_{m_J} , \tag{5}$$

where the total angular momentum $J = \frac{3}{2}$ and its z-component lies in the range $-\frac{3}{2} \leq m_J \leq \frac{3}{2}$. Let us consider again a single quantum well. The in-plane translational invariance implies that $\mathbf{k} = (k_x, k_y)$ is a good quantum number and Ψ factorises like

$$\Psi(\mathbf{r}) = A^{-1/2} \exp\left(\mathrm{i}\mathbf{k}\cdot\boldsymbol{\rho}\right)\Psi(z). \tag{6}$$

We assume in the following that growth is along the [001] axis and quantise $\mathbf{J}$ along this axis. Then the Hamiltonian becomes

$$H_v = \begin{matrix} \langle +\frac{3}{2}| \\ \langle -\frac{1}{2}| \\ \langle +\frac{1}{2}| \\ \langle -\frac{3}{2}| \end{matrix} \begin{pmatrix} H_{hh} & c & b & 0 \\ c^* & H_{lh} & 0 & b \\ b^* & 0 & H_{lh} & -c \\ 0 & b^* & -c^* & H_{hh} \end{pmatrix} . \tag{7}$$

The bras $\langle \pm\frac{3}{2}|$ and $\langle \pm\frac{1}{2}|$ have been inserted to indicate the basis, and the components of H_v are

$$H_{hh} = -\frac{\hbar^2}{2m_0}\frac{d}{dz}(\lambda_1 - 2\lambda_2)\frac{d}{dz} - \frac{\hbar^2 k^2}{2m_0}(\lambda_1 + \lambda_2) \tag{8}$$

$$H_{lh} = -\frac{\hbar^2}{2m_0}\frac{d}{dz}(\lambda_1 + 2\lambda_2)\frac{d}{dz} - \frac{\hbar^2 k^2}{2m_0}(\lambda_1 - \lambda_2) \tag{9}$$

$$b = -i\sqrt{3}\frac{\hbar^2}{2m_0}(k_x - ik_y)\left(\lambda_3\frac{d}{dz} + \frac{d}{dz}\lambda_3\right) \tag{10}$$

$$c = \sqrt{3}\frac{\hbar^2 k^2}{2m_0}\lambda_2\left(k_x^2 - k_y^2\right). \tag{11}$$

Here λ_1, λ_2 and λ_3 are Luttinger parameters which, in general, depend on z; non-commutating products have therefore been anti-symmetrised in Equations (8)–(11). For a quantum wire grown along the [010] direction by etching and/or patterning the [001] quantum well, k_y would still be a good quantum number but $\hbar k_x$ would have to be replaced by p_x in Equations (8)–(11) and suitable extra anti-commutators be introduced at any place where p_x appears near an x-dependent operator (as, for example, in Equation 11).

For a symmetrical quantum well, one where the confining barriers are of identical materials, the parity operator with respect to the centre of the well commutes with the Hamiltonian. Thus each of the components $\Psi_{m_J}(z)$ of the spinor Ψ should have a definite parity. Suppose we take the $\frac{3}{2}$-component as being even in z. Then, due to the fact that b is odd in z and c is even in z, the parity of the three other components of the spinor follows:

$$\Psi_+(z) = (A_e, B_e, C_o, D_o)\,; \tag{12}$$

the subscripts e and o stand for even and odd in z respectively. The functions A, B, C and D have to be calculated by explicit solution of Equation (4). Instead of taking the $\frac{3}{2}$-component as being even, one could instead have chosen it odd in z. This would generate another spinor which has the opposite parity to the that of the Ψ_+ state. A particular state of such an opposite parity is interesting. It is obtained by taking the $-\frac{3}{2}$ component as proportional to A_e^*, and is thus written

$$\Psi_-(z) = (-D_o^*, C_o^*, -B_e^*, A_e^*)\,. \tag{13}$$

This state is degenerate in energy with Ψ_+ (Uenoyama and Sham 1990, Ferreira and Bastard 1991). Both Ψ_+ and Ψ_- correspond to the *same* $\mathbf{k}$. It is tempting to identify Ψ_+ and Ψ_- with a 'spin' degeneracy (Ferreira and Bastard 1991). We put quotation marks to make it clear that of course neither Ψ_+ nor Ψ_- are eigenstates of σ_z or J_z. What is however true is

$$\begin{aligned}\langle\Psi_+|J_z|\Psi_+\rangle &= -\langle\Psi_-|J_z|\Psi_-\rangle;\\ \langle\Psi_+|J_z|\Psi_-\rangle &= 0.\end{aligned} \tag{14}$$

For the ground heavy hole state $\langle\Psi_\pm|J_z|\Psi_\pm\rangle$ converges towards $\pm\frac{3}{2}$ and the components B_e, C_o and D_o go to zero in this limit. Actually, the 'spin' quantum number $\pm$ can be identified (apart from a constant that depends on k) with the eigenvalue of the restriction of J_z to the single subband n:

$$J_z \rightarrow |\Psi_{n\mathbf{k}\eta}\rangle\langle\Psi_{n\mathbf{k}\eta}|J_z|\Psi_{n\mathbf{k}\eta}\rangle\langle\Psi_{n\mathbf{k}\eta}|, \quad \eta,\eta' = \pm.$$

Equation (14) shows that Ψ is an eigenstate of in this restricted operator. Note that matrix elements of the form $\langle\Psi_{n\mathbf{k}+}|J_z|\Psi_{m\mathbf{k}+}\rangle$ with $n \neq m$ do not vanish, otherwise J_z and H_v would truly commute. Note also that any matrix elements of the form $\langle\Psi_{n\mathbf{k}\eta}|J_z|\Psi_{m\mathbf{k}'\eta'}\rangle$ vanish by virtue of the orthogonality between two plane waves with different in-plane wave-vectors $\mathbf{k}$ and $\mathbf{k}'$.

Let us now discuss the eigenvalues. In the diagonal approximation, where the b and c terms of the Luttinger matrix are neglected, one finds an exact classification of the eigenstates in terms of m_J. The heavy hole states correspond to $m_J = \pm\frac{3}{2}$ while the

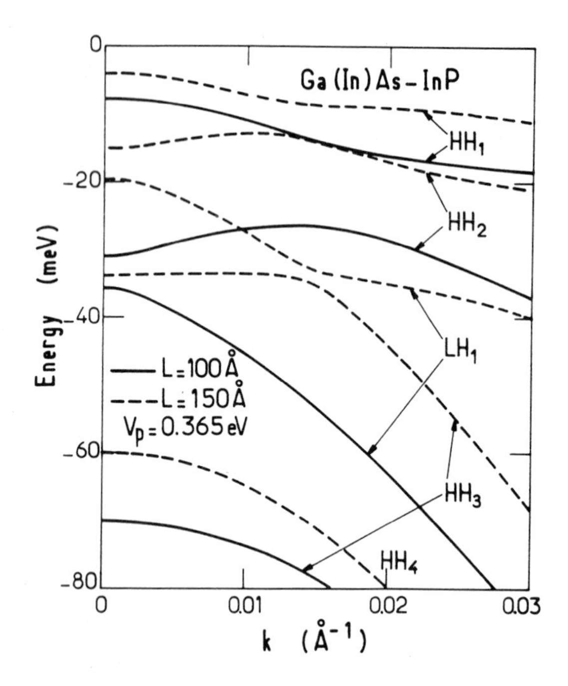

Figure 1: *Calculated in-plane dispersion relations of two* $Ga_{0.47}In_{0.53}As$-InP *quantum wells of different thicknesses.*

light hole states correspond to $m_J = \pm\frac{1}{2}$. The in-plane effective mass $m_0/(\lambda_1 + \lambda_2)$ of the heavy hole branches is lighter than that of the light hole branches, $m_0/(\lambda_1 - \lambda_2)$. One should therefore expect a crossing of the subbands upon increasing $\mathbf{k}$. The non-vanishing b and c terms replace these crossings by anti-crossings, and lead to strongly non-parabolic in-plane dispersion relations for the valence subbands. In particular, for unstrained systems like GaAs-AlGaAs quantum wells, one often finds a level ordering at $\mathbf{k} = 0$ of $HH_1, LH_1, HH_2, \ldots$. The ground subband HH_1 is always pushed upwards in energy by the other levels at $\mathbf{k} = 0$. On the other hand, the LH_1 subband experiences conflicting repulsions arising from the presence of HH_1 above it and all the other edges below it, so its actual dispersion at low $\mathbf{k}$ can be either hole-like or electron-like. [For results in GaAs quantum wells see, for example, Bastard (1988) and Bastard *et al.* (1991).] Note that a similar feature is found for HH_2 when the well depth, thickness, and material parameters are such that the ordering at $\mathbf{k} = 0$ is $HH_1, HH_2, LH_1, \ldots$. This feature is ilustrated in Figure 1 in the case of quantum wells in $Ga_{0.47}In_{0.53}As$-InP. Note finally that there is always a strong mixing between the heavy hole and light hole characters when an anti-crossing takes place.

3 Optical properties

Let us now turn our attention to the influence of the dispersion relations of the valence band and the admixture of light and heavy holes on the optical properties. The transition probability per unit time that an electromagnetic wave propagating along the z-axis induces a transition between a valence state $F_{v\mathbf{k}}$ to a conduction state $F_{c\mathbf{k}'}$ in the electric dipole approximation is

$$P_{\mathbf{k}\mathbf{k}'}(\omega) = \frac{\pi e^2 F^2}{2hm_0^2\omega^2} \left|\langle F_{c\mathbf{k}'}|\boldsymbol{\epsilon}\cdot\mathbf{p}|F_{v\mathbf{k}'}\rangle\right|^2 \delta\left(\varepsilon_{c\mathbf{k}'} - \varepsilon_{v\mathbf{k}'} - h\omega\right). \tag{15}$$

Here $\hbar\omega$ is the energy of the photon, F is the strength of the electric field in the electromagnetic wave and $\boldsymbol{\epsilon} = (\cos\beta, \sin\beta, 0)$ is its polarisation vector. The transition probability for any such transition to be excited is simply obtained by summing $P_{\mathbf{k}\mathbf{k}'}$ over all quantum numbers: $\mathbf{k}, \mathbf{k}'$, the spin and 'spin' variables, the subband index, *etc.*

Consider again the case of a single, symmetrical quantum well. The matrix element can be written

$$\langle F_{c\mathbf{k}'\sigma}|\boldsymbol{\epsilon}\cdot\mathbf{p}|F_{v\mathbf{k}\eta}\rangle \approx \sum_{m_J}\langle iS\sigma|\boldsymbol{\epsilon}\cdot\mathbf{p}|\tfrac{3}{2}, m_J\rangle\langle\Psi_{n\mathbf{k}'}|\Psi_{m_J\mathbf{k}}\rangle, \tag{16}$$

where $\eta = \pm$. It is clear from Equation (16) that $\mathbf{k}'$ should be equal to $\mathbf{k}$ (vertical transitions), since the momentum of the photon is neglected in the electric dipole approximation. The symmetry properties of the $|S\rangle$ and $|\frac{3}{2}, m_J\rangle$ Bloch functions imply that

$$\begin{aligned}
\langle iS\uparrow|\boldsymbol{\epsilon}\cdot\mathbf{p}|\tfrac{3}{2},\ \tfrac{3}{2}\rangle &= m_0P/\sqrt{2}\,e^{i\beta} \\
\langle iS\downarrow|\boldsymbol{\epsilon}\cdot\mathbf{p}|\tfrac{3}{2}, -\tfrac{3}{2}\rangle &= m_0P/\sqrt{2}\,e^{-i\beta} \qquad (17)\\
\langle iS\uparrow|\boldsymbol{\epsilon}\cdot\mathbf{p}|\tfrac{3}{2}, -\tfrac{1}{2}\rangle &= -m_0P/\sqrt{6}\,e^{-i\beta} \\
\langle iS\downarrow|\boldsymbol{\epsilon}\cdot\mathbf{p}|\tfrac{3}{2},\ \tfrac{1}{2}\rangle &= m_0P/\sqrt{6}\,e^{i\beta}. \qquad (18)
\end{aligned}$$

All other matrix elements vanish, and $P = -(i/m_0)\langle S|p_x|X\rangle$. Let us consider the transitions to the ground conduction subband, E_1. Then

$$\langle F_{c,n=1,\mathbf{k}'\uparrow}|\boldsymbol{\epsilon}\cdot\mathbf{p}|F_{v\mathbf{k}+}\rangle = \frac{m_0P}{\sqrt{2}}\left(e^{i\beta}\ \langle\chi_1|A_e\rangle - \frac{1}{\sqrt{3}}e^{-i\beta}\langle\chi_1|B_e\rangle\right)\delta_{\mathbf{k}\mathbf{k}'}; \tag{19}$$

$$\langle F_{c,n=1,\mathbf{k}'\downarrow}|\boldsymbol{\epsilon}\cdot\mathbf{p}|F_{v\mathbf{k}-}\rangle = -\frac{m_0P}{\sqrt{2}}\left(e^{-i\beta}\langle\chi_1|A_e^*\rangle - \frac{1}{\sqrt{3}}e^{i\beta}\ \langle\chi_1|B_e^*\rangle\right)\delta_{\mathbf{k}\mathbf{k}'}; \tag{20}$$

$$\langle F_{c,n=1,\mathbf{k}'\uparrow}|\boldsymbol{\epsilon}\cdot\mathbf{p}|F_{v\mathbf{k}-}\rangle = \langle F_{c,n=1,\mathbf{k}'\downarrow}|\boldsymbol{\epsilon}\cdot\mathbf{p}|F_{v\mathbf{k}+}\rangle = 0. \tag{21}$$

Thus it appears very clearly that a polarisation anisotropy, *i.e.* a dependence of the transition probability $P_{\mathbf{k}\mathbf{k}'}$ upon β, is possible only to the extent that there is a valence band mixing. Otherwise, either A_e or B_e (which respectively correspond to a heavy or a light hole component of the valence spinor) would vanish, which would lead to $P_{\mathbf{k}\mathbf{k}'}$ being independent of β. The polarisation anisotropy is expected to be weak for quantum wells grown on [001]. This is easily worked out numerically if one calculates the dispersions and wavefunctions by diagonalisation of the Luttinger matrix inside a small number of states at $\mathbf{k}=0$, for instance by retaining only the HH_1, HH_2 and LH_1 states (Ferreira and Bastard 1991, Twardowski and Hermann 1987). This simple approach remains acceptable for wells that are not too wide, and relatively small $\mathbf{k}$, and is very useful to estimate the oscillator strengths and optical properties near the absorption band edges for the quantum well (HH_1–E_1 and LH_1–E_1). The polarisation anisotropy is expected to be stronger for quantum wells grown along less symmetrical orientations, like [110] for instance. This can be verified by transforming the Luttinger matrix to comply with a z-direction which is now [110] instead of [001] as in Equation (7). A recent experiment by Gershoni *et al.* (1991) has clearly revealed this increased polarisation anisotropy.

4 Quantum wires

Other striking anisotropic effects of the light absorption were reported in quantum wires (Tsuchiya *et al.* 1989, Tanaka *et al.* 1990, Kohl *et al.* 1989). It is not difficult to understand the electronic origin of such anisotropy from the previous reasoning. Quantum wires are often made by etching and patterning a quantum well, usually grown along the [001] direction. The etching produces a new direction of quantisation in the plane of the quantum well. Usually this quantisation is weak compared to the one brought about by the epitaxy, *i.e.* the wire is much broader than deep ($L_x \gg L_z$). It can be shown that the Schrödinger equation is quasi-separable for the electron states under such circumstances (Bastard *et al.* 1991). It is exactly separable if the confining barriers along the x axis are infinite, and in this case we can write

$$F_{cn,m,k} = u_c(\mathbf{r})\Psi_{cnmk} = \chi_n(z)\sqrt{\frac{2}{L_x}}\sin\frac{m\pi x}{L_x}\sqrt{\frac{1}{L_y}}\exp\left(\mathrm{i}k_y y\right). \tag{22}$$

As for the valence subbands, one can no longer find any eigenstate which displays 100% heavy or light hole character, *even if* $k_y = 0$. This is because p_x now appears not only on the diagonal but also in the b and c terms of the Luttinger matrix. In other words, the localisation of the eigenstates along the x-direction (that is to say the loss of translational invariance along the x-direction) forces their wavefunctions to contain Fourier components with all possible values of k_x in the general case. With this remark in mind, and with the previous analysis of the physical origin of the polarisation anisotropy, it is clear that any quantum wire should display some anisotropy of electronic origin depending on whether the light, propagating along z, is polarised parallel or perpendicular to the wire. It is also clear that one should expect this anisotropy to increase with increasing quantisation along the x axis. A more complete analysis of the polarisation anisotropy in wires has reccently been undertaken by Bockelmann *et al.* (1991). Let us for simplicity keep the assumption of infinite barrier height along the x-direction. Then, if the underlying quantum well is symmetrical and not too wide, one can construct a valence state Ψ_v by retaining only the HH_1, HH_2 and LH_1 z-dependent subbands in the analysis. Thus

$$\Psi_{v+} = \sqrt{\frac{2}{L_x L_y}}\sum_{m=1}\sin\frac{m\pi x}{L_x}\exp\left(\mathrm{i}k_y y\right)\begin{pmatrix} g_m\,\chi_{hh1} \\ h_m\,\chi_{lh1} \\ 0 \\ l_m\,\chi_{hh2} \end{pmatrix}. \tag{23}$$

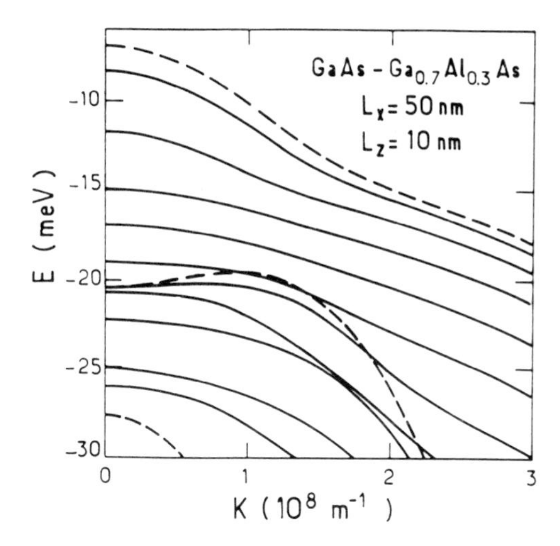

Figure 2: *Calculated dispersion relations along the wire axis of the valence subbands in a* GaAs-AlGaAs *quantum wire of cross-section* 500 nm × 100 nm. *The lateral confinement is assumed to be infinite. The dashed curves represent* $L_z = 10\,\mathrm{nm}$.

Since the parity operator with respect to z still commutes with the Hamiltonian, there exists another wavefunction Ψ_{v-} which is orthogonal to Ψ_{v+} and corresponds to the same k_y. It can be constructed from Ψ_{v+} in the same way as was done in the case of quantum wells.

In contrast with $\Psi_{cn,m,k}$, Ψ_v cannot correspond to a single x-dependent subband m since p_x appears not only on the diagonal but also in the b and c terms of the Luttinger matrix. Figure 2 shows the dependence on k_y of the valence subbands of a 50 nm × 10 nm quantum wire together with the 10 nm quantum dispersion upon k_y for $k_x = 0$. If the motion along (x, y) and z were decoupled, as for the conduction eigenstates, one would find that the wire dispersions upon k_y would be parallel to that of the quantum well but shifted by $(\hbar j\pi/L_x)^2/2m^*$, for $j = 1, 2, \ldots$. One sees from figure 2 that this is far from being the case except for the top-most valence level at small k_y. This is a result of the interactions brought by the b and c terms between the ladder attached to HH_1 and that attached to LH_1. Note in particular the increasingly heavier effective mass for the motion along the wire when the state approaches the onset of the LH_1 edge in the quantum well. These interactions can eventually give rise to camel-back shaped dispersions for some wire subbands. A similar signature of the interaction between heavy and light hole characters in the wire eigenstates can be found in the dependence on L_x of the energies of the edges of the valence subbands in the wire (Figure 3). The ground subbands loosely follow the L_x^{-2} trend when L_x is not too small. For the other edges, however, this simple trend is clearly offset by the heavy-to-light interactions with its tendency of increasing the apparent effective mass when one approaches the LH_1 edge of the quantum well. In particular some edges have practically no dependence on L_x when their energy corresponds to the camel-back portion of the LH_1 subband in

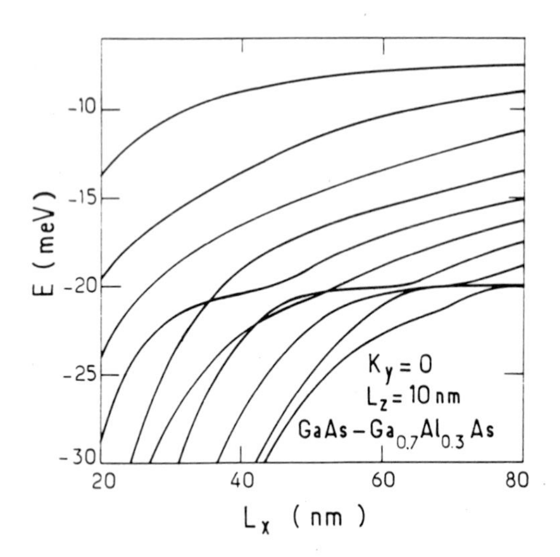

Figure 3: *Energies of the edges of the valence subbands in a* GaAs *quantum wire as a function of the lateral wire width* L_x *for a fixed* $L_z = 10\,\mathrm{nm}$.

the well. An intricate level pattern versus L_x also arises in this energy range due to the crossings and anti-crossings between wire subbands related to LH_1 and HH_1. The eigenstates in the diagonal approximation are even or odd with respect to $x = \frac{1}{2}L_x$, the centre of the wire along the x-axis, due to the parity along the x-direction. The b and c terms have different parities with respect to this inversion, but the parities of all components of Ψ_v are fixed if the parity in $(x - \frac{1}{2}L_x)$ of the $\frac{3}{2}$-component of the valence spinor is given. At a given L_x, it may well happen that the parities of the diagonal, energy degenerate, LH_1- and HH_1-related states are such that they are not coupled by the off-diagonal terms of the Luttinger matrix. In this case the two resulting states will cross. For instance, the $m = 6$ (odd) heavy hole and the $m' = 1$ (even) light hole at the diagonal approximation can cross at some value of L_x because the required coupling (the c term) is even in x. On the other hand the $m = 5$ (even) heavy hole and the $m' = 1$ (even) light hole at the diagonal approximation should anti-cross.

5 Polarisation effects in quantum wires

Let us now turn our attention to the optical polarisation effects of electronic origin in quantum wires (Bockelmann and Bastard 1991). We take the electric polarisation vector of the electromagnetic wave to be $\boldsymbol{\epsilon} = (\cos\varphi\sin\theta, \sin\varphi\sin\theta, \cos\theta)$ and assume that the final state corresponds to $n = 1$ and a given m in Equation (22), *i.e.* to a given wire conduction state which is derived from the ground subband of the quantum well. We obtain

$$|\langle F_c|\boldsymbol{\epsilon}\cdot\mathbf{p}|F_V\rangle|^2 = 2\sum_\sigma \left|\langle S\sigma|\boldsymbol{\epsilon}\cdot\mathbf{p}|\tfrac{3}{2}\rangle J_{h1} + \langle S\sigma|\boldsymbol{\epsilon}\cdot\mathbf{p}| -\tfrac{1}{2}\rangle J_{l1}\right|^2, \tag{24}$$

where

$$J_{h1} = g_m \langle \chi_1 | \chi_{h1} \rangle; \quad J_{l1} = h_m \langle \chi_1 | \chi_{l1} \rangle. \tag{25}$$

Due to parity there is no term involving J_{h2}. We again recognise in Equation (24) that the polarisation anisotropy of electronic origin arises from valence band mixing. Working out the $\boldsymbol{\epsilon} \cdot \mathbf{p}$ matrix elements we get

$$|\langle F_c | \boldsymbol{\epsilon} \cdot \mathbf{p} | F_v \rangle|^2 = \left(\frac{m_0 P}{\hbar} \right)^2 \left[\left(J_{h1}^2 - \frac{2}{\sqrt{3}} J_{h1} J_{l1} \cos(2\varphi) + \frac{1}{3} J_{l1}^2 \right) \sin^2\theta + \frac{4}{3} J_{l1}^2 \cos^2\theta \right]. \tag{26}$$

As expected, this reveals an anisotropy (a dependence on φ) due to the effects of band-mixing, demonstrated by the multiplicative factor $J_{h1} J_{l1}$. Finally, the absorption coefficient is given by the expression:

$$\alpha(\omega) = \left(\frac{2\pi e}{m_0} \right)^2 \frac{1}{n c \omega V} \sum_{m,k} |\langle F_{c1mk} | \boldsymbol{\epsilon} \cdot \mathbf{p} | F_{vk} \rangle|^2 \, \delta \left(E_{c1mk} - E_{vk} \right), \tag{27}$$

where n is the refractive index and V is the volume of the sample. In Equation (27) we have already accounted for the vertical nature of the inter-band transitions (translational invariance along the axis of the wire). Figure 4 displays the calculated absorption of the wire versus photon energy for the three light polarisations, $\boldsymbol{\epsilon}$ parallel to z, x and y. To compute such spectra and mimic the broadening effects, the δ-function in Equation (27) has been replaced by a Lorentzian curve with a width at half maximum of 2 meV. The first striking feature is the appearance of peaked structures in the calculated absorption despite the inclusion of broadening. They arise from the quasi-one-dimensional nature of the wire's eigenstates, *i.e.* from the restriction of the carrier freedom to the wire axis. It is well known (see Stern's lecture notes in this volume for a proof) that the density of states for such quasi-one-dimensional motion is singular at the edges of a subband, where the wave-vector of a carrier vanishes. For the z-polarisation (light propagating in the plane of the underlying quantum well) the heavy hole parts of the valence spinor do not contribute to the light absorption, since $\langle S | p_z | X \rangle = 0$. Thus, apart from band mixing and broadening effects, the absorption should be zero until one reaches an energy where the light hole component of the valence spinor is significant. Indeed, as seen from Figure 4 the first absorption peak in the z-polarisation involves the sixth valence state which has an energy right in the camel-back of the light hole in the quantum well. For a light propagating along the z-direction, the absorption at the edge is calculatd to display a *faint* anisotropy, depending on whether the light wave is polarised parallel or perpendicular to the wire axis. In spite of our assumption of a relatively narrow wire ($L_x = 50$ nm) for today's technology, it appears that the polarisation anisotropy should not be very large. Large anisotropy effects of electronic origin are *not* expected until the x-confinement is such that it pushes the ground hole level into a region where there is a large mixing of the bands in the underlying quantum well.

There is however another source of anisotropy. It is related to the dielectric mismatches encountered by the electromagnetic wave when it propagates through the array of wires. In practice, one usually grows a large number of parallel wires rather than a single wire in order to increase the absorption of light (like one grows multiple quantum wells instead of a single well in optical modulators). These wires are decoupled from

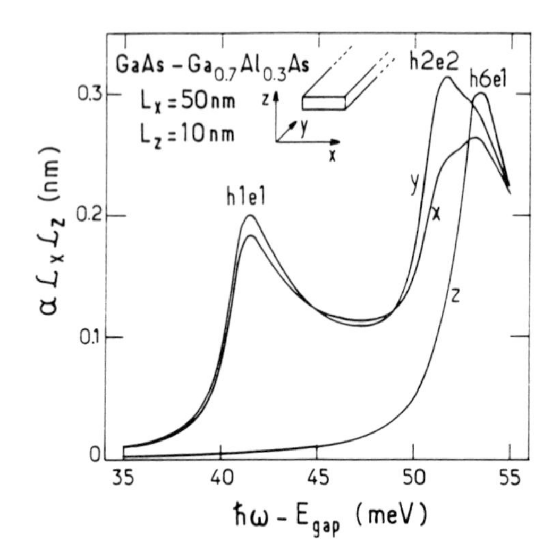

Figure 4: *Optical absorption of a quantum wire plotted versus photon energy for the three polarisations of light.* E_{gap} *is the band-gap of bulk* GaAs. *The wire has dimensions* $L_z = 10\,\mathrm{nm}$, $L_x = 50\,\mathrm{nm}$.

the electronic point of view: the usual separation between two wires is of the order of a few hundreds of nanometers. Thus our analysis of one wire is still adequate to treat the energy levels. On the other hand, the wires are not decoupled from the point of view of the electromagnetic wave since the distance between two wires is of the same order of magnitude as the wavelength of light. Thus the array of wires is a stratified medium from the optical point of view, and one may expect pronounced diffraction effects in these structures. A particular class of wires can give rise to large effects. They are those where the different wires are separated by the ambient because they have been produced by etching. The dielectric mismatch is larger compared to that when the wires are buried since the relative dielectric constant of a semiconductor is of the order of 12 while that of the ambient is 1. An analysis of electromagnetic propagation in wires has recently been undertaken by Bockelmann (1991). It was shown that large polarisation effects of electromagnetic origin can take place in wires where the polarisation anisotropy of electronic origin is negligible.

Acknowledgments

I would like to thank the organisers for their invitation to participate in the summer school and for their kind hospitality. This work was performed with my collaborators U Bockelmann and R Ferreira. U Bockelmann is grateful to the organisers for financial support. This work has been partly supported by CNRS (France), CAPES (Brazil), CNET (contract number 906B067 007909245) and a Procope contract.

References

Allan G, Bastard G, Boccara N, Lannoo M, and Voos M (eds), 1986. *Heterojunctions and Semiconductor Superlattices*, Berlin, Springer-Verlag.

Bastard G, Brum J A, and Ferreira R, 1991. *Solid State Phys.*, **44**, 229.

Bastard G, 1988. *Wave Mechanics Applied to Semiconductor Heterostructures*, Les Ulis, Les Editions de Physique.

Beenakker C W J, and van Houten H, 1991. *Solid State Phys.*, **44**, 1–288, and references cited therein.

Bockelmann U, and Bastard G, 1991. *Europhys. Lett.*, **15**, 215.

Bockelmann U, 1991. *Europhys. Lett.*, **16**, 601.

Ferreira R, and Bastard G, 1991. *Phys. Rev. B* **43**, 9687.

Gershoni D, Brener I, Baraff G A, Chu S N G, Pfeiffer L N, and West K,1991. *Phys. Rev. B*, **44**, 1930.

Kohl M, Heitmann D, Granbow P, and Ploog K, 1989. *Phys. Rev. Lett.*, **63**, 2124.

Tanaka M, Motohisa J, and Sakaki H, 1990. *Surface Sci.*, **228**, 408.

Tsuchiya M, Gaines J M, Ryan R H, Simes R J, Holtz P O, Coldren L A, and Petroff P, 1989. *Phys. Rev. Lett.*, **62**, 466.

Twardowski A, and Hermann C, 1987. *Phys. Rev. B*, **35**, 8144.

Uenoyama T, and Sham L J, 1990. *Phys. Rev. Lett.*, **64**, 3070; *Phys. Rev. B*, **42**, 7114 (1990).

Weisbuch C, and Vinter B, 1991. *Quantum Semiconductor Structures: Fundamentals and Applications*, New York, Academic Press, and references cited therein.

Theory of Coherent Quantum Transport

A D Stone

Yale University
New Haven, Connecticut, U.S.A.

1 Introduction

The importance of phase-coherence effects in normal electron transport has been appreciated for some time through the study of localisation, and particularly weak localisation, in macroscopic conductors (Anderson 1958, Lee and Ramakrishnan 1985a, Bergmann 1984). Nonetheless, the dramatic improvement in fabrication of ultrasmall (mesoscopic) conducting devices has increased interest in this subject because of the experimental accessibility of novel phenomena associated with quantum interference in such systems, and the potential applicability of these phenomena to new microelectronic devices in the (admittedly distant) future.

Quantum coherence in the diffusive limit

If we exclude from discussion phase-coherence phenomena directly related to interaction (*e.g.* superconductivity, or the Coulomb suppression in the electronic density of states), we are now able to identify three distinct classes of quantum coherence effects in conductors in the *diffusive* limit. Here 'diffusive' means that the *elastic* mean free path, l, is smaller than the dimensions of the device.

First there are the *weak localisation* (WL) effects on the average conductance, known since the work of Abrahams *et al.* (1979), which arise due to the coherent back-scattering of diffusing electrons in the presence of time-reversal symmetry. Because of this coherent back-scattering the *average* low temperature conductance of a film or wire of arbitrary size was shown to be sensitive to a weak magnetic field or weak spin-orbit scattering (Figure 1). The sensitivity to magnetic field was understood theoretically (Altshuler *et al.* 1980, 1981) to be a manifestation of the Aharonov-Bohm effect for multiply-scattered electrons, with the average over disorder leading to an effective doubling of

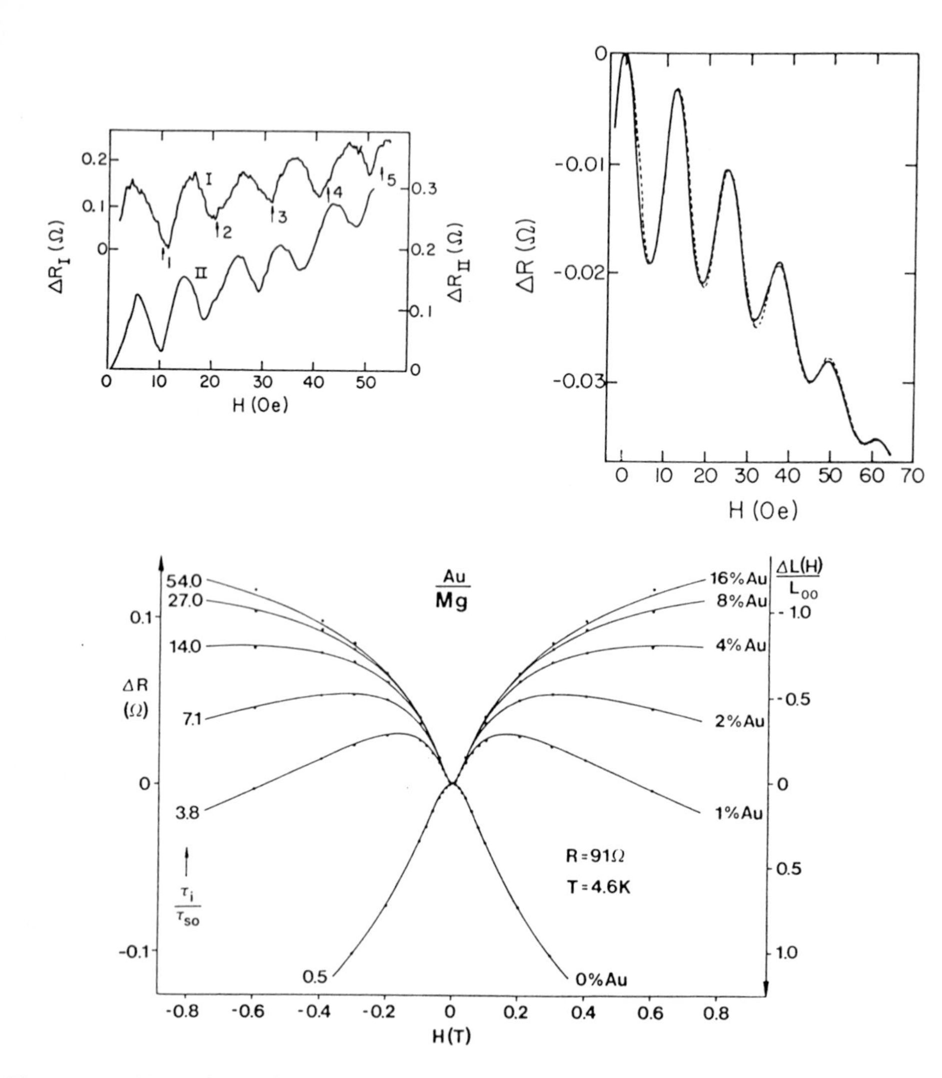

Figure 1: *Normal metal Aharonov-Bohm effect with period $h/2e$ in magnesium (top left) and lithium (top right) cylinders, from Aronov and Sharvin (1987). Weak localisation magnetoresistance of a magnesium film coated with gold to enhance spin-orbit scattering (bottom), from Bergmann (1984). Note the reversal of sign due to spin-orbit scattering.*

the electron charge ($e \to 2e$). We shall review this briefly below. Recently it has been shown (Mathur and Stone 1991) that the sensitivity to weak spin-orbit scattering is a manifestation of the Aharonov-Casher effect in which an electric field couples to the phase of the wavefunction by its influence on the spin magnetic moment (Aharonov and Casher 1984).

In both cases the WL effects arise from the influence of dynamically negligible per-

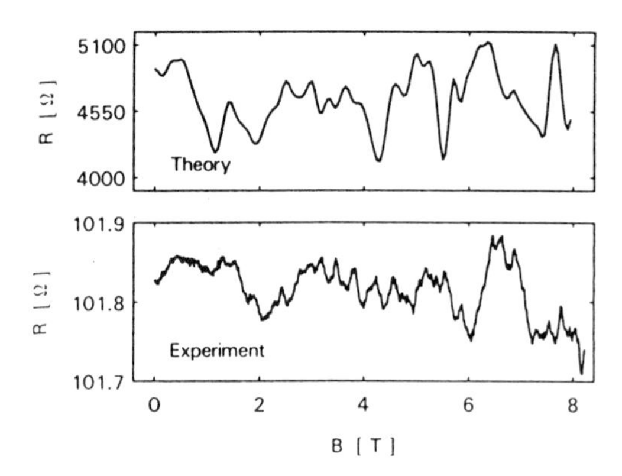

Figure 2: *Universal conductance fluctuations in a simulation of the Schrödinger equation (top) and in a gold-palladium wire of length* 790 nm *and width* 50 nm *(bottom); from Stone (1985).*

turbations on the phase of the wavefunctions, and hence are interference effects in the purest sense. The weak localisation effects depend on dimensionality and on the inelastic scattering length, l_{in}, but not on the the size of the sample. They are a manifestation of time-reversal symmetry, and are completely suppressed by a moderate magnetic field. Because they are corrections to the average conductance they are generic: two samples of the same material will show the same WL effect.

The second class of quantum coherence effects in disordered conductors are the sample-specific variations in the transport properties of mesoscopic devices at low temperature (Washburn and Webb 1986; see Figure 2). These are described statistically by the theory of universal conductance fluctuations (UCF) (Altshuler 1985, Lee and Stone 1985, Stone 1985, Altshuler and Shklovskii 1986, Lee *et al.* 1987). In this theory the conductance, g, of a phase-coherent device is shown to be sensitive to small changes in magnetic field, Fermi energy or impurity configuration. The maximum degree of sensitivity is given by the condition

$$\delta g \approx \frac{e^2}{h}, \tag{1}$$

independent of $\langle g \rangle$. The scale of variation of the control parameter needed to achieve this maximal fluctuation is simply that needed to alter the action (phase) along a typical diffusive path by order unity (see, for example, Stone and Imry 1986). Hence the conductance fluctuations and related effects show a sensitivity to weak magnetic field and spin-orbit scattering as do the WL effects. This sensitivity appears in two ways. First, in a given sample the resistance oscillates on a small field scale, typically around 100 G at 1 K (Figure 2); second, the magnitude of the variance of g depends on the time-reversal and spin symmetry of the scattering processes. The breaking of either symmetry leads to universal reduction factors for the variance of g (Altshuler and Shklovskii 1986, Stone 1989), but not (as in WL) to a complete suppression of the effect. Again the effects of a magnetic field arise completely from coupling to the phase of the wavefunction, and not from dynamical effects such as Landau quantisation, which was completely neglected in the standard theory. In fact studies of semiconductor

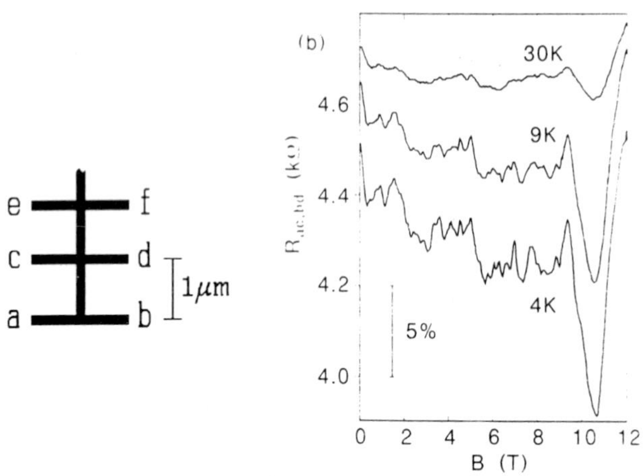

Figure 3: *Magnetoresistance traces showing co-existence of Shubnikov–de Haas oscillations with* UCF*; from Geim et al. (1991).*

microstructures show that UCF effects coexist with the Shubnikov–de Haas oscillations arising from Landau-level quantisation (Figure 3) and even in the transition region between quantum Hall steps. Below we shall show how to generalise UCF theory to the high-field limit within the self-consistent Born approximation.

In general, UCF effects depend on the ratio of the inelastic length to the sample dimensions and are not observable in macroscopic samples. However they do not depend on time-reversal or any other symmetry for their existence and are robust against magnetic field

A third class of phenomena which has recently received a great deal of theoretical attention relate to quantum coherence effects in the *thermodynamic* properties of mesoscopic samples, such as the persistent current or orbital magnetic response. The sample-to-sample fluctuations of these quantities are conceptually quite similar to UCF, but it has recently been predicted theoretically that there should be an *average* effect as well due to the constraint of fixed particle number in such isolated mesoscopic systems (Altshuler *et al.* 1991). This average effect is unlike UCF in that it should be observable in large *arrays* of isolated mesoscopic elements (*e.g.* quantum dots), but would also be unlike WL in that the amplitude will decay with the ratio of the size of individual elements to the inelastic mean free path. Although average persistent currents have been observed experimentally by Levy *et al.* (1990), it is unclear whether their origin is this new type of quantum coherence effect. Nonetheless I believe that the prediction of such effects opens a new and fascinating area for future work.

Quantum coherence in the ballistic limit

All of the above interference phenomena involve diffusing electrons and have their mathematical origin in the properties of the disorder-averaged two-particle Green function, and in particular in the dominance of the contributions in the cooperon or diffuson channels at long distances and low temperatures. A fourth class of phenomena at present outside of this type of description are coherent effects in *ballistic* transport, where the motion may not be diffusive and there is no obvious ensemble over which to average. Initially it was assumed that such devices would show rather elementary wave-guide interference effects, equivalent to microwave interferometers. However we have recently shown that fluctuating interference phenomena very similar to UCF occur in ballistic systems at low temperatures if the sample geometry is sufficiently complex so that the classical scattering from such a confining potential is chaotic (Jalabert *et al.* 1990); thus it remains an open question whether such simple interference effects are achievable in ballistic conductors.

In addition to *interference* effects, there are other novel transport phenomena in the ballistic limit which are essentially classical in origin. These include the quenching of the Hall resistance or four-probe bend resistance (see, for example, Beenakker and van Houten 1989), as well as effects related to mode quantisation such as the quantised conductance of a point contact (van Houten *et al.* 1990).

Finally one must mention the quantised Hall effect (Prange and Girvin 1987). Although not a phenomenon that requires phase coherence across the sample, we shall see below that the occurence of Hall quantisation and vanishing longitudinal resistance in a two-dimensional electron gas (2DEG) can be understood quite simply by the Landauer-Büttiker approach commonly used in studying phase-coherent transport. Moreover the stability of this phenomenon against disorder, and the occurence of plateaus, can only be understood by invoking localisation effects related to phase coherence. Unfortunately in this case a detailed physical picture has not yet been developed comparable with the theory of weak localisation.

Formalism for quantum transport

Since all the phenomena discussed above are directly related to interference of multiply-scattered electron waves, they are not described by the conventional approach to transport in solids based on the Boltzmann equation. Instead they require a formalism in which quantum coherence may be incorporated from the beginning. Linear response theory (Kubo 1957) has traditionally been employed to treat such situations, although often with some cost to physical intuition due to its complexity. An alternative approach which has proved extremely useful in the study of mesoscopic transport was pioneered by Landauer over thirty years ago (Landauer 1957, 1970). Using a counting argument and the Einstein relation, he related the resistance of a 1D conductor to the quantum transmission probability for the conductor (treated as a single composite scattering center); in principle the exact transmission coefficient will contain all of the interference effects associated with multiple scattering of the electrons. Landauer's original work did not consider in detail the nature of the resistance measurement, but experimental progress in the study of mesoscopic conductors rapidly uncovered the important role of the contacts and the measuring geometry in determining the results of transport measurements (Benoit *et al.* 1986, Skocpol *et al.* 1986). For example, a qualitative difference was found between two-probe and four-probe measurements of resistance with respect to their symmetry under reversing the magnetic field (Stone and Szafer 1988). This discovery motivated a crucial generalisation of the original Landauer approach, due to Büttiker (1986), in which all the measurement probes are treated on an equal footing and the correct reciprocity relations for multi-probe measurements in magnetic fields arise naturally. It was subsequently shown that Büttiker's multi-probe formula could be derived from a version of Kubo linear response theory (Stone and Szafer 1988, Baranger and Stone 1989). Although the Green function formulation is often more convenient for microscopic calculations, the Landauer-Büttiker (LB) formulation in terms of transmission coefficients has often proven useful and more congenial to physical intuition. We review briefly the physical assumptions of this approach and its derivation from linear response theory below.

2 Landauer-Büttiker approach to transport

2.1 Counting argument

The basic physical idea of the LB approach is to consider the sample whose resistance is being measured as a single phase-coherent unit attached to perfect reservoirs serving as current source and sink and as voltage probes. In an ideal *two-probe* measurement the sample is attached between two perfect reservoirs with electrochemical potentials, μ_1 and $\mu_2 = \mu_1 + eV$ respectively, where V is the applied voltage; these reservoirs serve both as current source and sink *and* as voltage terminals. In the energy interval eV between μ_2 and μ_1, electrons are injected into right-going states emerging from reservoir 1, but none are injected into left-going states emerging from reservoir 2. Thus there is a net right-going current proportional to the number of states in the interval $\mu_2 - \mu_1$, given by

$$I = e\sum_i^{N_c} v_i \frac{dn_i}{d\varepsilon}\, eV \sum_j^{N_c} T_{ij} = \left(\frac{e^2}{h}\sum_{i,j}^{N_c} T_{ij}\right) V, \tag{2}$$

where N_c is the number of propagating channels including spin *in the sample*, v_i is the the longitudinal velocity for the ith momentum channel at the Fermi surface, T_{ij} is the transmission probability from j to i, and we have used the fact that $dn_i/d\varepsilon = 1/hv_i$ for a quasi-1D density of states. Equation (2) yields an expression for the two-probe conductance in terms of the total transmission coefficient (normalised to N_c),

$$g = \frac{e^2}{h}\sum_{i,j}^{N_c} T_{ij} \equiv \frac{e^2}{h} T. \tag{3}$$

Many experimental measurements are not made in a two-probe configuration, but in a multi-probe configuration in which current and voltage probes are different. Büttiker (1986) generalised the above argument to calculate the current in a phase-coherent system connected to N_L reservoirs, where any two can serve as current source and sink, and a voltage can be applied (or induced) between any two. The currents and voltages on all the leads are now related by a matrix whose elements, g_{mn}, are known as conductance coefficients. Büttiker showed that an argument exactly analogous to that leading to Equation (3) yields the result

$$I_m = \sum_n^{N_L} g_{mn} V_n = \frac{e^2}{h}\sum_n^{N_L} (T_{mn} - N_c\delta_{mn}) V_n\,, \tag{4}$$

where I_m is the total current into lead m, V_n is the voltage applied at lead n, and T_{mn} is the total transmission coefficient (or reflection coefficient for the case $m = n$) at the Fermi energy for electrons injected at lead n to be collected at lead m (summed over all channel indices, which have been suppressed for clarity). In the case $N_L = 2$ this formula reduces exactly to Equation (3); for four or more probes it can be inverted to yield the Hall resistance if the T_{mn} are known. This formula is very appealing because it provides a simple Fermi-surface expression for the Hall resistance, R_H, which is valid in an arbitrary magnetic field. This property is noteworthy in view of the fact that the Hall conductance in the Kubo formulation is commonly expressed in terms of *all* the states below the Fermi surface. In fact until recently the conditions under which the LB approach (which is based on the above physical argument, and not on a derivation from an underlying Hamiltonian) and the Kubo approach were equivalent were unclear. Hence a derivation of the LB formulas from linear response theory became of interest.

2.2 Derivation from linear response theory

Derivations of Equation (3) for the two-probe conductance from linear response theory began with the work of Economou and Soukoulis (1981) and Fisher and Lee (1981), although the interpretation in terms of two-probe measurements was not understood until later (Imry 1986). Derivations of this type were generalised to the multi-probe case by Stone and Szafer (1988) for zero magnetic field, and by Baranger and Stone (1989) for arbitrary magnetic field. Here we sketch only the simpler version of the derivation for multi-probe systems for $B = 0$. One considers a non-interacting electron system under the influence of an arbitrary potential $V(\mathbf{r})$ in a finite region of space, defined to be the 'sample'. Electrons can flow out of the sample to infinity along N_L strips or bars of finite width which are translationally invariant in the longitudinal direction ('perfect leads'). The infinite perfect leads serve to make the spectrum continuous in energy, with eigenstates in the form of a wave approaching the sample from infinity in a single mode and a given lead and transmitted and reflected waves leaving the sample in all the leads and modes (the so-called scattering-wave states). Since the system is infinite a d.c. current can flow through the sample in response to a potential difference imposed between two edges of the sample. The perfect leads are assumed to be equipotentials out to infinity and unaffected by the current flow; these assumptions combined with the lack of back-scattering in the leads allows them to function like the perfect reservoirs in the LB argument.

Since we are treating our system as approximately non-interacting, the many-body eigenstates are Slater determinants of single-particle wavefunctions. The expectation value of any single-body operator $\hat{\mathcal{O}}_1$ evolves according to the time-dependent Schrödinger equation, and can be expressed as $\langle \mathcal{O}_1 \rangle = \mathrm{Tr}\{\rho(t)\hat{\mathcal{O}}_1\}$, where $\rho(t)$ is the *single-particle* density matrix satisfying the equation of motion $i\hbar(d\rho/dt) = [H, \rho]$. The unperturbed system is described by

$$\rho_0 = \int d\alpha\, f(\varepsilon_\alpha)\, |\psi_\alpha\rangle \langle \psi_\alpha| , \tag{5}$$

where $f(\varepsilon_\alpha)$ is the Fermi function, $|\psi_\alpha\rangle$ are the exact scattering-wave states of the equilibrium system with energy ε_α, and we have written an integral over α to emphasise that the energies are continuous.

We assume that the system is perturbed by an external scalar potential oscillating with frequency ω, which is turned on adiabatically from $t = -\infty$; the potential is arbitrary in the sample and approaches a different constant value in each lead. By solving the equation of motion for ρ to obtain the correction to ρ_0 to linear order in the perturbation, and then taking its trace with the current operator in the limit $\omega \to 0$, one obtains

$$\langle \mathbf{J}(\mathbf{r}) \rangle = \int d\mathbf{r}' \,\underline{\sigma}(\mathbf{r}, \mathbf{r}') \cdot \mathbf{E}(\mathbf{r}'), \tag{6}$$

where the local Kubo conductivity tensor is given by

$$\underline{\sigma}(\mathbf{r}, \mathbf{r}') = -\hbar \int d\alpha \, d\beta \left[f'(\varepsilon_\alpha) \pi \delta(\varepsilon_{\beta\alpha}) + i \frac{f_{\beta\alpha}}{\varepsilon_{\beta\alpha}} P\left(\frac{1}{\varepsilon_{\beta\alpha}} \right) \right] \mathbf{J}_{\beta\alpha}(\mathbf{r}) \, \mathbf{J}_{\alpha\beta}(\mathbf{r}'). \tag{7}$$

In Equation (7) $\varepsilon_{\beta\alpha} = \varepsilon_\beta - \varepsilon_\alpha$, $f' = \partial f / \partial \varepsilon$, $f_{\beta\alpha} = f(\varepsilon_\beta) - f(\varepsilon_\alpha)$ and P denotes the principal value of the integral. $\mathbf{J}_{\beta\alpha}(\mathbf{r})$ is the matrix element of the current operator between exact eigenstates,

$$\mathbf{J}_{\beta\alpha}(\mathbf{r}) = \frac{-ie\hbar}{2m} \left[\psi_\beta^*(\mathbf{r}) D \psi_\alpha(\mathbf{r}) - \psi_\alpha(\mathbf{r}) D^* \psi_\beta^*(\mathbf{r}) \right], \tag{8}$$

where $D = [\nabla - (ie/\hbar c)\mathbf{A}(\mathbf{r})]$ is the gauge-invariant derivative. It is convenient to introduce a double-sided derivative $\overleftrightarrow{D}$ defined such that

$$\mathbf{J}_{\beta\alpha}(\mathbf{r}) = \frac{-ie\hbar}{2m} \left[\psi_\beta^*(\mathbf{r}) \overleftrightarrow{D} \psi_\alpha(\mathbf{r}) \right].$$

Note that the δ-function term in Equation (7) only involves states at the Fermi surface as $T \to 0$, whereas the principal value term involves a sum over all states. Under time-reversal symmetry, $\mathcal{T}$, the matrix elements of current satisfy $\mathcal{T}[J_{\beta\alpha}(\mathbf{r}, B)] = J_{\alpha\beta}^*(\mathbf{r}, -B)$. By applying time-reversal symmetry and interchanging indices α, β in Equation (7) we see that the principal value term is anti-symmetric in magnetic field whereas the δ-function term is symmetric. At $B = 0$ this means that the principal value term must vanish.

The current I_m through probe m is obtained by integrating $\langle \mathbf{J}(\mathbf{r}) \rangle$ over a cross-section of lead m. One can then express $\mathbf{E}(\mathbf{r})$ as $\nabla\phi$, and use the divergence theorem to express I_m in terms only of the voltage at the boundaries as in Equation (4) [current conservation must be used in this step to eliminate additional terms involving volume integrals (Baranger and Stone 1989)]. One obtains the intuitive result

$$g_{mn} = \int dS_m \int dS_n \, \hat{\mathbf{m}} \cdot \underline{\sigma}(\mathbf{r}, \mathbf{r}') \cdot \hat{\mathbf{n}}, \tag{9}$$

where $\hat{\mathbf{m}}$ and $\hat{\mathbf{n}}$ are unit vectors normal to the cross-section. The task now is to show that this expression with $\underline{\sigma}$ given by Equation (7) is equivalent to $g_{mn} = (e^2/h)T_{mn}$. We only consider here the case of zero magnetic field in which the principal value term of Equation (7) vanishes. In this case one can immediately perform the integration over energy and take the limit $T \to 0$ in Equation (7) so that the integral over α, β becomes a sum over the discrete states at ε_F of the form

$$\frac{1}{2h} \sum_{\alpha,\beta} \int dS_m \, \hat{\mathbf{m}} \cdot \mathbf{J}_{\beta\alpha}(\mathbf{r}) \int dS_n \, \hat{\mathbf{n}} \cdot \mathbf{J}_{\alpha\beta}(\mathbf{r}'), \tag{10}$$

where we have included in Equation (10) the appropriate normalisation factor $(2\pi\hbar)^{-2}$ arising from the integrations over energy.

Thus we need to evaluate the current matrix elements of the exact eigenstates $\mathbf{J}_{\beta\alpha}(\mathbf{r})$ integrated over the cross-sections of leads m, n. Since their energy is now fixed at ε_F, states α, β are specified by a mode index a, b and a lead index p, q denoting the lead and mode which contains an incoming wave from infinity. Assume that $p, q \neq m, n$ for simplicity. In lead m, $\psi_\alpha(\mathbf{r}) = \sum_c t_{cm,ap}\, \phi_c^+(\mathbf{r})$, and $\psi_\beta^*(\mathbf{r}) = \sum_d t^*_{dm,bq}\, \phi_d^{+*}(\mathbf{r})$, where $t_{cm,ap}$ is the transmission amplitude for an incident wave in lead p and mode a to scatter to lead m and mode c, and $\phi_c^+(\mathbf{r})$ are the wavefunctions of the infinite perfect leads, consisting of a longitudinal plane wave traveling away from the sample multiplied by the transverse wavefunction for mode c; since the leads are translationally invariant we can always choose the longitudinal part to be a plane wave; the transverse part need not be specified. For $B = 0$ the transverse wavefunctions are orthogonal and we have $\int dS_m\, \hat{\mathbf{m}} \cdot J_{cd} = e\, \delta_{cd}$ (where we have normalised these states to unit flux); for $B \neq 0$ the same identity holds, but one needs to use current conservation to obtain it as the transverse wavefunctions are not orthogonal (Baranger and Stone 1989). Using this one finds

$$\int dS_m\, \hat{\mathbf{m}} \cdot \mathbf{J}_{\beta\alpha}(\mathbf{r}) = \sum_c t_{cm,ap}\, t^*_{cm,bq}. \tag{11}$$

The summation over α, β in Equation (10) will now be equivalent to summing Equation (11) over modes a, b and leads p, q except for the case $p = q = m$, which simply brings in an additional Kronecker delta term due to the incoming wave with unit flux. Hence we can express the conductance coefficient g_{mn} entirely in terms of summations over the transmission and reflection amplitudes for waves incident in all the leads and all the modes. This expression is then easily simplified using the unitarity relations for the scattering matrix to yield

$$g_{mn} = \frac{e^2}{2h}\left(T_{mn} + T_{nm}\right) = \frac{e^2}{h} T_{mn}, \tag{12}$$

where we have used fact that the scattering matrix is symmetric in the presence of time-reversal symmetry. In the presence of a magnetic field the derivation involving the δ-function term proceeds exactly as above except that $T_{mn} \neq T_{nm}$; however a much more involved argument (Baranger and Stone 1989) shows that the principal value term gives rise to a term $\left(T_{mn} - T_{nm}\right)/2$ which again yields Büttiker's result when added to the symmetric term. Hence the derivation from linear response theory shows that the LB equations are valid in an arbitrary magnetic field and that one can always think of phase-coherent quantum transport as a scattering problem between electronic states at the Fermi surface.

2.3 Two-probe conductance

The Kubo-type expression for $g_{12} \equiv g$ is simplified in the case of a two-probe measurement because the principal value term is zero even in the presence of a magnetic field. This may be seen by noting that time-reversal symmetry for the current matrix elements applied to Equation (7) implies that $g_{mn}(B) = g_{nm}(-B)$, hence $g_{mm}(B)$ is symmetric in field. But in the two-probe case $g_{12} = -g_{11}$ follows from the requirement that no current flow in response to zero voltage difference. Hence g is symmetric in field and the principal value term must vanish. In addition the total current flowing through the sample is independent of the cross-sections S_m, S_n in Equation (9) which can be chosen inside the sample as well. Since it is often more convenient in microscopic calculations to express g in terms of integrals over the entire volume, we may integrate Equation (9) over the positions of S_m, S_n and simply divide by L_x^2 (where henceforth we choose the x-axis to lie in the longitudinal direction). It is also convenient to express the exact current matrix elements in terms of a derivative of the advanced and retarded Green functions, $G^{\pm}(E,\mathbf{r},\mathbf{r}') = \sum_\alpha \psi_\alpha(\mathbf{r})\,\psi_\alpha^*(\mathbf{r}')\,(E-\varepsilon_\alpha \pm i\eta)^{-1}$. Define

$$\Delta G(E,\mathbf{r},\mathbf{r}') = G^+(\mathbf{r},\mathbf{r}') - G^-(\mathbf{r},\mathbf{r}') = -2\pi i \sum_\alpha \psi_\alpha(\mathbf{r})\psi_\alpha^*(\mathbf{r}')\delta(E-\varepsilon_\alpha). \tag{13}$$

Substitution of this relation into Equations (7) and (9) yields (at $T=0$)

$$g = -\frac{e^2\hbar^3}{16\pi m^2 L_x^2}\int d\mathbf{r}\, d\mathbf{r}'\, \Delta G(\varepsilon_F,\mathbf{r},\mathbf{r}')\overleftrightarrow{D}_x^*\ \overleftrightarrow{D}_{x'}\,\Delta G(\varepsilon_F,\mathbf{r}',\mathbf{r}). \tag{14}$$

Since we are now integrating over $\mathbf{r}$ and $\mathbf{r}'$, the double-sided derivatives in Equation (14) are equivalent by integration by parts to twice the velocity operator. Hence Equation (14) can be written in operator form as

$$g = -\frac{e^2\hbar}{4\pi L_x^2}\mathrm{Tr}\left\{v_x\,\Delta G\,v_x\,\Delta G\right\}, \tag{15}$$

an expression which dates back to the early work of Greenwood (1958) and Kubo (1965) where it is interpreted as the longitudinal conductivity. Thus we see that two-probe conductance is essentially equal to the spatially-averaged symmetric conductivity tensor, whereas the four-probe resistance is a more complicated quantity, which in general receives contributions from the symmetric and anti-symmetric part of the conductivity tensor.

3 Physical consequences of the LB formula

A number of physical consequences can be obtained without detailed calculations based on the LB formula.

3.1 Quantised contact resistance

Since the conductance coefficients are expressed as the fundamental quantum of conductance, e^2/h, multiplied by transmission coefficients, it is easily seen that quantisation of these transmission coefficients leads to quantisation of transport coefficients in units of e^2/h. The simplest example is the two-probe conductance in the ballistic limit. In the absence of scattering in the sample we will have $T_{ij} = \delta_{ij}$ in Equation (3), leading to

$$g = 2N_c\frac{e^2}{h}, \tag{16}$$

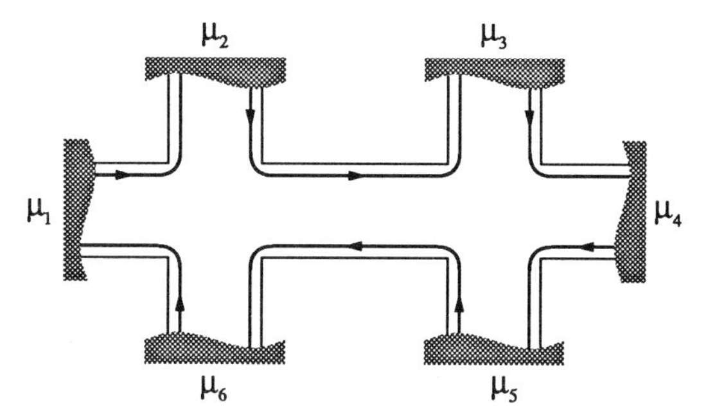

Figure 4: *Standard geometry for Hall effect measurements.*

where I have explicitly introduced a factor of two for spin degeneracy. This is the 'quantised contact resistance' which has now been widely observed in semiconductor point contacts in GaAs heterostructures (van Houten *et al.* 1990). Typically a 2DEG in such systems is divided into two regions by a constriction of variable width created by a split gate. As the width is varied, steps in g of height $2e^2/h$ are observed, separated by reasonably flat plateaus. The experimental observations indicate that the wide regions of the 2DEG behave remarkably like the perfect reservoirs invoked in the LB approach. In particular the existence of fairly sharp steps as a new channel opens is somewhat surprising in a real system where no attempt was made to eliminate impedance mismatch at the interface between the wide and narrow regions. A number of theoretical calculations have now been performed to understand this point (see, for example, Glazman *et al.* 1988, Szafer *et al.* 1989). Although robust with respect to the geometry of the interface, the quantised point contact resistance is quite sensitive to disorder and disappears for constrictions longer than few thousand Ångstroms (Timp *et al.* 1989).

3.2 Quantised Hall effect

A more spectacular and robust example of quantised transport coefficients which can be understood by the LB approach is the integer quantised Hall effect. Here one must of course use the multi-probe formula, Equation (4). The resistance is obtained from this formula by assuming that a current I is fed in from the source probe and withdrawn through the sink, and that the net current in the voltage probes is zero, and inverting Equation (4) to find the voltage difference of interest. Zero current in the voltage probes in steady state is achieved by adjusting their voltages to null any transient current.

Now, consider a two-dimensional electron gas in high magnetic field with the standard Hall geometry as shown in Figure 4. The high field causes the formation of Landau levels with large degeneracy and low velocity throughout the bulk of the 2DEG; however for each bulk Landau level (LL) there exists a current-carrying quantum state at the edge of the sample with (roughly speaking) the the same cyclotron energy and an additional kinetic energy due to its skipping motion along the edge of the sample (see, for example, Streda *et al.* 1987, Büttiker 1988). Assume that the Fermi energy is above the centre of the Nth LL so that a current $N(e/h)\mu_1$ is being carried into the sample by these states from reservoir 1 along the upper edge. There is disorder in the sample which in general can cause scattering; however *if* all the current continues to flow along that edge and into voltage probe 2 (*i.e.* $T_{21} = N$), then we must have $N(e/h)(\mu_1 - \mu_2) = 0$ in order for the total current into probe 2 to vanish, *i.e.* $\mu_1 = \mu_2 = \mu_{\mathbf{source}}$. By repeating this argument for μ_3 one immediately sees that all voltage probes on this side of the current path will adjust their chemical potentials to be equal to that of the source. The current sink, on the other hand, is maintained at a different chemical potential $\mu_{\mathbf{sink}}$, and by the same argument all probes on the lower side of the current path will be at the potential $\mu_{\mathbf{sink}}$. Thus we see that a net current $I = N(e/h)(\mu_{\mathbf{source}} - \mu_{\mathbf{sink}})$ can flow from source to sink with no voltage appearing between any two probes on the same side of the current path, *i.e.* the longitudinal resistance R_L vanishes. The Hall resistance is just the ratio of the voltage induced between any two probes separated by the current path, which is $(\mu_{\mathbf{source}} - \mu_{\mathbf{sink}})/e$, divided by the net current, giving $R_H = (h/Ne^2)$. Since N, the number of states at the Fermi level, is equal to the number of Landau levels below the Fermi surface, this formula yields the familiar quantised Hall resistance. This argument establishes that if the transmission coefficients $T_{n,n-1}$ are quantised to N, and if all other transmission coefficients are zero, then one obtains at the same time the quantised Hall resistance and the vanishing longitudinal resistance (which is normally not treated on the same footing by Laughlin's famous argument).

It is easy to see how these quantised transmission values can occur when all the states at the Fermi level are true edge states. Electrons are injected near one edge, and in order to back-scatter they must be removed at the other edge; as long as the cyclotron radius is short compared to the width of the sample, and the disordered potential is weak, it will be unable to scatter them across the sample (Streda *et al.* 1987, Jain and Kivelson 1988, Büttiker 1988). An arbitrary amount of forward scattering between edge states will not break the quantisation, since by unitarity the transmission coefficients must still sum to N. However it must be emphasised that the Fermi level is typically pinned near the center of the LL due to its high density of states in a macroscopic 2DEG of the type typically used for quantum Hall measurements, and the Fermi level resides in the region of edge states only very briefly at the center of a plateau. Hence the existence of plateaus cannot be explained by this simple argument; they arise due to localisation, which prevent states near the center of the LL from carrying current across the sample [see Stone *et al.* (1990) for a detailed discussion of this point].

3.3 Reciprocity symmetry of resistance

Although I derived Hall quantisation from a more detailed physical argument above, it is possible simply to solve Equation (4) in general for a four-probe measurement with the total current I through the current probes and zero current in the voltage probes. The conductance matrix g_{mn} is non-invertible in the whole vector space because its rows all sum to zero, but since the currents considered also always sum to zero it is invertible in the relevant subspace, and it is easy to confirm that voltage *differences* are always uniquely determined (Stone and Szafer 1988). If the current probes are designated 1, 3 and the voltage probes are 2, 4, one finds the resistance

$$R_{13,24} = \frac{h}{e^2}\left[\frac{T_{21}T_{43} - T_{41}T_{23}}{S}\right], \tag{17}$$

where S is any 3×3 sub-determinant of the conductance matrix. If one substitutes into Equation (17) the conditions $T_{nm} = N\delta_{m,n+1}$, considered in our physical argument above, it is easily seen to yield a quantised Hall resistance $h/(Ne^2)$.

Equation (17) is also useful in establishing the exact time-reversal symmetry of quantum transport measurements (Büttiker 1986). As noted above, the scattering matrix is not only unitary but symmetric in the presence of time-reversal symmetry, hence $g_{mn}(B) = g_{nm}(-B)$. It is then easily seen from Equation (17) that

$$R_{13,24}(-B) = R_{24,13}(B) \tag{18}$$

under field-reversal, *i.e.* time-reversal symmetry connects two *different* measurements in which current and voltage probes are interchanged and the field is reversed. This relation is referred to as *reciprocity symmetry* and was known prior to the work of Büttiker, although not on such a fundamental basis. The familiar symmetry of longitudinal resistance and antisymmetry of Hall resistance measurements under field reversal results not simply from time-reversal symmetry but from additional spatial symmetries of the scattering potential in the sample. Such symmetries exist on average in macroscopic samples, but are essentially always violated in mesoscopic samples at low temperatures where interference effects dominate. It was strikingly shown by Benoit *et al.* (1986) that there is a complete absence of symmetry under field-reversal for a fixed measuring configuration in samples with well-developed UCF effects, but that the reciprocity symmetry of Equation (18) is satisfied within experimental accuracy upon interchanging voltage and current probes.

3.4 Interference effects

The relationship between S-matrix coefficients and transport coefficients makes it clear that the latter will show significant interference effects in phase-coherent samples, even in the presence of static disorder. For example it was shown in the original work of Aharonov and Bohm that the scattering matrix of a doubly-connected system threaded by a flux will oscillate periodically with the enclosed flux due to the coupling between the vector potential and the phase of the quantum wavefunctions. It is straightforward then to show that for a one-dimensional ring-shaped conductor connected to leads one expects Aharonov-Bohm oscillations of order the average conductance even with additional scattering in the ring (Gefen *et al.* 1984). The crucial point is that elastic scattering only alters the phase of the wavefunction in a fixed, sample-specific manner, without averaging over that phase. An additional source of averaging, such as the energy fluctuations associated with inelastic scattering, is needed to eliminate the contributions due to interference. However in mesoscopic samples at low temperatures the inelastic scattering time can be much longer than the diffusive transit time across the sample, and the full sample-specific interference effects will show up in transport measurements. A physical analogy is to a laser beam transmitted through a static medium with randomly-varying dielectric constant; such a beam when detected will create a *speckle pattern* in which the spatial intensity varies rapidly due to the complex interference of the various light paths to a given point.

It is then natural to consider the electronic system using the analogue of ray optics, which is the semi-classical path integral formulation of quantum mechanics. To consider transmission and d.c. conductance we are interested in the Green function at fixed energy, ε_F, instead of at a fixed time, so we employ the formulation of Gutzwiller

(1990) in which the semiclassical (WKB) Green function is expressed as a sum over classical paths from $\mathbf{r}$ to $\mathbf{r}'$ at energy E. Since from Equation (14) we are interested in derivatives of the product of two Green functions, we use Gutzwiller's expression and define the electronic intensity as

$$\begin{aligned} I(\mathbf{r},\mathbf{r}') &= G(E,\mathbf{r},\mathbf{r}')G(E,\mathbf{r}',\mathbf{r}) \\ &= \frac{1}{\hbar^3}\sum_{p(\mathbf{r},\mathbf{r}')}\sum_{q(\mathbf{r},\mathbf{r}')}\sqrt{D_p}\sqrt{D_q}\exp\left\{\frac{i}{\hbar}\left[S_p(E)-S_q(E)\right]-i\frac{\pi}{2}\mu_{pq}\right\}, \end{aligned} \quad (19)$$

where S_p is the action integral along classical path p at energy E, D_p is a positive amplitude given by the stability of the path with respect to variations in its initial momentum and μ_{pq} is the difference of the Maslov indices of the paths p and q, equal to the number of conjugate points along each trajectory (the details of this formulation will not be relevant for the qualitative discussion here, so we need not go into them in depth). The terms in the double sum for which $p \neq q$ represent the interference of the different paths between the points $\mathbf{r}$ and $\mathbf{r}'$.

To get a feeling for how important such interference might be in a disordered but weak potential in which many paths may exist between any two points, assume that there are N_p paths with equal amplitudes (which we take to be unity) and equal Maslov indices, so that

$$I(\mathbf{r},\mathbf{r}') = \sum_{p(\mathbf{r},\mathbf{r}')}\sum_{q(\mathbf{r},\mathbf{r}')}\exp\left\{\frac{i}{\hbar}\left[S_p(E)-S_q(E)\right]\right\}. \quad (20)$$

Assume that the phases of all terms with $p \neq q$ are uncorrelated and vanish upon averaging over disorder. It is then a trivial exercise to show that

$$\frac{\langle(\delta I)^2\rangle}{\langle I\rangle^2} = 1 + \frac{1}{N_p}; \quad (21)$$

the local relative intensity fluctuations are order unity. In fact many speckle patterns obey Equation (21) to good accuracy even though the argument was over-simplified, and we expect similarly complex local transmission fluctuations in electronic systems.

As noted above, the particular interference pattern should be sensitive to external parameters which couple to the phase, such as energy or magnetic field, and of course to the particular realisation of the random scattering potential. To estimate the scale of sensitivity to E and B one need only consider the change in phase of a given path as these parameters are varied. Let Λ represent an external parameter upon which the action depends, and assume that the classical path smoothly evolves as a function of this parameter over the scale of interest (so that its identification remains unambiguous). Then

$$\Delta\phi = \frac{1}{\hbar}\left[S_p(\Lambda+\Delta\Lambda)-S_p(\Lambda)\right] \approx \frac{1}{\hbar}\frac{\partial S_p}{\partial\Lambda}(\Delta\Lambda). \quad (22)$$

When $\Lambda = E$, we have from classical mechanics the relation $\partial S_p/\partial\Lambda = T_p$, the time required to traverse the path. The interference pattern will fluctuate when most paths which traverse a sample of length L have changed their phase by order unity, so estimating the correlation scale E_c by $\Delta\phi(\Delta E = E_c) \approx 1$ yields

$$E_c \approx \frac{\hbar}{\langle T_p\rangle}, \quad (23)$$

where $\langle T_p \rangle$ is the mean traversal time. The mean time is of order D/L^2 in the diffusive case, where D is the diffusion constant, so we have

$$(E_c)_{\text{diff}} \approx \frac{\hbar D}{L^2}. \tag{24}$$

When the parameter Λ is a uniform field B perpendicular to the plane of motion, the action depends on the field through a term $(e/c) \int \mathbf{A} \cdot d\mathbf{l}$ where the line integral is along the path. If we assume that change in field ΔB has a negligible effect on the path, we have

$$\Delta\phi(\Delta B) = \frac{e}{\hbar c} \left\langle \int \Delta\mathbf{A} \cdot d\mathbf{l} \right\rangle. \tag{25}$$

For a diffusive trajectory which intersects itself many times, the line integral will just give ΔB times the area enclosed, which for a path crossing the sample will be of order the sample area. Hence

$$(B_c)_{\text{diff}} \approx \frac{L^2}{(hc/e)} \Delta B. \tag{26}$$

A slightly more complicated argument along these lines can be used in the diffusive limit to estimate the sensitivity of the interference pattern to changes in the scattering potential (Feng *et al.* 1986). Without reproducing the detailed argument here, let me note that it is clear from the above analysis that a change in the scattering phase shift of order unity anywhere in a given path will alter its phase sufficiently. Since in diffusion a number of scattering event of order L^2 occurs in traversing the sample, a change of this magnitude in the scattering potential anywhere in the sample will be sufficient to alter the phase of a fraction of order unity of all the paths in 2D and 1D. Hence a phase-coherent 2D conductor will be sensitive to the motion of a single impurity in a manner which is independent of the size of the system. This extreme sensitivity to small changes in the sample leads to a new kind of quantum low-frequency noise (Feng *et al.* 1986) which has now been convincingly observed in a number of experiments (see, for example, Birge *et al.* 1989). I will not discuss this aspect of the theory further.

Finally, it is worth noting that these general ideas concerning the sensitivity of interference effects to changing parameters have recently been shown to apply to the ballistic as well as the diffusive regime (Jalabert *et al.* 1990). It is only necessary to evaluate the average of $\partial S/\partial\Lambda$ for the particular classical dynamics relevant to the system of interest.

In addition to the fluctuations arising from the interference of unrelated classical paths between $\mathbf{r}$ and $\mathbf{r}'$ in Equation (19), there are interference effects associated with paths related by time-reversal symmetry, which lead to weak localisation. By a well-known argument which I shall not repeat here (Bergmann 1984), these effects double the intensity $I(\mathbf{r}, \mathbf{r})$ in the absence of magnetic field and spin-orbit scattering. The suppression of these effects by a magnetic field gives the distinctive negative magnetoresistance associated with weak localisation. A quantitative theory of weak localisation based on the semiclassical approach has been developed by Chakravarty and Schmid (1986).

Although it is possible to estimate the local fluctuations in intensity and the correlation lengths from elementary considerations using the semiclassical approach, one needs to evaluate the amplitudes in Equation (19), and most importantly to understand the spatial correlations in the intensity, to evaluate the interference effects in the total transmission coefficient (*i.e.* the conductance). This is a difficult task which has not been accomplished so far using the semiclassical technique. Hence we leave this approach at this point and employ the impurity-averaged Green function technique, which provides a systematic perturbation theory for the quantities of interest.

4 Impurity-average technique in real space

Below I will review the aspects of the impurity-average Green function technique required to develop the theory of universal conductance fluctuations. Almost identical calculations arise in the theory of weak localisation and persistent currents, so the techniques are of wide applicability. I choose to use the less common formulation in real space primarily because we have recently shown (Xiong and Stone 1991) that this approach can be generalised to arbitrary magnetic fields which satisfy $N \gg 1$ (where N is the Landau level index) within the self-consistent Born approximation, whereas the previous theory only applied at fields for which the cyclotron radius was much larger than the elastic mean free path. The condition $r_c \gg l$ is violated at moderate fields in most two-dimensional electron gas systems, so the generalisation of the theory is of some importance.

4.1 Historical background

The impurity-averaging formalism dates back to Edwards (1958), having been proposed and used to derive the Drude conductivity only a few months after Greenwood's paper in which he first derived Equation (15) for the conductivity. At zero magnetic field and weak disorder the technique provides a systematic expansion in the small parameter $(\varepsilon_F \tau/\hbar)^{-1}$, where τ is the elastic mean free time (see, for example, Abrikosov *et al.* 1965). The Drude conductivity can be obtained by treating a particular contribution known as the ladder diagrams. The technique was used in the study of superconductivity during the 60's, and Langer and Neal (1966) discovered that the perturbation theory for the conductivity was formally divergent in dimensions $d \leq 2$, in the sense that there existed contributions of lower order in $(\varepsilon_F \tau)^{-1}$ which depended upon the lower momentum cut-off and scaled with the system size. However it was not until the work of Abrahams *et al.* (1979) that this divergence was shown to indicate the nonexistence of extended states in the infinite system at $T = 0$ in 2D. The divergent contribution studied by Langer and Neal, referred to either as the maximally-crossed graphs or as the *cooperon* contribution because of its role in the study of superconductivity, then became the basis of the theory of weak localisation. The divergence of such contributions is always ultimately cut off by finite temperature effects, and thus only leads to a small correction to g in good 2D conductors at any practical temperature.

In 1985 Altshuler, and Lee and Stone, independently discovered that the ladder or *diffuson* contribution, which was well-behaved for the average conductivity, was divergent in $d \leq 4$ for the variance of g. This led to conductance fluctuations in phase-coherent metals that are anomalously large from the classical point of view, and formed the basis of the theory of universal conductance fluctuations. The theories of WL and UCF only included the effect of a magnetic field through its coupling to the phase of the wavefunction, as discussed in Section 3.4. A version of the impurity-averaging technique appropriate for a 2DEG in a high field where Landau level (LL) quantisation is important was developed by Ando (1974, 1975) based on the *self-consistent Born approximation* (SCBA). Carra, Chalker and Benedict (1989) showed that this was a systematic expansion in $1/N$ in the limit of a short-ranged potential, where N is the LL index. Again the expansion for the conductivity was shown to contain divergences in terms of lower order, indicating the importance of the localisation effects in such systems which are essential to the quantum Hall effect as discussed above. Hence this theory is also not expected to apply to the infinite system at $T = 0$; nonetheless for systems at finite temperature the perturbation theory appears to have a reasonable range of validity near the center of the LL. For example the SCBA predicts that the height of the peaks in σ_{xx} is linear in the Landau index, N, in the limit of a short-ranged potential, and this prediction is often quantitatively satisfied in 2DEG's (see, for example, Luo *et al.* 1989). The perturbation theory should be even better near the center of the LL in mesoscopic systems which are of course far from the limit of an infinite volume. Below we will show for the first time that the there exists a generalisation of UCF theory to arbitrary magnetic field, valid to order $(\varepsilon_F\tau)^{-1}$ when the cyclotron radius $r_c > l$, and valid to order $1/N$ when $r_c < l$.

4.2 White noise model for average Green function

The basic principles of the impurity-averaging technique are as follows.

1. To express quantities of physical interest in terms of the electronic Green functions for a given configuration of the random impurity potential $V(\mathbf{r})$. If the system is assumed non-interacting, all such quantities will be expressible as products of the advanced and retarded one-particle Green functions (1PGF), $G^{\pm}$, introduced above before Equation (15).

2. To express the Green functions as a perturbation expansion in $V(\mathbf{r})$ using the Dyson equation for the 1PGF.

3. To average the quantities of interest over these realisations; this averaging is equivalent to introducing a special kind of static two-body interaction between electrons which leads to a non-trivial perturbation expansion.

Since our system is non-interacting the 1PGF is just the inverse of the one-body hamiltonian which we take to be $H = H_0 + V(\mathbf{r})$, with

$$H_0 = \frac{1}{2m}\left(\mathbf{P} - \frac{e}{c}\mathbf{A}\right)^2, \tag{27}$$

with m the effective mass and $\mathbf{A}(\mathbf{r})$ the vector potential of a uniform magnetic field perpendicular to the transport direction. Let $G_0^{\pm} = [E - H_0 \pm i\eta]^{-1}$ where the superscript $\pm$ gives the sign of the infinitesimal η which fixes the analytic properties of G, but will be suppressed henceforth except where it is needed to resolve an ambiguity.

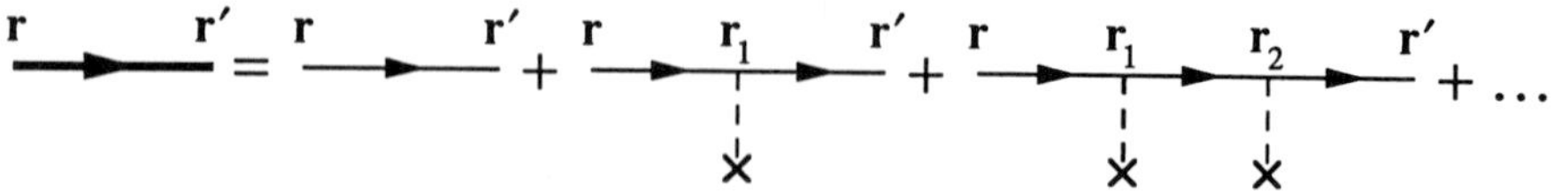

Figure 5: *Dyson equation for the one-particle Green function for a given configuration of impurities, before averaging.*

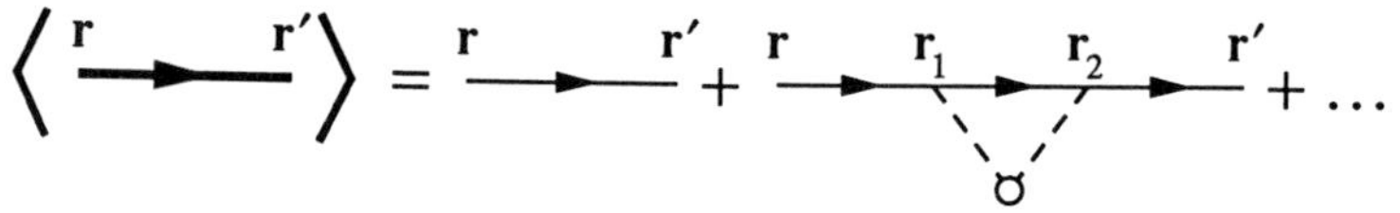

Figure 6: *Dyson equation for the one-particle Green function after averaging; dashed lines connected by a circle denote the effective interaction.*

The operator form of the Dyson equation for $G^{\pm}$ before averaging is

$$\begin{aligned} G(E) &= [E - H_0 - V]^{-1} \\ &= \left[G_0^{-1}(1 - G_0 V)\right]^{-1} \\ &= \left[\sum_{j=0}^{\infty}(G_0 V)^j\right] G_0 \\ &= G_0 + G_0 V G. \end{aligned} \tag{28}$$

This equation is represented by the sequence of diagrams in Figure 5, where the crosses represent interactions with the random potential. Consider the third term in Figure 6, representing the term with $j = 2$ in the third line of Equation (28). Upon averaging this over impurity configurations in a real space representation this takes the form

$$\int d\mathbf{r}_1\, d\mathbf{r}_2\, G_0(\mathbf{r}, \mathbf{r}_1)\, G_0(\mathbf{r}_1, \mathbf{r}_2)\, G_0(\mathbf{r}_2, \mathbf{r}')\langle V(\mathbf{r}_1) V(\mathbf{r}_2)\rangle, \tag{29}$$

where we see that the statistical average (denoted either by angle brackets or an overbar below) of the random potential at two points in space enters as an effective elastic two-body interaction (since the energy is unchanged along each line). Diagrammatically we represent this by joining the two crosses to make a single dashed line as shown in Figure 6. Different statistical models for the correlation of moments of the potential are known to give essentially identical results in the low-field limit, so we use the simplest one, the white noise model, which yields a particularly simple theory in the limit of high B. In the WN model

$$\langle V(\mathbf{r})\rangle = 0 \tag{30}$$

$$\langle V(\mathbf{r}) V(\mathbf{r}')\rangle = c_i u^2 \delta(\mathbf{r} - \mathbf{r}'). \tag{31}$$

All odd higher moments are zero, and all even higher moments are pairwise decompositions in terms of the second moment; c_i is the number density of impurities and u^2 may be regarded as the mean-squared strength of the scattering potential in Fourier space.

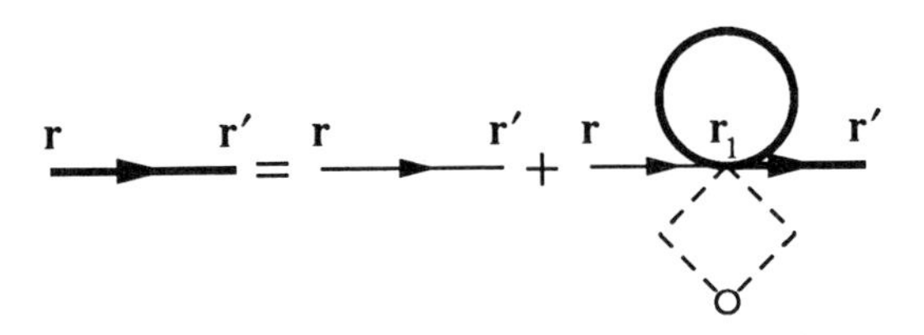

Figure 7: *Diagrammatic representation of the self-consistent equation for the Green function in the self-consistent Born approximation.*

5 Average Green function in SCBA

In the WN model the disorder average is identical to a static two-body interaction for all terms in the perturbation expansion for the disorder-averaged 1PGF, $\overline{G}(\mathbf{r},\mathbf{r}')$. Hence $\overline{G}(\mathbf{r},\mathbf{r}')$ satisfies an integral equation in terms of the proper self-energy insertion of the usual type:

$$\overline{G}(\mathbf{r},\mathbf{r}') = G_0(\mathbf{r},\mathbf{r}') + \int d\mathbf{r}_1\, d\mathbf{r}_2\, G_0(\mathbf{r},\mathbf{r}_1)\,\Sigma(\mathbf{r}_1,\mathbf{r}_2)\,\overline{G}(\mathbf{r}_2,\mathbf{r}'). \tag{32}$$

The self-consistent Born approximation for the self-energy is to approximate Σ as follows:

$$\Sigma(\mathbf{r},\mathbf{r}') \approx c_i u^2\, \overline{G}(\mathbf{r},\mathbf{r}')\,\delta(\mathbf{r}-\mathbf{r}'); \tag{33}$$

this is shown diagramatically in Figure 7. When inserted into Equation (32) the approximation yields the self-consistent equation for $\overline{G}(\mathbf{r},\mathbf{r}')$,

$$\overline{G}(\mathbf{r},\mathbf{r}') = G_0(\mathbf{r},\mathbf{r}') + \int d\mathbf{r}_1\, G_0(\mathbf{r},\mathbf{r}_1)\, c_i u^2\, \overline{G}(\mathbf{r}_1,\mathbf{r}_1)\,\overline{G}(\mathbf{r}_1,\mathbf{r}'). \tag{34}$$

The significance of this equation can be understood by applying $(E - H_0)$ to both sides and using $\langle \mathbf{r}|(E-H_0)G_0|\mathbf{r}'\rangle = \delta(\mathbf{r}-\mathbf{r}')$ to obtain

$$\left[E - c_i u^2\, \overline{G}(\mathbf{r},\mathbf{r}) - H_0\right] \overline{G}(\mathbf{r},\mathbf{r}') = \delta(\mathbf{r}-\mathbf{r}'). \tag{35}$$

Since the system is translationally invariant on average, $\overline{G}(\mathbf{r},\mathbf{r},E)$ is just a complex function of energy, independent of $\mathbf{r}$, and we shall denote this function simply by $\overline{G}(E)$ henceforth. It follows that $\overline{G}(\mathbf{r},\mathbf{r}')$ satisfies exactly the same equation as $G_0(\mathbf{r},\mathbf{r}')$ except that the energy E is replaced by the complex energy $z(E) = E - c_i u^2\, \overline{G}(\mathbf{r},\mathbf{r})$. Hence

$$\overline{G}(\mathbf{r},\mathbf{r}',E) = G_0(\mathbf{r},\mathbf{r}',z). \tag{36}$$

This is the fundamental result of the SCBA for the 1PGF in the WN limit.

From now on we specialise to the two-dimensional case; generalisation to 3D is straightforward. Using Equation (35) we can immediately write an expression for $\overline{G}(\mathbf{r},\mathbf{r}')$ in the SCBA valid for arbitrary field by transcription into the expression for G_0:

$$\overline{G}(\mathbf{r},\mathbf{r}',E) = \sum_{n=0}^{\infty} \frac{P_n(\mathbf{r},\mathbf{r}')}{E - E_n - c_i u^2\, \overline{G}(E)}, \tag{37}$$

where $\overline{G}(E) = \overline{G}(\mathbf{r},\mathbf{r},E)$, and $P_n(\mathbf{r},\mathbf{r}')$ is the projection operator onto the nth LL in real space, and $E_n = (n+\frac{1}{2})\hbar\omega_c$ is the energy of the nth LL. It is possible to calculate P_n exactly in 2D [see the article by Haldane in Prange (1986)] and one finds

$$P_n(\mathbf{r},\mathbf{r}') = \frac{1}{2\pi\lambda^2}\exp\left(-\frac{R^2}{4\lambda^2}\right) L_n\left(\frac{R^2}{2\lambda^2}\right)\exp\left(\frac{ie}{\hbar c}\int_{\mathbf{r}'}^{\mathbf{r}}\mathbf{A}\cdot d\mathbf{l}\right), \tag{38}$$

where $R = |\mathbf{r}-\mathbf{r}'|$, $\lambda^2 = (\hbar c/eB)$ is the square of the magnetic length and L_n is the nth Laguerre polynomial. The line integral in the phase factor is taken along the straight line from $\mathbf{r}'$ to $\mathbf{r}$. Note that $P_n(0) = 1/(2\pi\lambda^2)$, since $L_n(0) = 1$ for all n. Using this fact and setting $\mathbf{r} = \mathbf{r}'$ in Equation (37) we obtain the self-consistent equation for $\overline{G}(E)$,

$$2\pi\lambda^2\overline{G}(E) = \sum_{n=0}^{\infty}\frac{1}{E - E_n - c_i u^2\,\overline{G}(E)}. \tag{39}$$

To obtain $\overline{G}(\mathbf{r},\mathbf{r}')$ in the limit of high fields, in which the broadening of the LLs by disorder is much less than their spacing $\hbar\omega_c$, we note that in this case $E = \varepsilon_F$ will be pinned near a particular LL with energy $E_N = \left(N+\frac{1}{2}\right)\hbar\omega_c$, and $\overline{G}(E)$ in Equation (37) will be dominated by the term with $n = N$ in the sum:

$$2\pi\lambda^2\overline{G}(0, E\approx E_N) \approx 2\pi\lambda^2\overline{G}_N(E) \approx \frac{1}{E - E_N - c_i u^2\,\overline{G}_N(E)}. \tag{40}$$

Solving this quadratic equation for $G_N(E)$ gives the explicit result

$$G_N^{\pm}(E) = \frac{e^{\pm i\theta(E)}}{2\pi\lambda^2\nu}, \tag{41}$$

where $\nu^2 = c_i u^2/2\pi\lambda^2$ and $\cos\theta(E) = (E - E_N)/2\nu$. Substitution into Equation (37) with terms $n \neq N$ neglected then yields

$$G_N^{\pm}(\mathbf{r},\mathbf{r}',E) \approx P_N(\mathbf{r},\mathbf{r}')G_N^{\pm}(E). \tag{42}$$

The spatial range of $P_N(\mathbf{r},\mathbf{r}')$ is obtained by maximising a polynomial of degree N against the gaussian fall-off with range λ. Just as for the wavefunctions this gives a maximum at the cyclotron radius $r_c = (N+\frac{1}{2})^{1/2}\lambda$. This approximation for $\overline{G}(\mathbf{r},\mathbf{r}')$ is valid when $r_c \ll l$ and shows explicitly that the range of $\overline{G}(\mathbf{r},\mathbf{r}')$ is of order r_c and not l in this limit. From the imaginary part of this expression one can obtain the shape of the broadened LL in the WN limit, which is simply a semi-circle of radius $\nu \approx (\hbar^2\omega_c/\tau)^{1/2}$ (Ando *et al.* 1975). In the high field limit the self-consistent calculation of the Green function is crucial because the LL is infinitely degenerate without disorder, and substitution of G_0 on the LHS of Equation (35) would give an infinite spike at $\varepsilon_F = E_N$ and zero elsewhere.

In the zero-field limit the mean density of states is not altered by disorder to leading order, and it is permissible to replace $\overline{G}(E)$ by $G_0(E)$ on the LHS of Equation (35) and neglect the real part of G_0. Since $\mathrm{Im}G_0^{\pm} = \mp\pi\rho_F$, and by definition the elastic scattering rate in Born approximation is $1/\tau = 2\pi c_i u^2\rho_F$, where ρ_F is the density of states at the Fermi energy, we have

$$\overline{G^{\pm}}(\mathbf{r},\mathbf{r}',\varepsilon_F, B=0) = G_0^{\pm}(\mathbf{r},\mathbf{r}',\varepsilon_F \pm i/2\tau, B=0). \tag{43}$$

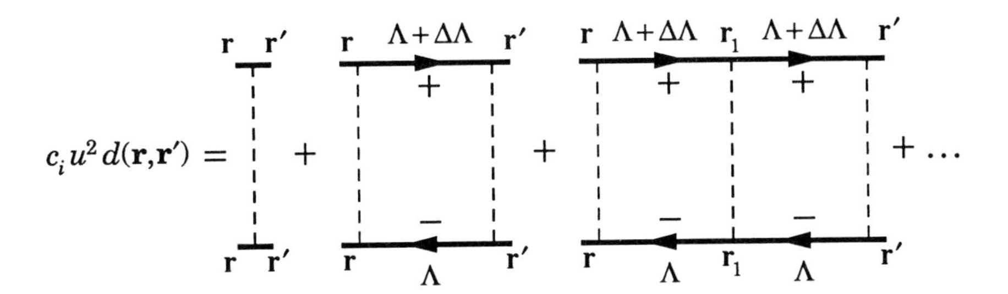

Figure 8: *Class of polarisation diagrams in real space whose sum gives the diffuson contribution.*

The rapidly-varying part of $G_0^{\pm}(\mathbf{r},\mathbf{r}') \sim \exp(\pm ik_F|\mathbf{r}-\mathbf{r}'|)$, where $k_F = (2m\varepsilon_F/\hbar^2)^{1/2}$ is the Fermi wave-vector, so substituting $\varepsilon_F \to \varepsilon_F \pm i/2\tau$ and expanding to leading order in $(\varepsilon_F\tau)^{-1}$ one obtains the familiar result

$$\overline{G^{\pm}}(\mathbf{r},\mathbf{r}',\varepsilon_F,B=0) = G_0^{\pm}(\mathbf{r},\mathbf{r}',\varepsilon_F,B=0)\exp(-|\mathbf{r}-\mathbf{r}'|/2l). \tag{44}$$

The decay of $\overline{G}$ on a scale l is due to averaging over the random phase-shift of G for a given impurity configuration; it does not mean that the modulus of G decays this rapidly for a given configuration. Nonetheless we see that the average Green function is short-ranged for arbitrary field within the SCBA, with a range given by the shorter of r_c and l which we shall refer to as $l_{\min}$.

5.1 Diffuson correlator

Referring back to Equation (14) one sees that the averages of products of two Green functions are needed in order to calculate the average conductance. This is done by expanding each factor $G(\mathbf{r},\mathbf{r}')$ in a power series in $V(\mathbf{r})$ using Equation (28), and then averaging the product of these two series. Two types of terms result. The first type only involve impurity interactions along each line separately; this corresponds to the factorisation $\overline{GG} = \overline{G}\,\overline{G}$, and all such contributions can be accounted for by making the appropriate self-energy insertion in each line separately in the manner described above. The second type of term arises from the creation of interaction lines that join the two Green functions as shown in Figure 8, and correspond to a non-factorisable contribution. These vertex or polarisation corrections can introduce qualitatively new features because now it is possible to correlate Green functions with different energy, magnetic field and analyticity. These correlators can also be long-ranged in space compared to $\overline{G}\,\overline{G}$, each factor of which must decay over a distance of $l_{\min}$. In particular, the correlator $\overline{G^+G^-}$ can have long-ranged behaviour because of the relation $G^+(\mathbf{r},\mathbf{r}') = [G^-(\mathbf{r}',\mathbf{r})]^*$, which makes it possible for the phase shift at each scattering event to cancel.

This is true for the particular infinite class of diagrams known as the *diffuson* diagrams shown in Figure 8 which contribute to the correlator $\overline{G^+(\mathbf{r},\mathbf{r}')\,G^-(\mathbf{r}',\mathbf{r})}$. Note that in the WN model each impurity interaction line has a single point in space associated with it, so it is tempting to associate a given polarisation diagram (before integrating over intermediate positions) with a particular pair of trajectories in space which visit the same set of impurity sites in some order. With this interpretation the diffuson contribution is precisely the interference of a given trajectory with itself (complex conjugated) and corresponds to the diagonal approximation discussed in Section 3.4 above. Define the contribution of these diagrams to be $c_i u^2 d(\mathbf{r},\mathbf{r}')$; then it is clear that d satisfies the integral equation

$$d(\mathbf{r},\mathbf{r}') = \delta(\mathbf{r}-\mathbf{r}') + \int d\mathbf{r}_1\, d_0(\mathbf{r},\mathbf{r}_1)\, d(\mathbf{r}_1,\mathbf{r}'), \tag{45}$$

where

$$d_0(\mathbf{r},\mathbf{r}') = c_i u^2\, \overline{G}^+(\mathbf{r},\mathbf{r}',\Lambda+\Delta\Lambda)\, \overline{G}^-(\mathbf{r}',\mathbf{r},\Lambda), \tag{46}$$

and Λ represent the external parameters such as Fermi energy, frequency, or magnetic field which may be different along each line. Equation (45) is just a geometric series in the operator d_0 and may be solved easily in operator form

$$d = 1 + d_0 d; \quad d = [1-d_0]^{-1}. \tag{47}$$

Introduce the eigenfunctions, $\chi_j(\mathbf{r})$, and eigenvalues, $1-\xi_j^2$, of the integral operator d_0 defined by

$$\int d\mathbf{r}'\, d_0(\mathbf{r},\mathbf{r}')\chi_j(\mathbf{r}') = (1-\xi_j^2)\chi_j(\mathbf{r}). \tag{48}$$

Then we can write a spectral representation of d,

$$d(\mathbf{r},\mathbf{r}') = \sum_{j=1}^{\infty} \frac{\chi_j(\mathbf{r})\chi_j^*(\mathbf{r}')}{\xi_j^2}. \tag{49}$$

First consider the case where $\Delta\Lambda = 0$ so d_0 is given by the real positive function $\left|\overline{G^+}(|\mathbf{r}-\mathbf{r}'|)\right|^2$ whose range is l_{min}. We follow a technique introduced by Altshuler *et al.* (1980) in order to convert the integral equation (45) into a diffusion equation. Since d_0 is short-ranged we can regard the eigenfunction $\chi(\mathbf{r})$ in Equation (45) as slowly-varying and expand it around $\mathbf{r}$:

$$\chi(\mathbf{r}') \approx \chi(\mathbf{r}) + \nabla\chi\cdot(\mathbf{r}-\mathbf{r}') + \tfrac{1}{2}\sum_{a,b} \nabla_a\nabla_b\chi\,(\mathbf{r}-\mathbf{r}')_a(\mathbf{r}-\mathbf{r}')_b + \dots. \tag{50}$$

Substitution of this expansion into Equation (45) yields a differential equation of the form

$$\left(C_2\nabla^2 + C_1\right)\chi_j(\mathbf{r}) = (1-\xi_j^2)\chi_j(\mathbf{r}), \tag{51}$$

where C_2 and C_1 are constants and we have used the fact that d_0 is even in $\mathbf{r}-\mathbf{r}'$ to eliminate terms linear in $\mathbf{r}-\mathbf{r}'$ and simplify the quadratic terms. The constant C_1 is, from Equations (46) and (48),

$$C_1 = c_i u^2 \int d\mathbf{r}' \left|\overline{G^+}(|\mathbf{r}-\mathbf{r}'|)\right|^2. \tag{52}$$

Equation (37) shows that in general the spatial dependence of $\overline{G}(\mathbf{r},\mathbf{r}')$ comes only from the projection operators $P_n(\mathbf{r},\mathbf{r}')$, which satisfy

$$\int d\mathbf{r}' P_n(\mathbf{r},\mathbf{r}')P_m(\mathbf{r}',\mathbf{r}) = \delta_{mn}P_n(\mathbf{r},\mathbf{r}) = (2\pi\lambda^2)^{-1}\delta_{mn}.$$

Thus one can simplify Equation (52) to

$$C_1 = \frac{c_i u^2}{2\pi\lambda^2} \sum_n^\infty \frac{1}{\left|E - E_n - c_i u^2 \overline{G}^+(E)\right|^2}. \tag{53}$$

If we now use the identity

$$\begin{aligned}
&\sum_n^\infty \frac{1}{\left|E - E_n - c_i u^2 \overline{G}^+(E)\right|^2} \\
&= \frac{1}{c_i u^2 \left[\overline{G}^+(E) - \overline{G}^-(E)\right]} \sum_n^\infty \left[\frac{1}{E - E_n - c_i u^2 \overline{G}^+(E)} - \frac{1}{E - E_n - c_i u^2 \overline{G}^-(E)}\right] \\
&= \frac{1}{c_i u^2 \left[\overline{G}^+(E) - \overline{G}^-(E)\right]} 2\pi\lambda^2 \left[\overline{G}^+(E) - \overline{G}^-(E)\right],
\end{aligned} \tag{54}$$

we find that $C_1 = 1$. In deriving this result we have used the self-consistent Equation (39) relating the Green function $\overline{G}(E)$ and the self-energy $c_i u^2 \overline{G}(E)$, valid for an arbitrary magnetic field. The exact cancellation giving $C_1 = 1$ is a manifestation of a general Ward identity relating the polarisation vertex (here represented just by the prefactor $c_i u^2$) and the Green function that will be valid even in the presence of interactions if a consistent approximation is made for all quantities. We shall see below that $C_1 = 1$ is required for there to be a diffusion pole in $d(\mathbf{r}, \mathbf{r}')$; the robustness of this condition with respect to interactions ensures that diffusive behaviour occurs at long wavelengths even in the presence of interactions as we expect for a Fermi liquid (neglecting localisation effects).

The constant C_2 has the dimensions of length squared so we define $C_2 = l_0^2$, where

$$\begin{aligned}
l_0^2 &= \frac{c_i u^2}{2d} \int d\mathbf{r}' \, (\mathbf{r} - \mathbf{r}')^2 \left|\overline{G^+}(|\mathbf{r} - \mathbf{r}'|)\right|^2 \\
&= \frac{c_i u^2}{2d} \int d\mathbf{R} \, R^2 \left|\overline{G^+}(R)\right|^2.
\end{aligned} \tag{55}$$

From this we see that the length l_0 is just the spatial range of the average Green function. Hence the differential equation that we need to solve to obtain the diffuson with the same external parameters on each line is simply

$$-\, l_0^2 \nabla^2 \chi_j(\mathbf{r}) = \xi_j^2 \chi_j(\mathbf{r}). \tag{56}$$

It is easily found from Equations (41) and (44) that

$$l_0^2 = \begin{cases} l^2/2, & l \ll r_c \\ r_c^2, & r_c \gg l, \end{cases} \tag{57}$$

consistent with our earlier explicit results for $\overline{G}(\mathbf{r}, \mathbf{r}')$.

Consider a rectangular 2D sample of length L_x and width L_y. The appropriate boundary conditions are that the diffuson must vanish on the surfaces normal to the direction of current flow (which we take to be the x-direction) since no excess density can build up in the leads (reservoirs), and the derivative of the diffuson must vanish at the walls because the current there (which is proportional to the derivative of the excess density) must vanish. Solving Equation (56) explicitly and substituting into Equation (49) yields

$$d(\mathbf{r},\mathbf{r}') = \sum_{m=1,n=0}^{\infty} \frac{\chi_{mn}(\mathbf{r})\,\chi^*_{mn}(\mathbf{r}')}{l_0^2(k_m^2+q_n^2)}, \tag{58}$$

where $q_n = n\pi/L_y$, $k_m = m\pi/L_x$, and the eigenfunctions χ_{mn} are appropriately normalised products of $\sin(k_m x)\cos(q_n y)$. Hence we have the familiar diffusion pole diverging as Q^{-2} for arbitrary magnetic field. If we take the $B = 0$ limit $l_0^2 \to D\tau$ (where $D = l^2/2\tau$ is the elastic diffusion constant), and at high field we find $l_0^2 \to \hbar D_N/\nu$, where ν is the LL broadening mentioned above and $D_N = r_c^2\nu/\hbar$ is the diffusion constant for a single LL in SCBA (Ando 1975). Note that the Q^{-2} divergence is automatically cut off by the sample size due to the boundary conditions imposed on Equation (56), which meant that the summation over k_m starts from $m = 1$ and not $m = 0$.

5.2 Generalised diffuson

For the theory of UCF we need to consider the generalised kernel d_0 of Equation (45) with the factor $\overline{G}^+$ taken at shifted values of the energy and magnetic field. The generalisation to finite ΔE is straightforward. The analysis leading to Equation (51) is unchanged, except that the expressions for C_1 and C_2 now contain the generalised kernel. It is easy to see that the correction to C_2 is negligible in the limit of interest, small Q, so we need only consider C_1 explicitly. The calculation of C_1 proceeds as before up to Equation (53) with the only difference that instead of the factor $\left|E - E_n - c_i u^2 \overline{G}^+(E)\right|^2$ in the denominator one has the factor $\left[E + \Delta E - E_n - c_i u^2 \overline{G}^+(E+\Delta E)\right]\left[E - E_n - c_i u^2 \overline{G}^-(E)\right]$ which requires the use of a slight generalisation of the identity used to derive C_1 above. Again using the self-consistent equation for $\overline{G}$ one finds

$$\begin{aligned} C_1(\Delta E) &= \frac{c_i u^2}{2\pi\lambda^2}\sum_n^{\infty} \frac{1}{\left[E+\Delta E - E_n - c_i u^2 \overline{G}^+(E+\Delta E)\right]\left[E - E_n - c_i u^2 \overline{G}^-(E)\right]} \\ &= \frac{\overline{G}^+(E+\Delta E) - \overline{G}^-(E)}{\overline{G}^+(E+\Delta E) - \overline{G}^-(E) - (\Delta E/c_i u^2)} \\ &\approx 1 - i(\Delta E \tau_0/\hbar), \end{aligned} \tag{59}$$

where we defined the generalised scattering time

$$\frac{i\hbar}{\tau_0} = c_i u^2\left[\overline{G}^+(E) - \overline{G}^-(E)\right], \tag{60}$$

and we can now ignore the energy difference in $\overline{G}^+$ because as usual we are assuming $\Delta E \ll \hbar/\tau_0$. It is easy to check from our above results that $\tau_0 \to \tau$ for $l \ll r_c$, and $\tau_0 \to \hbar/2\nu$ for $r_c \ll l$ and $\varepsilon_F = E_N$.

To generalise the diffusion equation (51) to finite ΔB at arbitrary field we note that each term in Equation (37) for $\overline{G}(\mathbf{r},\mathbf{r}')$ contains an overall phase factor from P_n,

$$\exp\left[\frac{ie}{\hbar c}\int_{\mathbf{r}'}^{\mathbf{r}} \mathbf{A}\cdot d\mathbf{l}\right] = \exp\left[\frac{ie}{\hbar c}\Delta\mathbf{A}(\mathbf{r})\cdot(\mathbf{r}-\mathbf{r}')\right], \tag{61}$$

where we have evaluated the line integral explicitly assuming a uniform field in the z-direction. This phase factor is the same for $\overline{G}^{\pm}(\mathbf{r}, \mathbf{r}')$ but of course changes sign when $\mathbf{r}$ and $\mathbf{r}'$ are interchanged as they are in the product defining d_0; hence $d_0(\mathbf{r}, \mathbf{r}', B + \Delta B)$ involves the difference of this phase (which cancels when $\Delta B = 0$). If we assume that the most rapid variation of d_0 with ΔB comes from this factor then we have

$$\begin{aligned} d_0(\mathbf{r}, \mathbf{r}', \Delta B) &= c_i u^2 \overline{G}^{+}(\mathbf{r}, \mathbf{r}', B + \Delta B) \overline{G}^{-}(\mathbf{r}', \mathbf{r}, B) \\ &= d_0(\mathbf{r}, \mathbf{r}') \exp\left[\frac{ie}{\hbar c} \Delta \mathbf{A}(\mathbf{r}) \cdot (\mathbf{r} - \mathbf{r}')\right]. \end{aligned} \tag{62}$$

To check that this is indeed the leading dependence on ΔB we recall that at low field the only dependence on B comes through this phase (Section 3), while at high field this phase is of order $\Delta B r_c^2/(\hbar c/e) \approx N \Delta B/B$. Since $N \gg 1$ by assumption, this phase can be of order unity for $\Delta B \ll B$, whereas the denominators in Equation (37) only vary on the scale $\Delta B \approx B$ and may be treated as constant on this scale.

We shall see that the contribution of this phase term to the diffusion equation will become important for $\Delta B r_c^2/(\hbar c/e) \ll 1$, so we can assume that the phase is small in Equation (62), expand it,

$$d_0(\mathbf{r}, \mathbf{r}', \Delta B) \approx d_0(\mathbf{r}, \mathbf{r}') \left\{1 + \frac{ie}{\hbar c} \Delta \mathbf{A}(\mathbf{r}) \cdot (\mathbf{r} - \mathbf{r}') + \frac{1}{2}\left[\frac{ie}{\hbar c} \Delta \mathbf{A}(\mathbf{r}) \cdot (\mathbf{r} - \mathbf{r}')\right]^2 + \ldots\right\}, \tag{63}$$

and multiply this expansion with the Taylor expansion of the $\chi(\mathbf{r})$ in Equation (48). As before we use the symmetry of $d_0(\mathbf{r}, \mathbf{r}')$ to eliminate terms odd in $\mathbf{r} - \mathbf{r}'$ and simplify the quadratic terms, and we find unsurprisingly that the effect on the diffusion equation is just to make the minimal substitution, $-i\boldsymbol{\nabla} \rightarrow [-i\boldsymbol{\nabla} - (e/\hbar c)\Delta \mathbf{A}(\mathbf{r})]$.

The final generalisation of the equation for the diffuson relevant to UCF concerns the effect of inelastic scattering, which can be taken into account by dressing the Green functions with appropriate interaction corrections to the self-energy and polarisation vertex. Unfortunately there is no general answer to the effect of inelastic processes on the diffuson, since this effect depends on the physical quantity which is being averaged. For the average diffusion constant we noted above that the Ward identity used to obtain $C_1 = 1$ and hence diffusive behaviour was valid in the presence of interactions if the self-energy and vertex are treated consistently. However, when calculating the mesoscopic fluctuations of g and other quantities, the diffuson typically represents the statistical correlation of different *measurements*, and the two Green functions involved cannot be connected by interaction lines. Hence interaction corrections appear in the imaginary part of the self-energy which do not cancel with the undressed vertex, and i/τ_{in} appears as an imaginary energy shift for the diffuson (Lee *et al.* 1987).

Thus the generalisation of Equation (56) valid in an arbitrary magnetic field, with non-zero ΔE, ΔB and τ_{in}, is

$$\left\{l_0^2\left[-i\boldsymbol{\nabla} - \frac{e}{\hbar c} \Delta \mathbf{A}(\mathbf{r})\right]^2 + \frac{\tau_0}{\tau_{\text{in}}} - i\frac{\Delta E \tau_0}{\hbar}\right\} \chi_j(\mathbf{r}) = \xi_j^2 \chi_j(\mathbf{r}). \tag{64}$$

The generalised diffuson is then given by the spectral representation of Equation (49) with the eigenvalues and eigenfunction obtained from this equation.

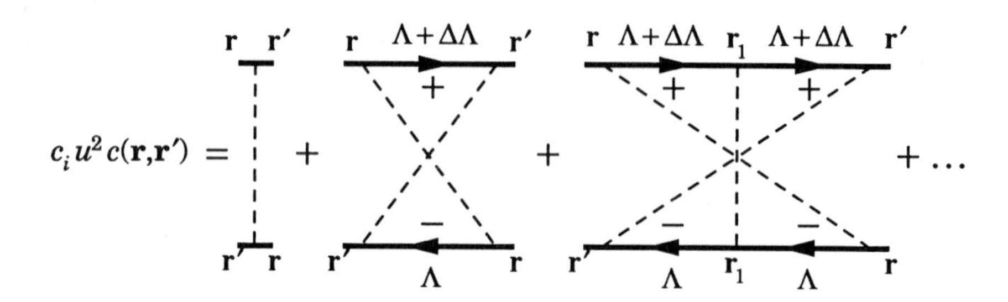

Figure 9: *Class of polarisation diagrams in real space whose sum gives the cooperon contribution.*

6 Cooperon correlator

The other important contribution to the average of the product of two Green functions comes from the *cooperon* diagrams shown in Figure 9. These contribute to $\overline{G^+(\mathbf{r},\mathbf{r}')\,G^-(\mathbf{r},\mathbf{r}')}$ (note that the second factor is no longer the complex conjugate of the first in general). If we define this contribution as $c_i u^2 c(\mathbf{r},\mathbf{r}')$, it satisfies an integral equation of exactly the same form as the diffuson,

$$c(\mathbf{r},\mathbf{r}') = \delta(\mathbf{r},\mathbf{r}') + \int d\mathbf{r}_1\, c_0(\mathbf{r},\mathbf{r}_1)\, c(\mathbf{r}_1,\mathbf{r}'), \tag{65}$$

where

$$c_0(\mathbf{r},\mathbf{r}') = c_i u^2\, \overline{G}^+(\mathbf{r},\mathbf{r}',\Lambda+\Delta\Lambda)\, \overline{G}^-(\mathbf{r},\mathbf{r}',\Lambda). \tag{66}$$

Equation (65) is again solved in operator form by $c = [1-c_0]^{-1}$ and if we introduce the eigenfunctions $\eta_j(\mathbf{r})$ and eigenvalues $1-\zeta_j^2$ of the integral operator c_0, defined by

$$\int d\mathbf{r}'\, c_0(\mathbf{r},\mathbf{r}')\, \eta_j(\mathbf{r}') = (1-\zeta_j^2)\eta_j(\mathbf{r}), \tag{67}$$

then we can write a spectral representation of c,

$$c(\mathbf{r},\mathbf{r}') = \sum_{j=1}^{\infty} \frac{\eta_j(\mathbf{r})\eta_j^*(\mathbf{r}')}{\zeta_j^2}. \tag{68}$$

Consider $c_0(\mathbf{r},\mathbf{r}',B=0)$. Since we have time-reversal symmetry at $B=0$ the wavefunctions can be chosen real, and it follows from their spectral representation that the Green functions are symmetric in their spatial arguments, $\overline{G}^\pm(\mathbf{r},\mathbf{r}') = \overline{G}^\pm(\mathbf{r}',\mathbf{r})$. Hence $\overline{G}^-(\mathbf{r},\mathbf{r}')$ *is* the complex conjugate of $\overline{G}^+(\mathbf{r},\mathbf{r}')$, and we have $c_0(\mathbf{r},\mathbf{r}',B=0) = d_0(\mathbf{r},\mathbf{r}',B=0)$; the same conclusion can be reached by noting that $\overline{G}^\pm$ is a function only of $|\mathbf{r}-\mathbf{r}'|$ at $B=0$. Thus all the steps used above for d_0 (including those involving a finite energy or frequency difference) go through to yield the differential equation

$$\left(-D\tau\boldsymbol{\nabla}^2 + \frac{\tau}{\tau_{\text{in}}} - i\frac{\Delta E\tau}{\hbar}\right)\eta_j(\mathbf{r}) = \zeta_j^2\eta_j(\mathbf{r}), \tag{69}$$

where we have used $\tau_0 = \tau$ since this equation is only valid for $B = 0$ unlike that for the diffuson. This equation will then yield a Q^{-2} pole which does not cancel in the conductivity and gives the weak localisation corrections.

However, as soon as one considers non-zero B the analysis differs importantly from that leading to Equations (56) and (64) for the diffuson. For $B \neq 0$ the phase factors [Equation (61)] in the product $\overline{G}^{+}(\mathbf{r},\mathbf{r}')\overline{G}^{-}(\mathbf{r},\mathbf{r}')$ do not cancel but rather add to give

$$\begin{aligned} c_0(\mathbf{r},\mathbf{r}',B,\Delta B) &= c_i u^2 \overline{G}^{+}(\mathbf{r},\mathbf{r}',B+\Delta B)\overline{G}^{-}(\mathbf{r},\mathbf{r}',B) \\ &= c_0(\mathbf{r},\mathbf{r}',B=0)\exp\left\{\frac{ie}{\hbar c}\left[2\mathbf{A}(\mathbf{r})+\Delta\mathbf{A}(\mathbf{r})\right]\cdot(\mathbf{r}-\mathbf{r}')\right\}. \end{aligned} \quad (70)$$

As before we are interested in points separated by at most the spatial range of c_0 which is approximately equal to l for small B, and we would like to expand this phase factor in the integral equation to yield a differential equation of the diffusion type. However the the typical phase in c_0 even for $\Delta B = 0$ is of order $Bl^2/(\hbar c/2e)$, and is only small compared to unity when $l^2 \ll \lambda^2$. Since $\lambda \approx 1000\,\text{Å}$ at $B = 1\,\text{kG}$, this condition will usually not be satisfied in a 2DEG at fields larger than 1 T and fails for much weaker fields in almost all GaAs heterostructures. It is well known (and we shall see below) that the main effect of the low-temperature divergence in the cooperon channel is removed by even weaker fields such that $Bl_{\text{in}}^2 \approx (\hbar c/2e)$, where $l_{\text{in}}^2 = (D\tau_{\text{in}})$ is the inelastic scattering length. Hence the cooperon contribution is always negligible at the fields where Landau level quantisation becomes important, and there is no high-field ($r_c < l$) analogue of Equations (56) and (64) for the diffuson. However, if we restrict ourselves to fields where $(B+\Delta B)l^2 \ll (\hbar c/2e)$, then the analysis does go through exactly as for the generalised diffuson with $\Delta\mathbf{A} \to 2\mathbf{A} + \Delta\mathbf{A}$ to yield

$$\left\{D\tau\left[-i\nabla - \frac{e}{\hbar c}(2\mathbf{A}+\Delta\mathbf{A})\right]^2 + \frac{\tau}{\tau_{\text{in}}} - i\frac{\Delta E\tau}{\hbar}\right\}\eta_j(\mathbf{r}) = \zeta_j^2\,\eta_j(\mathbf{r}). \quad (71)$$

This is the famous diffusion equation for the cooperon first derived by Altshuler *et al.* (1980), except that usually $\Delta\mathbf{A}$ is taken to be zero and ΔE is replaced by the external frequency of the electric field. Note that when $\Delta\mathbf{A} = 0$ we have a Schrödinger-type equation for a particle of charge $2e$; hence in a doubly-connected geometry with an Aharonov-Bohm flux the solutions are periodic with period $hc/2e$, half the period of the conventional electronic Aharonov-Bohm effect. Thus the weak-localisation correction due to the cooperon oscillates with flux period $hc/2e$ and not hc/e, a dramatic effect first observed by Sharvin and Sharvin (1981).

7 Weak localisation magnetoresistance

Having developed this generalised real-space formulation of impurity-averaged perturbation theory, we now apply it to transport. We review the WL effects before moving on to the more novel UCF effects. Although well known, the WL calculation in real-space will have immediate applications to UCF.

The WL magnetoresistance (neglecting spin effects) is obtained by calculating the impurity-averaged conductance with the cooperon vertex correction. Equation (14)

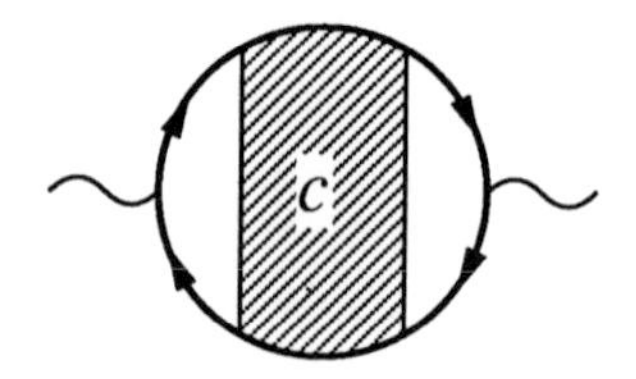

Figure 10: *Conductance diagram with a cooperon polarisation insertion which leads to weak localisation. Wavy lines at the end of the bubble represent the velocity operators.*

for the conductance is the most convenient starting point, and the set of diagrams included is indicated in real-space formulation in Figure 10. Although in principle each of the two Green function lines can correspond to G^+ or G^- due to the factors ΔG in Equation (14), the cooperon divergence only occurs when averaging $\overline{G}^{\pm}\overline{G}^{\mp}$ terms and not $\overline{G}^{\pm}\overline{G}^{\pm}$, so we may drop these from the outset. The two remaining terms from Equation (14) are complex conjugates of each other so we may take only twice the real part of one term. And finally, since the contribution from one term turns out to be real (for $\omega = 0$), we may omit taking the real part for simplicity of notation. With these simplifications the weak localisation magnetoconductance correction $\Delta g_{\mathrm{WL}}(B)$ is given by

$$\Delta g_{\mathrm{WL}}(B) = \frac{-e^2\hbar}{2\pi L_x^2}\int d\mathbf{r}\, d\mathbf{r}'\, J(\mathbf{r},\mathbf{r}')\, c(\mathbf{r},\mathbf{r}',B), \tag{72}$$

where

$$J(\mathbf{r},\mathbf{r}') = c_i u^2 \int d\mathbf{r}_1\, d\mathbf{r}_2\, v_x(\mathbf{r}_1)\,\overline{G}^-(\mathbf{r}',\mathbf{r}_1)\,\overline{G}^+(\mathbf{r}_1,\mathbf{r})\, v_x(\mathbf{r}_2)\,\overline{G}^-(\mathbf{r}_2,\mathbf{r}), \tag{73}$$

and I have already integrated the double-sided derivatives by parts to replace them by velocity operators acting only to the right.

The factor $J_0(\mathbf{r},\mathbf{r}')$ arises from the external portions of the diagrams and the crucial point here is that it is short-ranged since it only involves factors of $\overline{G}$, and will be small for points $\mathbf{r},\mathbf{r}'$ separated by more than l. Thus to leading order in $(k_F l)^{-1}$ we can make the approximation

$$J(\mathbf{r},\mathbf{r}') \approx J_0\delta(\mathbf{r}-\mathbf{r}') \tag{74}$$

with the constant J_0 given by integrating $J(\mathbf{r},\mathbf{r}')$ over both arguments and dividing by the sample area:

$$J_0 = \frac{c_i u^2}{A}\int d\mathbf{r}\, d\mathbf{r}'\, d\mathbf{r}_1\, d\mathbf{r}_2\, v_x(\mathbf{r}_1)\,\overline{G}^-(\mathbf{r}',\mathbf{r}_1)\,\overline{G}^+(\mathbf{r}_1,\mathbf{r})\,\overline{G}^+(\mathbf{r}',\mathbf{r}_2)\, v_x(\mathbf{r}_2)\,\overline{G}^-(\mathbf{r}_2,\mathbf{r}). \tag{75}$$

This expression for J_0 can be put in a more useful general form by noting that the phase factors from the field, $(2ie/\hbar c)\mathbf{A}\cdot(\mathbf{r}-\mathbf{r}')$, will be small for points separated by less than l and can be neglected, so that all factors $\overline{G}$ are symmetric in their spatial arguments. By interchanging spatial arguments in the factors $\overline{G}^+(\mathbf{r}_1,\mathbf{r})$ and $\overline{G}^-(\mathbf{r}',\mathbf{r}_1)$, J_0 can be written as the trace of an operator

$$J_0 = \frac{c_i u^2}{A}\mathrm{Tr}\left\{\overline{G}^+ v_x \overline{G}^-\overline{G}^+ v_x \overline{G}^-\right\} = \frac{c_i u^2}{A}\mathrm{Tr}\left\{\overline{G}^+ v_x \overline{G}^+\overline{G}^- v_x \overline{G}^-\right\}, \tag{76}$$

where in the second step we have used the fact that $\overline{G}^{\pm}$ commute.

This expression can then be simplified using the identity $v_x = (-i/\hbar)[H_0, x]$ for the velocity operator, and the operator equation $\left[E - c_i u^2 \overline{G}^{\pm}(E) - H_0\right] \overline{G} \equiv (Z_0^{\pm})\overline{G} = 1$. Because Z_0 is just a constant shift of H_0 which cancels in the commutator we have

$$v_x = \frac{i}{\hbar}[Z_0^{\pm}, x]. \tag{77}$$

If we substitute this identity with the choice Z_0^+ for the first occurence of v_x in Equation (76), and Z_0^- for the second occurence, we easily obtain

$$\begin{aligned} J_0 &= \frac{-c_i u^2}{A\hbar^2} \mathrm{Tr}\left\{\left(x\overline{G}^+ - \overline{G}^+ x\right)\left(x\overline{G}^- - \overline{G}^- x\right)\right\} \\ &= \frac{c_i u^2}{A\hbar^2} \int d\mathbf{r}\, d\mathbf{r}'\, (x - x')^2 \overline{G}^+(\mathbf{r}, \mathbf{r}') \overline{G}^-(\mathbf{r}', \mathbf{r}) \\ &= \frac{c_i u^2}{dA\hbar^2} \int d\mathbf{r}\, d\mathbf{r}'\, (\mathbf{r} - \mathbf{r}')^2 \overline{G}^+(\mathbf{r}, \mathbf{r}') \overline{G}^-(\mathbf{r}', \mathbf{r}) \\ &= \frac{2l_0^2}{\hbar^2}, \end{aligned} \tag{78}$$

where we have used the definition of l_0^2, Equation (55). Note that the derivation is completely general once expressed as a trace of this form, and applies to an arbitrary field. For the cooperon contribution to be important, however, we need to consider the weak-field limit as discussed above, in which case $2l_0^2 = l^2$.

Having evaluated J_0 we immediately find from Equations (68) and (72)

$$\begin{aligned} \Delta g_{\mathrm{WL}}(B) &= \frac{-e^2 l^2}{2\pi\hbar L_x^2} \int d\mathbf{r}\, c(\mathbf{r}, \mathbf{r}, B) \\ &= \frac{-e^2 l^2}{2\pi\hbar L_x^2} \sum_j \frac{1}{\zeta_j^2}. \end{aligned} \tag{79}$$

If one sets $B = \Delta B = 0$ in Equation (71) the solutions are just those discussed for the diffuson after Equation (58) with $\zeta_j^2 = D\tau(Q^2 + l_{\mathrm{in}}^{-2})$, and changing the sum to an integral leads to the familiar logarithmically divergent WL correction,

$$\Delta g_{\mathrm{WL}}(B = 0) = \frac{-e^2}{2\pi^2\hbar} \int_L^l \frac{Q\, dQ}{Q^2 + l_{\mathrm{in}}^{-2}}. \tag{80}$$

For $B \neq 0$, $\Delta B = 0$ in 2D the eigenvalues of Equation (71) correspond to Landau levels of a particle with charge $2e$ and mass $\hbar^2/2D\tau$. Expressing the sum over j in terms of these eigenvalues, including the appropriate degeneracy factor per LL, yields the well-known result (Altshuler *et al.* 1980)

$$\Delta g_{\mathrm{WL}}(B) = \frac{-e^2}{4\pi^2\hbar} \sum_{n=0} \frac{1}{n + \frac{1}{2} + \gamma_{\mathrm{in}}}, \tag{81}$$

where $\gamma_{\mathrm{in}} = (\hbar c/4eBl_{\mathrm{in}}^2)$. This expression with the appropriate upper cut-off [determined by the range of validity of the expansion of the phase factor in Equation (70)] tends correctly to the limit at $B = 0$ and tends to zero for $Bl_{\rm in}^2 \gg \hbar c/e$. This shows that a magnetic field large enough to alter substantially the phase of classical paths within a phase-coherent region is sufficient to destroy the WL effect, which relies on the exact cancelation of phases due to time-reversal symmetry. We need not engage in further analysis of the well-studied WL effects here; the effort paid to them is primarily to emphasise their close affinity to the UCF effects treated below.

8 Universal conductance fluctuations

Having laid the groundwork carefully above, we are now able to derive the microscopic theory of conductance fluctuations in a few lines.

8.1 Definition of correlation function

As discussed in Section (3.4), the elastic transmission coefficient through a multiple-scattering medium is expected to have sizeable fluctuations due to random interference effects, and these interference effects can be modulated by changing external parameters which couple to the phase of the scattered wave. In the case of a mesoscopic solid the two most natural external parameters to consider are the magnetic field and Fermi energy. One needs to determine the scale over which the interference effects vary as a function of these parameters, and the typical amplitude of the fluctuating interference effects. In the conventional low-field theory of UCF Lee and Stone (1985) introduced the *ergodic hypothesis* stating that the amplitude of the fluctuations as a function of external parameters B, ε_F was equal to the variance of g (averaged over impurity configurations) at fixed external parameters. This hypothesis was found to hold quite well numerically for the fluctuations as a function of magnetic field in the low-field limit (Lee and Stone 1985). However in the high-field limit this hypothesis most certainly fails over a sufficiently large interval of field and will have to be modified in a manner to be discussed below. Nonetheless we shall see that typically the scale of the fluctuations is significantly less than the scale over which the ergodic hypothesis breaks down, allowing its use over this interval.

If the ergodic hypothesis holds, the typical amplitude and scale of the sample-specific fluctuations may be obtained from the conductance correlation function

$$F(\Delta E, \Delta B) = \langle \delta g(\varepsilon_F + \Delta E, B + \Delta B)\, \delta g(\varepsilon_F, B) \rangle\,, \tag{82}$$

where $\delta g(\varepsilon_F, B) = g(\varepsilon_F, B) - \overline{g}(\varepsilon_F, B)$. The statistical variance of g is given by $F(0,0) = \langle \delta g^2 \rangle \equiv \mathrm{var}(g)$, which by the ergodic hypothesis is the mean-squared variation as a function of ε_F, B. The decay width of $F(\Delta E, 0)$ gives the correlation range, E_c, of the fluctuations with Fermi energy; similarly the width of $F(0, \Delta B)$ gives the correlation range, B_c, of the fluctuations with magnetic field.

8.2 Evaluation of diagrams

Diagrammatically the correlation function is obtained by considering two conductance bubbles with different values of the external parameters $\Lambda = B$, ε_F (corresponding

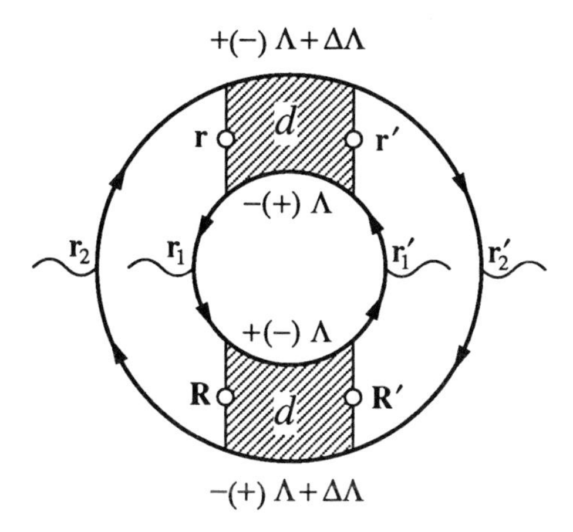

Figure 11: *Diagram representing one of the two diffuson contributions to the conductance correlation function.*

to two different conductance measurements) connected by impurity-averaging lines as shown in Figure 11. All diagrams in which the two bubbles are not connected are contributions to the square of the average conductance and will cancel by definition of var(g). Not surprisingly the leading contribution to F comes from diagrams where the two bubbles are connected by diffuson ladders or cooperon ladders as shown in Figure 11. When B is greater than a few hundred Gauss the cooperon contribution is negligible for the reasons discussed above, whereas for $B = 0$ it gives a contribution identical to that of the diffuson. Its detailed treatment involves a trivial extension of the analysis of the diffuson contribution, and we will not discuss it in detail here. For the diffuson contribution there are only two types of diagrams, one of which is shown in Figure 11. The two types give very similar contributions to F and so we only consider the simpler kind shown in Figure 11, which we denote by F_1.

This diagram consists of two diffuson ladders and two external vertices on each side of the diagram involving two factors $\overline{G}^+, \overline{G}^-$ and two velocity vertices. As before, each Green function line can be $\overline{G}^+$ or $\overline{G}^-$ but they must be chosen so that the diffuson connects $\overline{G}^+\overline{G}^-$. Diffusons with different signs of the term $i\Delta E\tau_0$ result from different pairing, but this is irrelevant in the external vertices J_0 which only involve $\overline{G}$ and are short-ranged, so that the small energy or field difference may be neglected. Hence all the external vertices are equal and can be approximated as δ-functions connecting $\mathbf{r}, \mathbf{R}$ and $\mathbf{r}', \mathbf{R}'$. This observation leads to the exhilarating realisation that each of these vertices is equivalent to the vertex $J(\mathbf{r}, \mathbf{r}') = J_0\,\delta(\mathbf{r} - \mathbf{r}')$ considered above, where because of the definition of the diffuson one need not appeal to time-reversal symmetry to express it in the trace form of Equation (76). Then by inspection

$$F_1 = \left(\frac{e^2\hbar}{4\pi L_x^2}\right)^2 J_0^2 \int d\mathbf{r}\, d\mathbf{r}' \left\{2\left|d(\mathbf{r}, \mathbf{r}', \Lambda + \delta\Lambda)\right|^2 + 2\,\mathrm{Re}\left[d(\mathbf{r}, \mathbf{r}', \Lambda + \delta\Lambda)\, d(\mathbf{r}', \mathbf{r}, \Lambda + \delta\Lambda)\right]\right\} \tag{83}$$

Recalling that $J_0 = 2l_0^2/\hbar^2$ for arbitrary field, and using the spectral representation of

$d(\mathbf{r}', \mathbf{r}, \Lambda + \delta\Lambda)$ [Equation (49)] to perform the integrations over $\mathbf{r}, \mathbf{r}'$, yields

$$F_1 = 2\left(\frac{e^2}{h}\right)^2 \frac{l_0^4}{L_x^4} \sum_j \left[\frac{1}{\xi_j^4} + \mathrm{Re}\left(\frac{1}{\xi_j^4}\right)\right], \tag{84}$$

where we have used the orthonormality of the $\chi_j(\mathbf{r})$. The contribution of the other type of diagram to F is exactly equal to the second term in the square brackets, hence

$$F(\Delta E, \Delta B) = 2\left(\frac{e^2}{h}\right)^2 \frac{l_0^4}{L_x^4} \sum_j \left[\frac{1}{\xi_j^4} + 2\,\mathrm{Re}\left(\frac{1}{\xi_j^4}\right)\right], \tag{85}$$

where the dependence on ΔE, ΔB comes through the dependence of the eigenvalues ξ_j^2 of the diffusion equation (64) on these quantities. Equation (85) gives the full correlation function for $T = 0$. It is valid at arbitrary magnetic field (except for fields near $B = 0$ where it is straightforward to include the cooperon contribution) as long as the SCBA is reasonable. This expression is identical to the conventional theory which neglects Landau quantisation effects. Thus we have obtained the striking new result (Xiong and Stone 1991) that the conventional theory goes over unchanged to the high-field limit except for the dependence of the constants l_0^2 and τ_0 on the magnetic field, which we now show does not appear in $\mathrm{var}(g)$ just as at low fields.

8.3 Variance of conductance

The variance of g is obtained from Equation (85) simply by setting $\Delta E = \Delta B = 0$ in the evaluation of ξ_j^2, which we have already found above gives $\xi_j^2 = l_0^2(k_m^2 + q_n^2) = l_0^2(m\pi/L_x^2 + n\pi/L_y^2)$. Inserting this into Equation (85) we immediately see that the factor l_0^4/L_x^4 cancels to give

$$\mathrm{var}(g) = \frac{6}{\pi^4}\left(\frac{e^2}{h}\right)^2 \sum_{m=1,n=0}^{\infty} \frac{1}{[m^2 + n^2(L_x/L_y)^2 + (L_x/\pi l_{\mathrm{in}})^2]^2}, \tag{86}$$

where the generalised inelastic diffusion length is defined as

$$l_{\mathrm{in}}^2 = \frac{l_0^2}{\tau_0}\tau_{\mathrm{in}} \equiv D_0(B)\tau_{\mathrm{in}}. \tag{87}$$

In the limit $l_{\mathrm{in}} \to \infty$ $(T \to 0)$ this equation implies that $\mathrm{var}(g) \approx (e^2/h)^2$, independent of the size of the sample, degree of disorder and magnetic field (insofar as the ergodic hypothesis is satisfied), hence the term 'universal conductance fluctuations'. At $B = 0$ a factor of two is needed to include the cooperon contribution (which again is equal to the diffuson by time-reversal symmetry), and a factor of 4 is needed to account for spin degeneracy in the absence of spin-orbit interactions or Zeeman splitting larger than k_BT. Spin-orbit interactions reduce both diffuson and cooperon contributions by a factor of 4 due to suppression of the triplet channels (Altshuler and Shklovskii 1986); we are able to show that this suppression also occurs at high fields by an extension of the analysis presented above. The specific value of $\mathrm{var}(g)$ at $T = 0$ can easily be evaluated from Equation (86) but in typical experiments, which are done in multi-probe geometries, this value depends on the probe configuration and it is not possible to make precise comparisons with Equation (86). When $l_{\rm in} \ll L$ the variance of g depends on their ratio (we assume $L_x = L_y$ here), and since many terms contribute to the sum in Equation (86) we may accurately convert it to an integral giving in 2D

$$\mathrm{var}(g) = \frac{24}{\pi}\left(\frac{l_{\rm in}}{L}\right)^2\left(\frac{e^2}{h}\right)^2. \tag{88}$$

Note that inelastic scattering does reduce the fluctuations as a power of the system size, making the system self-averaging as the size goes to ∞.

8.4 Correlation ranges E_c and B_c

The functions $F(\Delta E, 0)$ and $F(0, \Delta B)$ can be obtained quantitatively from Equations (64) and (85) and plotted to find their precise half-widths in terms of the system parameters. Detailed evaluations in various limits have been given elsewhere (Lee and Stone 1985, Lee *et al.* 1987). However the parametric scale of E_c can be determined for the energy correlation function quite easily because ΔE appears as only a constant shift of the eigenvalue ξ_j^2. With no inelastic scattering the sum in Equation (85) is very rapidly convergent and E_c can be determined simply by looking at its largest term, which is the smallest eigenvalue $|\xi_0^2|^2 = \pi^4 l_0^4/L_x^4 + (\Delta E \tau_0/\hbar)^2$. The sum will decay substantially when the second term is comparable to the first, *i.e.* $\Delta E \approx \hbar l_0^2/\tau_0 L_x^2$. Thus

$$E_c(B) \approx \frac{\hbar D_0(B)}{L_x^2}, \tag{89}$$

which is the generalisation to arbitrary field of the well-known low-field result which was discussed in Section 3.4. By inserting the high and low-field limits of l_0 and τ_0 one finds

$$E_c = \begin{cases} \hbar l^2/(2\tau L_x^2), & l \ll r_c \\ (r_c^2 \nu \sin\theta(E))/L_x^2, & l \gg r_c. \end{cases} \tag{90}$$

It is easily shown from Equation (85) that $l_{\rm in}$ simply replaces L_x in these relations when $l_{\rm in} \ll L_x$.

The determination of the magnetic field correlation length from Equation (85) is slightly more involved as ΔB does not enter only as an eigenvalue shift, and one must solve the differential equation (64) for $\xi_j^2(\Delta B)$. As noted above in the discussion of WL effects in 2D, the solutions will give the analogue of Landau levels when $l_{\rm in} \ll L$, in this case with the field replaced by ΔB and mass $m = \hbar^2/2l_0^2$, but with the conventional charge e. The correlation length can again be obtained by looking at the lowest eigenvalue, $\xi_0^2(\Delta B) = (\Delta B l_0^2)/(\hbar c/e) + \tau_0/\tau_{\rm in}$, and finding the scale of ΔB at which the two terms become comparable. Thus

$$B_c(B) \approx \frac{(\hbar c/e)}{D_0 \tau_{\rm in}}, \tag{91}$$

which again generalises the low field result discussed in Section 3.4 to arbitrary field. In generalising the theory we obtain a field-dependent B_c as recently observed in low-mobility GaAs by Geim *et al.* (1991); see Figure 3 and the chapter by Main. If $\tau_{\rm in}$ is only weakly field-dependent on a scale B_c, then our theory predicts that when $r_c \ll l$

$$B_c \propto \frac{1}{D_0(B)} \propto \frac{1}{r_c^2 \nu} \propto B^{3/2}, \tag{92}$$

where it must be noted that the WN limit for the random potential has been assumed, and this may have a limited region of applicability to GaAs.

9 Summary and conclusions

I have discussed in detail the two types of experimentally observed quantum interference effects in disordered conductors, weak localisation and universal conductance fluctuations. A third type of interference effect for non-interacting electrons is associated with thermodynamic properties of mesoscopic systems such as persistent currents and can be treated by the same techniques, but has yet to be unambiguously observed. I have shown that both weak localisation and universal conductance fluctuations can be treated in a real-space formulation based on effective diffusion equations for the diffuson and cooperon contributions, and for the case of the diffuson (which does not rely on time-reversal symmetry) the approach can be generalised within the self-consistent Born approximation to the limit where the Landau level spacing is much greater than the disorder broadening. Since this generalisation is new a few comments about its expected range of validity are in order.

It is well known from the study of the quantum Hall effect that localisation effects become important rapidly as a function of magnetic field even in high (zero-field) mobility two-dimensional electron gases. From the perturbative point of view taken above this may be seen as due to the fact that no matter how large is $k_F l$, it is the parameter $1/N$ which measures the localisation effects in high field, and this parameter approaches unity in the quantum Hall regime. Even when $N \approx 10$ the beginning of Hall plateaus are observed and the SCBA breaks down in these intervals of field and the theory presented here does not apply. Conversely, at the centres of the LLs one expects extended states to exist even in the infinite system; however they are believed to be described by a strong-coupling fixed point whose properties are different in general from those of the perturbative unstable fixed point described by the SCBA. Nonetheless, by the universality hypothesis one expects var(g) to be independent of size and disorder at this fixed point as we find in the SCBA, so at most the precise constant and correlation scales could be different. Moreover, the mesoscopic systems of interest are likely to be far from the infinite volume behaviour (at least in the conducting region) and may reasonably be described by perturbation theory. Thus, in summary, we expect the theory presented here to give a reasonable description of mesoscopic transport fluctuations in a Landau quantised 2DEG even in the quantised Hall regime in the transition region between the plateaus, with the major *caveat* that the disordered potential is short-ranged compared to the cyclotron radius.

The theory presented here highlights further the universality of the transport fluctuation phenomena. The details of the bare quantum states, as long as they are extended, are unimportant. This is because at long wavelengths the only coherent scattering (represented by the diffuson) will satisfy the same diffusion equation; only the diffusion constant reflects the nature of the underlying states. Since the diffusion constant cancels in the variance of the conductance (and in other physical properties such as the persistent current (Altshuler *et al.* 1991) one finds remarkably general behaviour. Recently a theoretical study of ballistic conductors by Jalabert *et al.* (1990) has also found fluctuation effects identical to those of disordered conductors with correlation ranges given by expressions similar to those obtained above. Here the multiple scattering comes from the geometry of the device which generates classically chaotic scattering for the cases considered. I have conjectured, but not proven, that a diffuson-type approximation to the modulus of the semi-classical propagator can be used to describe this limit also. Thus it seems that complex quantum scattering of almost any type leads to a single type of coherent fluctuation phenomena in normal electron transport.

Acknowledgments

I would particularly like to acknowledge the major contributions of my student S Xiong to the generalisation of the UCF theory to high magnetic field and for helpful comments on the manuscript. This work was supported by NSF grant DMR-8658135, by the AT&T Foundation, and by the Alfred P Sloan Foundation.

References

Abrahams E, Anderson P W, Licciardello D C, and Ramakrishnan T V, 1979. *Phys. Rev. Lett.*, **42**, 673.
Abrikosov A A, Gorkov L P, Dyzaloshinski I E, 1965. *Methods of Quantum Field Theory in Statistical Physics*, New York, Pergamon.
Aharonov Y, and Casher A, 1984. *Phys. Rev. Lett.*, **53**, 19.
Al'tshuler B L, Khmelnitskii D E, Larkin A I, and Lee P A, 1980. *Phys. Rev. B*, **22**, 5142.
Al'tshuler, B L, Aronov A G, and Spivak B Z, 1981. *JETP Lett.*, **33**, 94.
Al'tshuler, B L, 1985. *JETP Lett.*, **41**, 648
Al'tshuler B L, and Shklovskii B I, 1986. *Sov. Phys. JETP*, **64**, 127.
Altshuler B L, Gefen Y, and Imry Y, 1991. *Phys. Rev. Lett.*, **66**, 88.
Anderson, P W, 1958. *Phys. Rev.*, **102**, 1008.
Ando T, Matsumoto Y, and Umera Y, 1974. *J. Phys. Soc. Jpn.*, **36**, 959.
Ando T, Matsumoto Y, and Uemura Y, 1975. *J Phys. Soc. Jpn.*, **39**, 279.
Aronov A G, and Sharvin Yu V, 1987. *Rev. Mod. Phys.*, **59**, 755.
Baranger H U, and Stone A D, 1989. *Phys. Rev. B*, **40**, 8169.
Beenakker C W J, and van Houten H, 1989. *Phys. Rev. Lett.*, **63**, 1857.
Benoit A D, Washburn S, Umbach C P, Laibowitz R B, and Webb R A, 1986. *Phys. Rev. Lett.*, **57**, 1765.
Bergmann G, 1984. *Phys. Rep.*, **107**, 11.
Birge N O, Golding B, and Haemmerle W H, 1989. *Phys. Rev. Lett.*, **62**, 195.
Büttiker M, 1986. *Phys. Rev. Lett.*, **57**, 1761.
Büttiker, M, 1988. *Phys. Rev. B*, **38**, 9375.
Carra P, Chalker J T, and Benedict K A, 1989. *Ann. Phys. (NY)*, **194**, 1.
Chakravarty S, and Schmid A, 1986. *Phys. Rep.*, **140**, 193.
Economou E N, and Soukoulis C M, 1981. *Phys. Rev. Lett.*, **46**, 618.
Edwards, S F, 1958. *Phil. Mag.*, **3**, 1020.
Feng S, Lee P A, and Stone A D, 1986. *Phys. Rev. Lett.*, **56**, 1960.

Fisher D S, and Lee P A, 1981. *Phys. Rev. B*, **23**, 6851.
Geim A K, Main P C, Beton P H, Streda P, Eaves L, Wilkinson C D W, and Beaumont S P, 1991. *Phys. Rev. Lett.*, **67**, 3014.
Gefen Y, Imry Y, and Azbel M Ya, 1984. *Phys. Rev. Lett.*, **52**, 129.
Glazman L I, Lesovick G B, Kmelnitskii D E, and Shekhter R I, 1988. *JETP Lett.*, **48**, 218.
Greenwood, D A, 1958. *Proc. Phys. Soc. (London)*, **71**, 585.
Gutzwiller M C, 1990. *Chaos in Classical and Quantum Mechanics*, Berlin, Springer-Verlag, pp. 184–190, 283–287.
Imry Y, 1986. In *Directions in Condensed Matter Physics*, Eds. Grinstein G, and Mazenko G, Singapore, World Scientific, p. 101.
Jain J K, and Kivelson S A, 1988a. *Phys. Rev. Lett.*, **60**, 1542.
Jain J K, and Kivelson S A, 1988b. *Phys. Rev. B*, **37**, 4276.
Jalabert, R A, Baranger H U, and Stone A D, 1990. *Phys. Rev. Lett.*, **65**, 2442.
Kubo, R, 1956, *Canad. J Phys.*, **34**, 1274.
Kubo R, Miyake S I, and Hashitsume N, 1900. In *Solid State Physics*, Eds. Seitz F, and Turnbull D, New York, Academic Press, **17**, 288.
Landauer R, 1957. *IBM J. Res. Dev.*, **1**, 233.
Landauer R, 1970. *Phil. Mag.*, **21**, 863.
Langer J S, and Neal T, 1966. *Phys. Rev. Lett.*, **16**, 984.
Lee P A, and Ramakrishnan T V, 1985a. *Rev. Mod. Phys.*, **57**, 287.
Lee P A, and Stone A D, 1985b. *Phys. Rev. Lett.*, **55**, 1622.
Lee P A, Stone A D, and Fukuyama H, 1987. *Phys. Rev. B*, **35**, 1039.
Levy L P, Dolan G, Dunsmuir J, and Bouchiat H, 1990. *Phys. Rev. Lett.*, **64**, 2094.
Luo J K, Ohno H, Matsuzaki K, Umeda T, Nakahara J, and Hasegawa H, 1989. *Phys. Rev. B*, **40**, 3461.
Mathur H, and Stone A D, 1991. In preparation.
Prange R E, and Girvin S M (Eds.), 1987. *The Quantum Hall Effect*, New York, Springer-Verlag.
Sanquer M, Mailly D, Pichard J-L, and Pari P, 1989. *Europhys. Lett.*, **8**, 471.
Skocpol W J, Mankiewich P M, Howard R E, Jackel L D, Tennant D M, and Stone A D, 1987. *Phys. Rev. Lett.*, **58**, 2347.
Sharvin D Yu, and Sharvin Yu V, 1981. *JETP Lett.*, **34**, 272.
Stone A D, 1985. *Phys. Rev. Lett.*, **54**, 2692.
Stone A D, and Imry Y, 1986. *Phys. Rev. Lett.*, **56**, 189.
Stone A D, and Szafer A, 1988. *IBM J. Res. Dev.*, **32**, 384.
Stone A D, 1989. *Phys. Rev. B*, **39**, 10736.
Stone A D, Szafer A, McEuen P L, and Jain J K, 1990. *Ann. N Y Acad. Sci.*, **581**, 21.
Streda P, Kucera J, and MacDonald A H, 1987. *Phys. Rev. Lett.*, **59**, 1973.
Szafer A, and Stone A D, 1989. *Phys. Rev. Lett.*, **62**, 300.
Timp G, Behringer R, Sampere S, Cunningham J E, and Howard R E, 1989. In *Nanostructure Physics and Fabrication*, Eds. Reed M A, and Kirk W P, Boston, Academic Press, p. 000.
van Houten H, Beenakker C W J, and van Wees B J, 1990. In *Semiconductors and Semiconductors*, Ed. Reed M A, New York, Academic Press.
van Wees B J, Van Houten N, Beenakker C W J, Williamson J G W, Kouwenhoven L, Van Der Marel D, and Foxon C T, 1988. *Phys. Rev. Lett.*, **60**, 848.
Washburn S, and Webb R A, 1986. *Adv. Phys.*, **35**, 375.
Xiong S, and Stone A D, 1991. In preparation.

Is Atomically Precise Lithography Necessary for Nanoelectronics?

G Timp

AT&T Bell Laboratories
Holmdel, New Jersey, U.S.A.

1 Introduction

As the active area of an electronic device shrinks, the transport characteristics become acutely sensitive to atomic variables. If the transport is incoherent, this sensitivity develops from statistical variations in the doping density, while for coherent transport it develops from quantum interference of the electronic wavefunction due to scattering from the dopant configuration. For example, to avoid a change of more than 20% in the threshold voltage of a MOSFET in which the transport is incoherent, statistical fluctuations in the channel doping must not exceed 40% because the threshold depends on the square root of the doping concentration. Doping fluctuations do not exceed 40% when the the length of the channel of the transistor $L_{\text{eff}} > 150\,\text{nm}$ if the doping of the substrate $N_a > 4 \times 10^{17}\,\text{cm}^{-3}$ (Hoeneisen and Mead 1972). In such a transistor, there is on average only 170 dopant atoms in a volume determined by $(150\,\text{nm}/2)^3$. But if the transport in such a MOSFET is coherent, *i.e.* if $L_{\text{eff}} < L_\phi$ where L_ϕ is the phase coherence length, then the conductance between the drain and source contacts exhibits interference effects: universal conductance fluctuations (UCF) (Lee and Stone 1985, Alt'shuler 1985; see also the chapter by Stone). These are random, but reproducible, fluctuations in the conductance as a function of gate voltage or as a function of the impurity configuration. Under these conditions, a change in the position of a *single* impurity comparable to the electron wavelength, $\lambda_F \approx 40\,\text{nm}$, changes the conductance by e^2/h (Feng *et al.* 1986, Alt'shuler and Spivak 1986).

The acute sensitivity of very small devices to the effects of impurities will confound our ability to manufacture them reproducibly. Atomic scale control over the position of dopants may ultimately be required for practical use of very small devices. Conventional microfabrication techniques such as electron beam lithography and molecular beam epitaxy, which have provided the springboard for the study of mesoscopic phenomenon, have not yet attained such control.

In this chapter, we first briefly review some of the effects of impurities on very small electronic devices, and then demonstrate a new technique which may ultimately be used to position dopants with atomic precision (see Timp and Howard 1991 and Timp *et al.* 1991a). To demonstrate unequivocally the effect of impurities on transport in very small electronic devices, we use an electron waveguide as a paradigm of a very small electronic device. An electron waveguide is a wire that is so clean and so small that electron waves can propagate coherently in guided modes, which are characteristic of the geometry, with minimal scattering. An electron waveguide is supposed to be reminiscent of an optical or microwave waveguide but, unlike an electromagnetic wave, an electron wave is sensitive to an applied electric or magnetic field because it possesses a charge. In response to an applied electric field (or to an applied current), an electron waveguide has a resistance which is related to the quantum mechanical transmission through the wire (Landauer 1985). Through numerical computations of the transmission probabilities in realistic structures, the theoretical effects of impurities are compared with the results of our measurements under a variety of circumstances. Even in an electron waveguide, the effect of impurities or dopants can be discerned.

The transport characteristics of an electron waveguide demonstrate a need for atomically precise, three-dimensional lithography and crystal growth. In the third section of this chapter we demonstrate a maskless method of lithography and thin film deposition which potentially could be used to position dopants economically and quickly with atomic precision. We show the effect of a light field interposed between a collimated neutral atomic beam and a flat substrate on the morphology of a thin film deposited on the substrate. The light field deflects atoms according to its spatial structure, intensity and frequency, and so the spatial structure of the light is transferred into the film deposited on the substrate. For low intensity light, we interpret our observations in terms of the action of a conservative, velocity-independent dipole force on the neutral atom. Since indirect measurements have indicated that similar forces can be used to localise neutral atoms on the scale of an optical wavelength and smaller (Salamon *et al.* 1987, Prentiss and Ezekiel 1986, Westbrook *et al.* 1990), we assert that this method can be extended to control the morphology of a film with nearly atomic precision.

2 Transport in an electron waveguide

It is now possible to make a wire that is reminiscent of an electron waveguide by forming a narrow constriction in a two-dimensional electron gas (2DEG). Following Pepper (Berggren *et al.* 1986), we have used the electrostatic potential provided by a split-gate geometry to laterally constrain the electron gas to the region within the gap between the gates, and make a one-dimensional (1D) constriction in a 2DEG. The fabrication procedures are described in detail elsewhere (Timp 1990). The device is made on top of the wide leads (100 μm wide) of a Hall bar that have been etched into a GaAs-AlGaAs heterostructure. A split gate is fabricated on top of the Hall bar using electron beam lithography. Electron beam lithography is used to prepare a mask for lift-off, then a Ti-Au or Ti-AuPd film approximately 7.5/50 nm thick is evaporated and

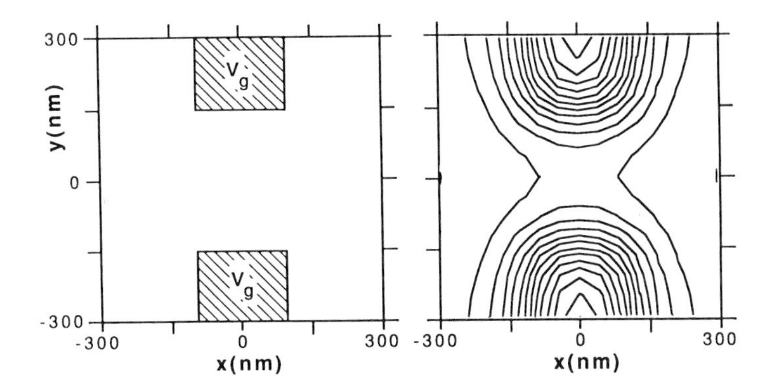

Figure 1: *Graphical representation of the potential in the heterostructure near the* GaAs-AlGaAs *interface defined by the split-gate electrodes on the surface* 70 nm *above the* 2DEG. *It is assumed that* −1.0 V *is applied to each electrode. The left plot represents the disposition of the gate electrodes on the surface of the heterostructure. The split gate is* 200 nm *long, with a* 300 nm *gap between electrodes. A contour plot of the potential in the plane of the* 2DEG *is shown at the right of the figure. The equipotential contours are at intervals of* 0.066 V.

the mask is removed to give the electrodes of the split gate. Typically, the split-gate electrodes are 200 nm wide with an intervening gap of 300 nm.

By applying a negative voltage to the split gates, the 2DEG gas at the GaAs-AlGaAs interface immediately beneath the gate electrodes is depleted and so the 2DEG is laterally constrained within the gap between the electrodes. The electrostatic potential due to split-gate electrodes, not accounting for the δ-doped impurity layer or the charge density associated with the 2DEG or the dielectric constant, is represented in Figure 1. This electrostatic potential, ϕ, is given by

$$\begin{aligned}\phi(x,y,z) &= V_g\,[f(x+L/2,y+w/2,z)-f(x-L/2,y+w/2,z)\\ &\quad -f(x+L/2,y-w/2,z)+f(x-L/2,y-w/2,z)]\,,\end{aligned}$$

where

$$f(X,Y,Z)=\frac{1}{2\pi}\left[\arctan\frac{X}{Z}-\arctan\frac{XY}{\sqrt{X^2+Y^2+Z^2}}\right],$$

and $2l$ is the lithographic length, and $2w$ the gap, between the gate electrodes. We assume that the gate electrodes are a distance $z=d$ above the 2DEG, extending along y from $w/2$ to ∞ and from $-w/2$ to $-\infty$, and along x from $-L/2$ to $L/2$ with the origin in the centre of the gap (Davies 1989).

The two salient features of the electrostatic confinement potential are the smooth taper from the 2DEG to the 1D constriction, and the potential barrier between the constriction and the 2DEG along the z-axis. The contact to the 2DEG is not abrupt because the constriction is formed by depletion using gate electrodes that are between 60 nm–200 nm away from the 2DEG, depending on the bias voltage and the heterostructure. A self-consistent calculation of the potential produces a smooth taper even if there are imperfections in the gate electrodes, for this reason (Kumar *et al.* 1989). The potential is relatively flat near the center of the constriction, but rises abruptly near the edges. When the gate voltage is large and negative, the constriction gradually widens to the 2D contact over approximately 250 nm which is about 5–10 λ_F. Since we assume that the Fermi energy is constant throughout the device, the higher potential within the constriction means that the carrier density is lower there.

Electron micrographs taken from a top view of the actual split-gate electrodes are shown in top portion of Figure 2, whose lower part shows the two-terminal resistance (conductance), $R_{12,12}$ ($G_{12,12}$), as a function of the applied gate voltage that we found at 280 mK in devices like that shown schematically in Figure 1. The gate voltage is applied between the split-gate electrodes and a grounded contact to the 2DEG. The notation $R_{kl,mn} \equiv 1/G_{kl,mn}$ denotes a measurement in which there is a positive current from lead k to l, and a positive voltage is detected between leads m and n. (A series resistance found at $V_g = 0$ V was subtracted from the measured resistance obtained as a function of gate voltage to give the data shown.) The depletion of the 2DEG immediately beneath the gate electrodes occurs near $V_p = -0.325$ V for the devices represented in Figure 2, and is not shown. As the gate voltage decreases below V_p, the constriction within the gap between the electrodes narrows, the carrier density within the constriction decreases, and plateaus are observed in the resistance. The two-terminal conductance, obtained by inverting the resistance versus gate voltage, is also shown in Figure 2. The *average* conductance (minus the series resistance) is approximately $(2e^2/h)N$, with N an integer ranging from one to six, and is evidently quantised in steps of height $2e^2/h$ with about one to five percent accuracy as an increasingly negative gate voltage makes the constriction narrower (van Wees *et al.* 1988, Wharam *et al.* 1988). The length of the gate electrode (along the x-axis in Figure 1), L, is about 200 nm for the device of Figure 2(a). The mobility of the 2DEG used to fabricate the devices of Figure 2 corresponds to a low temperature mean free path of about $L_e = \hbar k_F \mu / e \approx 7\,\mu$m. Since the 300 nm gap in each device pinches off for large negative gate voltages, the depletion around the gate must be less than 150 nm under conditions where the constriction conducts, and so we estimate the length of the constriction to be less than 500 nm in Figure 2(a) — much less than the mean free path deduced from the mobility of the 2DEG. At the temperature of the measurement, we estimate that $L_\phi \geq L$ [the phase coherence length was estimated using the Aharonov-Bohm effect as discussed in Timp *et al.* (1988)].

The resistance measured in the constriction is due to the redistribution of the current among the electronic states in the wide 2D contact and the 1D constriction, *i.e.* $R_{12,12}$ is a contact resistance between the 2DEG and the 1D constriction. In a conventional two-terminal measurement, two wider conductors make connection to each end of a constriction. If the contacts are wide enough, they become reservoirs in which the electronic motion approaches thermal equilibrium. The contacts can then be characterised by chemical potentials μ_1 and μ_2. When $\mu_1 > \mu_2$, there is a net current through the constriction. Generally, only a fraction of the flux incident in a particular subband is transmitted through the constriction, and the current carried by subband j is given by

$$I_j = ev_j \Delta n = ev_j \left(\frac{dg}{dE}\right)_j \sum_{k=1}^{N} t_{jk} \Delta\mu,$$

where v_j and (dg/dE) are the Fermi velocity and the density of states at the Fermi energy for subband j; t_{jk} is the probability intensity for transmission from subband k into subband j; N is the number of subbands in the constriction; and $\Delta\mu$ is the difference in chemical potential between the two wide contacts on either side of the constriction, *i.e.* $\Delta\mu = \mu_1 - \mu_2$. Thus the current is just the product of the number of excess electrons in the constriction and their velocity. In 1D, the density of states is inversely proportional to the velocity, so

$$I_j = \left(\frac{2e}{h}\right) \sum_{k=1}^{N} t_{jk} \Delta\mu.$$

Summing the contributions from each of the subbands gives the total current, and the voltage difference between the wide reservoirs is $V = \Delta\mu/e$, so the two-terminal conductance is:

$$G_{12,12} = \left(\frac{2e^2}{h}\right) \sum_{j,k=1}^{N} t_{jk}. \tag{1}$$

This is the Landauer formula for the two-terminal conductance (Economou and Soukoulis 1981). From Landauer's perspective the resistance develops from changes in quantum mechanical transmission probabilities. The dissipation associated with resistance occurs in the reservoirs, which are in thermal equilibrium and separate from the wire.

If there is no scattering, *i.e.* if $t_{jk} = \delta_{jk}$, then the current injected into subband j is $I_j = (2e/h)\Delta\mu$, independent of the specific 1D subband under consideration, and the total current is $I = N(2e/h)\Delta\mu$. The two-terminal conductance of a constriction with N-occupied subbands is exactly

$$G_{12,12} = \left(\frac{2e^2}{h}\right) N \tag{2}$$

in this limit. Thus the two-terminal conductance is ideally a quantised function that measures the number of occupied 1D subbands. The correspondence between Equation (2) and the data of Figure 2 is surprising, because to obtain Equation (2) it was assumed that (i) $T = 0$, and (ii) the transmission through the contacts into the constriction and through the constriction is perfect. The correspondence implies that the resistance of the constriction, minus the contact resistance, nearly vanishes. Moreover, the observation of an accurately quantised two-terminal conductance implies (i) that the energy separation between subbands is much larger than 0.1 meV; (ii) that the constriction is ballistic for $L = 200$ nm; and (iii) that the reflections associated with contacts to the constriction are negligible.

2.1 Effect of impurity scattering on transport in a waveguide

The quantisation of the two-terminal resistance of a 1D constriction was not anticipated theoretically because it was widely held that the nature of the contact, used to measure the resistance of the constriction, was not ideal. While Imry (1989) observed that, theoretically, the two-terminal conductance is a quantised function of the number of occupied 1D subbands, he concluded that fluctuations of magnitude e^2/h in the measured potential due to scattering at the contacts would obscure the effect. Subsequently, Glazman *et al.* (1988) developed an adiabatic model for conductance

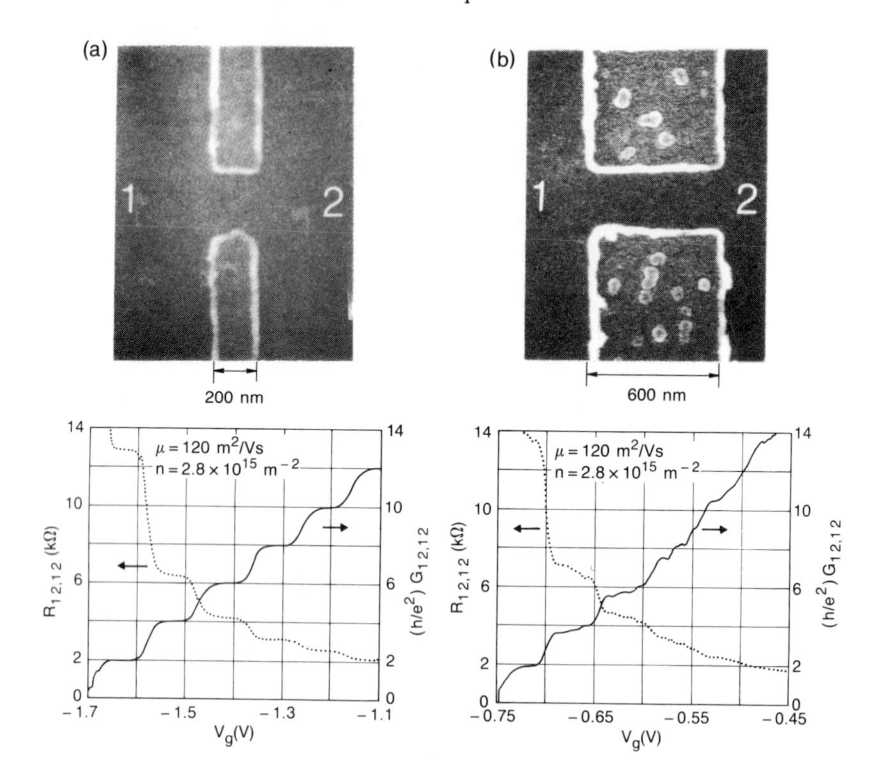

Figure 2: *Two-terminal resistances (conductances),* $R_{12,12}$ $(G_{12,12})$, *of two constrictions in a* 2DEG *as a function of gate voltage,* V_g, *obtained at* 280 mK *for a heterostructure with a mean free path, estimated from the mobility, of about* 7 μm. *The split gate in (a) is* 200 nm *long with a* 300 nm *gap between the electrodes; the conductance* $G_{12,12}$ *is quantised in steps of* $2e^2/h$ *as a function of* V_g. *In (b) the split gate is* 600 nm *long, again with a* 300 nm *gap. Although the mean free path in the* 2DEG *at low temperature is much longer than the length of the constriction, (b) shows deterioration of the quantisation of the two-terminal resistance.*

of a constriction that relies on a gradual taper from the 1D constriction to the 2D contact. Following Glazman, we imagine an electron waveguide with a width that is a function of the longitudinal position, $W(x)$, yielding an adiabatically smooth constriction of width W_i tapering from a width W_f. If the variation along x is slow, the Hamiltonian separates into terms in the longitudinal variable, x, and the transverse variable y. The y-dependent part of the Hamiltonian is taken as a hard-wall problem with solution $E_n(x) = \hbar^2\pi^2n^2/2m^*W^2(x)$, where $n = 1, 2, \ldots$. The x-dependent part of the Hamiltonian represents scattering from a barrier. For a given Fermi energy, E_F, only a finite number of modes, N, will be above the barrier. If tunnelling below and reflections above the barrier are negligible, then N modes with $t_{jk} = \delta_{jk}$ carry current while the transmission of all other modes vanishes exponentially, and so Equation (2) is recovered from Equation (1).

Actually, it is unlikely that the constriction is entirely adiabatic. It is not necessary that the taper be completely adiabatic for accurate quantisation; rather it is sufficient that the adiabaticity holds only up to a certain width beyond which reflections occur, or that the reflections that do occur are small (Imry 1989, Yacoby and Imry 1990, Payne 1989). As Szafer and Stone (1989) have shown, the resistance may be well quantised even if the constriction is not adiabatic at all (Laughton *et al.* 1991).

The accuracy of quantisation in an adiabatic model is an exponential function of the length of the constriction; the transmission coefficient is $t(\epsilon) = [1 + \exp(-\epsilon/\Delta E)]$ where $\Delta E = \hbar^2/m\sqrt{2RW^3}$, W is the minimum width of the constriction and R is the radius of curvature associated with the taper (Glazman *et al.* 1988). The accuracy is determined by evanescent modes which carry current over the finite length of the constriction. An exponential improvement in the quantisation is expected with increasing length due to the decay of the current in these modes. However, we find that the quantisation deteriorates as the length of the constriction increases.

Figure 2(b) shows the resistance and conductance determined as above for a constriction with gate electrodes of length $L = 600\,\mathrm{nm}$, spaced with a 300 nm gap on the same two heterostructures as the shorter constriction. For this geometry, the effective length of the constriction is estimated to be less than 900 nm which is still a factor of eight less than the mean free path estimated from the 2D mobility and carrier density, yet the quantisation has deteriorated dramatically from that shown in Figure 2(a). While quantisation of the resistance is still apparent in the higher mobility device represented in Figure 2(b), the accuracy of the quantisation is only about 90% for the first three steps with no quantisation observed for higher N. We attribute the deterioration of the quantisation in the long constrictions to disorder or impurity scattering within the constriction (van der Marel and Haanappel 1989, Chu and Sorbello 1989).

The transmission probability, t_{jk}, through a coherent device is affected by elastic scattering. Elastic scattering, such as might occur at an impurity, changes the distribution of the electrons between the 1D subbands in the constriction. If a defect in the constriction back-scatters or reflects an electron from the constriction (Haanappel and van der Marel 1989), the transmission probabilities t_{jj} in Equation (1) fall below unity and the conductance is reduced from the quantised value.

We assume that the scattering originates with ionised donors intentionally placed deep in the AlGaAs layer of the heterostructure, set back from the GaAs-AlGaAs interface and the 2DEG. The Coulomb potential associated with ionised dopants beneath the gate will be screened by the metal electrodes, but the dopants in the gap between the electrodes are not. The top of Figure 3 shows the calculated electron density obtained by Nixon *et al.* (1991a,b) in the 2DEG for a 1.5 μm square around idealised point contacts corresponding to the heterostructure and geometry of the devices of Figure 2, including the effect of the impurities in the δ-doped impurity layer in the AlGaAs set back 42 nm from the 2DEG. The smooth contours shown in Figure 1, due to the electrostatic potential of the gate electrodes, are grossly distorted by the remote ionised impurity layer in the more realistic appraisal shown in Figure 3. The distortions in the contours give rise to nonadiabatic transport and scattering in the constriction. The bottom of Figure 3 shows the calculated conductance as a function of gate voltage associated with profiles like the ones represented in the top portion of Figure 2. According to Davies and Nixon (1988), the 2DEG effectively screens the potential fluctuations due to Coulombic impurities in the doped layer, but as the constriction in the 2DEG

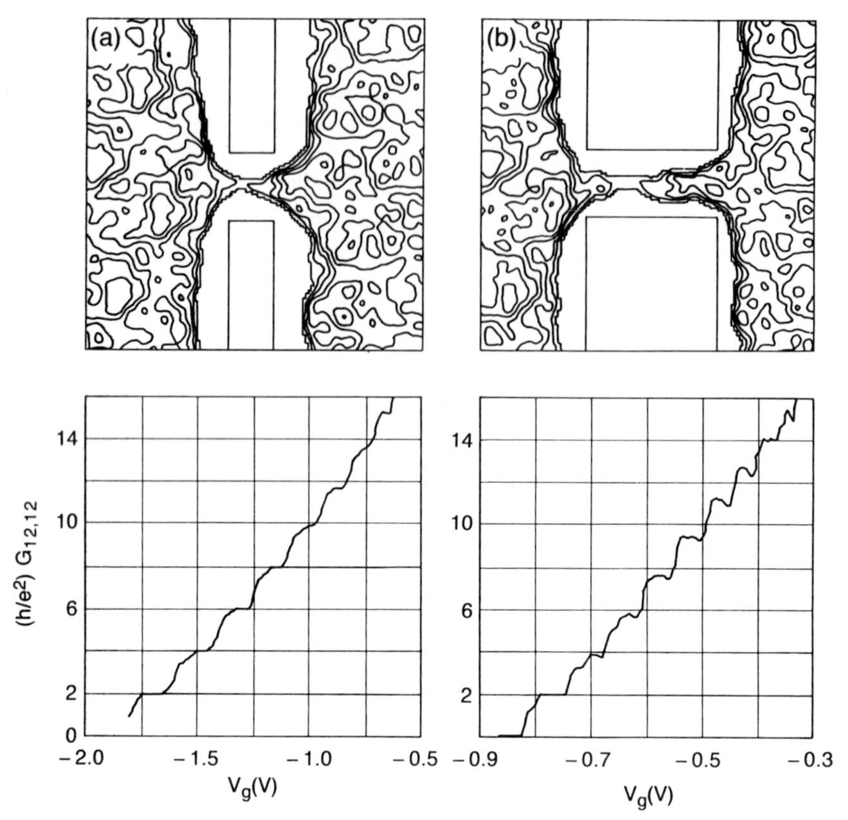

Figure 3: *Calculations by Nixon et al. (1991a,b) of the charge density and the conductance of two configurations of gate electrodes on the surface of a heterostructure, corresponding to the experimental parameters. In (a) the split gate is* 200 nm *long with a* 300 nm *gap between the electrodes, while in (b) the gate is* 600 nm *long. The gate electrodes are* 72 nm *above the* 2DEG. *The contours at the top of the figure represent the electron density in the* 2DEG *for a* 1.5 μm *square are around the point contacts and are* $6\times10^{14}\,\mathrm{m}^{-2}$ *apart starting at zero. In the top of (a) the bias is* $V_g = -1.8\,\mathrm{V}$*, and it is* $V_g = -0.8\,\mathrm{V}$ *in the top of (b). Below the contours, numerical calculations of the two-terminal conductance as a function of* V_g *for the two different devices are shown.*

becomes narrower with more negative gate voltage and the carrier density is reduced, the Fermi wave-vector k_F becomes smaller, and the screening of the impurities in the doped layer by the electron gas becomes less effective.

Figure 3 successfully mimics the effects found experimentally, although a smaller number of plateaus are found in the calculation of the conductance of the 200 nm long constriction than observed experimentally. The deterioration of the quantisation of the conductance of a 600 nm long constriction is supposed to be due either to the effect of small-angle scattering on the highest indexed 1D subbands (Glazman and Jonson 1991, Nixon *et al.* 1991a,b), or to indirect back-scattering processes which develop from small-angle forward scattering and coherent reflections near modal cutoff in the 'bulges' of

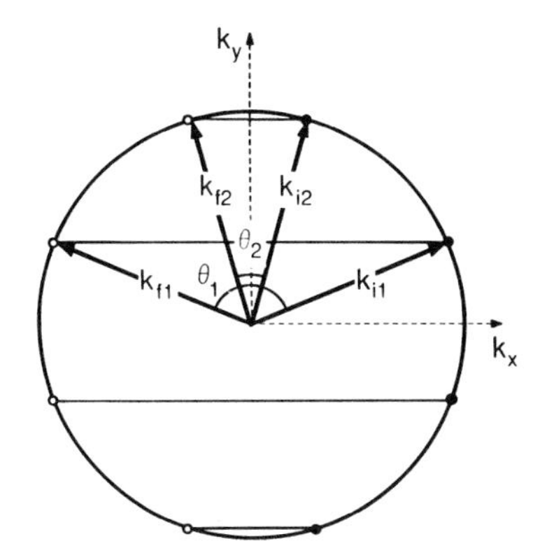

Figure 4: *Schematic representation of the Fermi surface of a narrow constriction in a 2DEG. The circle represents the Fermi surface of a 2DEG in reciprocal space; the horizontal lines reflect the allowed transverse wave-vectors in the 1D constriction. Current is applied along the k_x direction. Two scattering events occurring at the Fermi energy are depicted in the figure. In the first (unlikely) event, k_{i1} associated with an occupied subband is scattered through a large angle θ to k_{f1}, resulting in deterioration of quantisation. In the second (more probable) event, k_{i2} is back-scattered through a small angle $< 60°$ to k_{f2}, and the quantisation deteriorates.*

the real potential which defines the constriction (Laughton *et al.* 1991). Figure 4 shows two scattering events that lower the transmission probability through the constriction: (i) a large angle back-scattering event away from threshold; and (ii) a small angle back-scattering event near threshold for occupation of a subband. As illustrated in Figure 4, the quantisation is sensitive to small angle scattering only near the threshold for each subband. Small-angle scattering is prevalent in modulation-doped GaAs-AlGaAs heterostructures (Das Sarma and Stern 1985, Fang *et al.* 1988), and it is especially deleterious to the quantisation of the resistance when N is large because many subbands are available within a small angle, θ. The discrepancy between theory and experiment may be due to an overestimate of the small-angle scattering rate (Laughton *et al.* 1991). According to Laughton *et al.* there is no region of the taper between the constriction and the contact in which the transport is adiabatic!

While it is widely used as an estimate for the mean free path and a measure of the extent of impurity scattering, L_e inferred from the mobility of the 2DEG is not necessarily an appropriate estimate for the elastic scattering length in the constriction, especially for gate voltages near pinch-off. For example, a quantised conductance has been observed in constrictions with a gate length of 200 nm with a 2DEG density of $n = 8 \times 10^{15}\,\mathrm{m}^{-2}$ where $L_e \approx 450\,\mathrm{nm}$ is comparable to the length of the constriction (Timp *et al.* 1991b). Perhaps a more appropriate estimate of the elastic mean free path can be derived from the correlation length associated with potential fluctuations in the real constriction, estimated to be about 300 nm for the heterostructure of Figures 2 and 3 (Nixon *et al.* 1991a,b, Laughton *et al.* 1991).

2.2 Effect of a single impurity on transport in a waveguide

According to our interpretation, Figure 5 illustrates the extent to which the transport in the electron waveguides we fabricate is adiabatic. Moreover, Figure 5 also illustrates that the effect of impurity scattering can be nullified in a 1D constriction without eliminating the impurities altogether. Figure 5 shows the four-terminal conductance $G_{12,43}$ found when electrode pairs A-C of the device shown in Figure 6 are used to define a constriction in a high mobility 2DEG. Figure 6 shows an electron micrograph of two split-gate electrodes cascaded in series on top of a GaAs-AlGaAs heterostructure. Each set of electrodes is 200 nm long with a gap of 300 nm between the electrodes. The two sets are separated by 300 nm to make the device symmetric. Associated with each port of the device there are contacts to the 2DEG that are not shown. The convention for numbering the contacts is given in the figure. By applying a negative voltage to any pair of split-gate electrodes, (A-B), (C-D), (A-D), (B-C), (A-C) or (B-D), the 2DEG at the GaAs-AlGaAs interface immediately beneath the gate electrodes is depleted and laterally constrained within the gap between the electrodes. The two-terminal conductance through a single constriction is quantised in steps of height $2e^2/h$ with an accuracy of about 5% (van Wees *et al.* 1988, Wharam *et al.* 1988).

The depletion of the 2DEG immediately beneath gate electrodes A and C in the device of Figure 6 occurs at $V_g = V_A = V_C \approx -0.35\,\mathrm{V} \equiv V_p$, and is not shown in Figure 5. As the gate voltage decreases beyond V_p, the constriction within the gap between the electrodes narrows and plateaus are observed in the resistance. $G_{12,43}$ is evidently quantised in steps of $2e^2/h$ with about 1 to 5% accuracy as V_g decreases. (Note that there is no series resistance in this measurement when $V_A = V_C = 0$.)

The conductance is again well represented by $G_{12,43} = (2e^2/h)N$, with N an integer ranging from 1 to 7 depending on V_g. The conductance of the constriction formed by every other pair of electrodes in Figure 6 is quantised similarly.

When electrode pairs A-C or B-D are used to define a constriction, discrete time-dependent changes are observed in the conductance near the threshold for a step, that resemble a random telegraph signal (RTS) (Timp *et al.* 1990, Dekker *et al.* 1991). The RTS of Figure 5 is supposed to be due to the modulation of the scattering cross section of a conglomerate of a few defects or perhaps a single defect. RTS has been found only in a few of the devices we have fabricated, and then only for gate voltages $V_g < -1.5\,\mathrm{V}$, presumably because of the position of the defect in the AlGaAs (Timp *et al.* 1991b, Timp *et al.* 1990). Although the discrete changes in the conductance are time-dependent, the statistics associated with the fluctuations at a particular V_g are not (Kirton and Uren 1989).

As shown in Figure 5, the conductance as a function of time, measured when $V_g = V_A = V_C = -2.250\,\mathrm{V}$, $-2.103\,\mathrm{V}$ and $-1.950\,\mathrm{V}$, fluctuates with an amplitude of $\Delta G_{12,43} \approx e^2/h$. The largest change found in conductance versus time is $1.2\,e^2/h$. In this example, the conductance generally fluctuates between three different values. Each value is supposed to correspond to a particular configuration of the defect (Kirton and Uren 1989). The time spent at each value varies, depending on V_g. The variation of the frequency of the switching with V_g has been attributed to a change in the Fermi energy in the constriction relative to the energy of the defect. Near the threshold for the occupation of $N = 1$, the time-dependent fluctuations in the resistance are so large

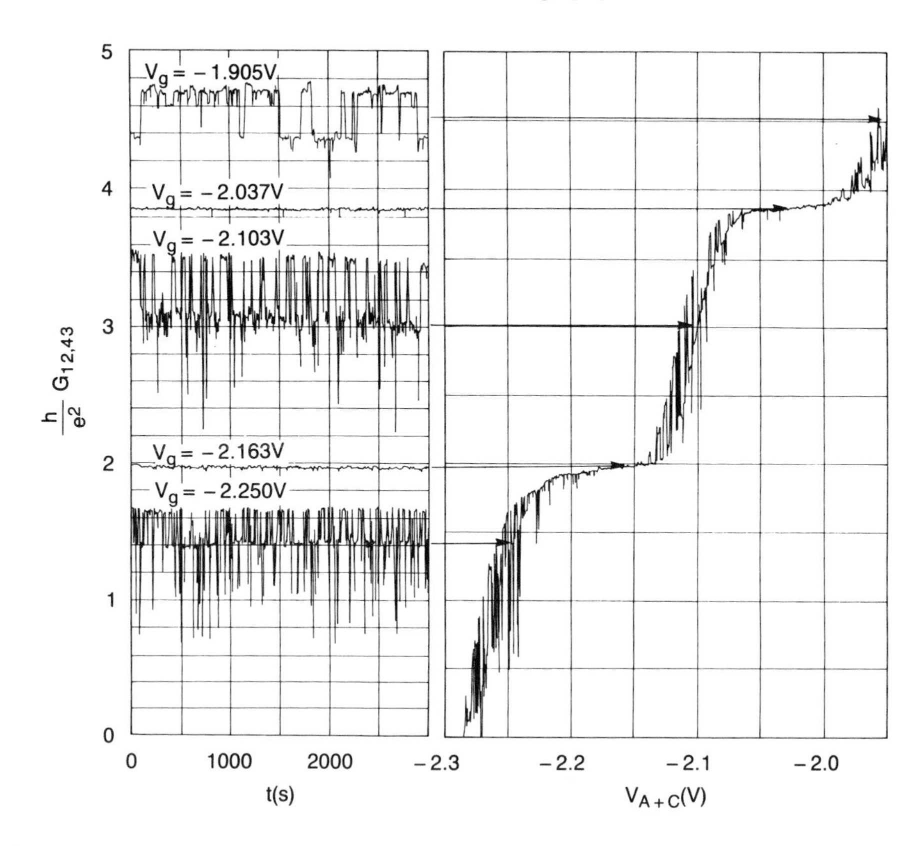

Figure 5: *The four-terminal conductance, $G_{12,43}$, of a constriction in a 2DEG as a function of gate voltage (right), $V_g = V_A = V_C$, and as a function of time t (left), found at a temperature of 280 mK. The resistance is quantised as a function of V_g corresponding to the number of 1D subbands occupied within the constriction. However, the resistance fluctuates wildly at the threshold of a step which is indicative of a random telegraph signal (RTS).*

that the device turns off. The fluctuations as a function of gate voltage provide an estimate of the effect of a few defects on the threshold of a device. From Figure 5 we deduce that defects gives rise to an uncertainty in the threshold voltage of about 50 mV at $T = 300$ mK in a waveguide approximately 100 nm wide.

In stark contrast with the results obtained near the threshold for a step, RTS is completely inhibited at the center of the $N = 1$, 2, 3 and 4 plateaus in the conductance. The value of $G_{12,43}$ is approximately $2e^2/h$ and $4e^2/h$ independent of time for the $N = 1$ and $N = 2$ plateaus, when $V_g = V_A = V_C = -2.163$ V and -2.037 V, as shown in Figure 5 (Timp *et al.* 1990, Dekker *et al.* 1991).

The suppression of RTS is due to the suppression of small-angle scattering in a 1D constriction (Sakaki 1980). The two-dimensional Fourier components of the interaction between a Coulomb potential in the δ-doped layer and an electron in the 2DEG are $(2\pi/Q)\exp(-Q|z_0|)$, where $\mathbf{Q} = \mathbf{k} - \mathbf{k}'$ is the momentum transferred in scattering, *i.e.*

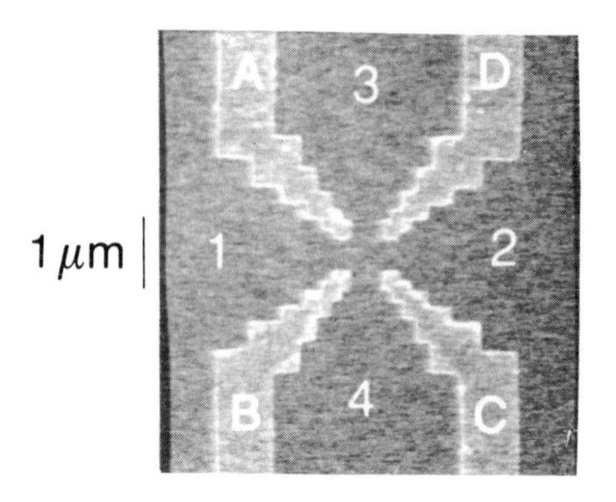

Figure 6: *Two sets of split-gate electrodes in series comprising a junction between waveguides. The disposition of the gate electrodes and the convention used to denote the leads and the gate electrodes are shown.*

$Q^2 = 2k^2(1 - \cos\theta)$, and z_0 is the thickness of the AlGaAs spacer. Thus the Fourier components are exponentially small when $|Q| > 1/z_0$ (Price 1984). If an impurity in the δ-doped layer gives rise to a potential fluctuation at the GaAs-AlGaAs interface responsible for the RTS, then back-scattering through a large angle from the potential fluctuation is improbable because the fluctuation only has spatial frequencies less than $z_0^{-1} = (42\,\mathrm{nm})^{-1}$. However, back-scattering through a small angle is not only possible but likely in GaAs (Fang *et al.* 1988), and could destroy the quantisation near the threshold for occupation of the subband of highest index. If an electron is reflected through a small angle, the change in the conductance associated with the reflection is constrained by unitarity to be less than $2e^2/h$ because the transmission probability associated with the 1D subband will either be 0 or 1. Thus, the amplitude of the RTS must be less than $2e^2/h$. Away from threshold, the spatial frequencies required to back-scatter an electron in the highest subband are not available since the potential is too smooth. Consequently scattering is suppressed and RTS is inhibited.

From another perspective (Fowler 1991), the transport through the constriction is non-adiabatic near threshold because the wavelength of the electrons is long relative to the potential fluctuations. Consequently the conductance is not quantised (Glazman *et al.* 1988), but the transport becomes adiabatic and the conductance is quantised when the subband is completely occupied and the wavelength becomes short relative to the potential fluctuations.

We contend that the defect associated with the RTS is near the constriction, somewhere in the gap between the electrodes *A-C* and *B-D*, because it is observed when the pairs *A-C* or *B-D* are biased independently or all together (Timp *et al.* 1990). Furthermore, we contend that a defect in the AlGaAs, set back vertically from the constriction along the z-direction, gives rise to the scattering since the RTS depends on V_g as shown in Figure 5, and is suppressed when an integral number of 1D subbands are occupied independent of the lateral position of the constriction within the gap between *A-C*. In an attempt to ascertain the position of the defect, we laterally changed the position of the constriction within the gap between *A-C* by independently biasing the split gates. Independent control of the electrodes provides the opportunity to change the position of

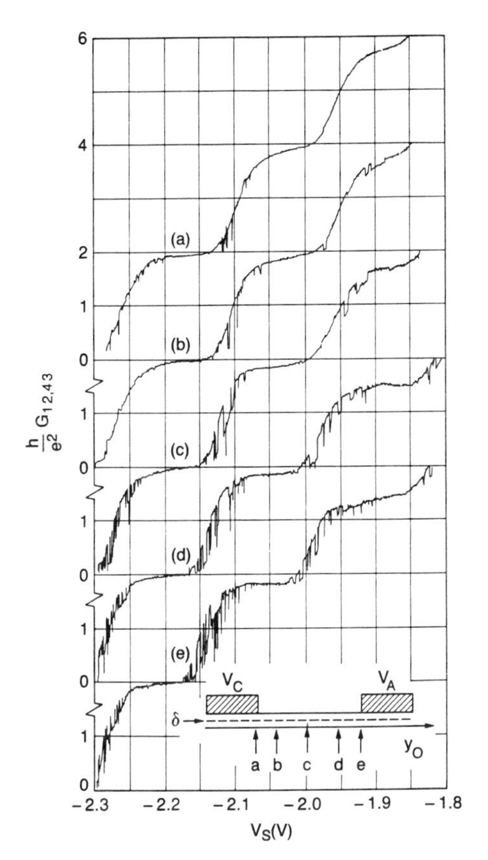

Figure 7: *The conductance, $G_{12,43} = 1/R_{12,43}$, versus gate voltage, $V_s = \frac{1}{2}(V_A + V_C)$, for (a) $\Delta V = V_A - V_C = -0.425\,\text{V}$, (b) $-0.275\,\text{V}$, (c) $0\,\text{V}$, (d) $0.275\,\text{V}$, and (e) $0.425\,\text{V}$. The position of the 1D constriction within the gap between the electrodes of Figure 6 is supposed to be linearly related to ΔV.*

the constriction laterally while maintaining the same number of 1D subbands (Glazman and Larkin 1991). For a split-gate configuration similar to Figure 6, the maximal value of the potential due to the electrodes, measured from the center of the gap between A-C, corresponds to the position $y_0 = D\Delta V/2V_p$ where $\Delta V = V_A - V_C$, $D = 425\,\text{nm}$ is the width of the gap between A-C and $V_p \approx -0.35\,\text{V}$ is the pinch-off voltage (Glazman and Larkin 1991). Thus the lateral position of the constriction within the gap depends linearly on ΔV. The changes in the threshold for each step in the conductance as a function of $V_s = \frac{1}{2}(V_A + V_C)$ depend on $(\Delta V/V_p)^2$ and are supposed to be small (Glazman and Larkin 1991).

Figure 7 depicts $G_{12,43}$ as a function of V_s for $\Delta V = -0.425\,\text{V}$, $-0.275\,\text{V}$, $0\,\text{V}$, $0.275\,\text{V}$, $0.425\,\text{V}$, ideally corresponding to $y_0 = -260\,\text{nm}$, $-170\,\text{nm}$, $0\,\text{nm}$, $170\,\text{nm}$ and $260\,\text{nm}$. The quantisation of the conductance is preserved when $\Delta V \neq 0$, and the thresholds are not very sensitive to changes in ΔV, as anticipated. We observe RTS at the threshold for occupation of a subband for each ΔV, but the magnitude and characteristic frequency of the switching noise change dramatically with the lateral position of the constriction. RTS is reduced when $\Delta V < 0$ compared to $\Delta V \geq 0$ which indicates that the trap is in the AlGaAs near electrode A.

We can estimate the vertical separation between the defect and the constriction from the width of the transition between plateaus where RTS is observed. For example, at the threshold for occupation of the $N = 2$ subband, where RTS is first observed ($V_g = -2.14\,\mathrm{V}$), the longitudinal component of the wave-vector, k_x in Figure 6, vanishes. The RTS vanishes on the $N = 2$ plateau at $V_g = -2.06\,\mathrm{V}$ where the Fermi wave-vector $k_F = 7.9 \times 10^7\,\mathrm{m}^{-1}$ (k_F is determined by the carrier density in the constriction, $n = 1 \times 10^{15}\,\mathrm{m}^{-2}$, which we deduced from magnetoresistance measurements). At the same gate voltage we deduce that $W \approx 100\,\mathrm{nm}$ using a hard wall confinement potential. Thus, $k_x = \sqrt{k_F^2 - (2\pi/W)^2} \approx 5 \times 10^7\,\mathrm{m}^{-1}$ when RTS is suppressed at the $N = 2$ plateau in conductance. The absence of scattering for $-2.06\,\mathrm{V} < V_g < -1.98\,\mathrm{V}$, where $k_x > 5 \times 10^7\,\mathrm{m}^{-1}$, implies a vertical separation between defect and constriction of about $k_x^{-1} > 21\,\mathrm{nm}$, a distance comparable to the spacer layer thickness of 42 nm.

Our data are consistent with an impurity in the AlGaAs layer switching (but it is not inconsistent with an impurity deep in the GaAs layer switching either). If the RTS found as a function of ΔV is all due to the same trap, that trap affects a region that is comparable to the size of the gap. This deduction is consistent with the long range effect that a poorly screened impurity potential in the doped layer would have at the GaAs-AlGaAs interface (Nixon *et al.* 1991a,b). Alternatively, if different traps are activated with each ΔV, the relevant trap must still be set back from the constriction into the AlGaAs because RTS is always suppressed whenever an integral number of subbands is occupied.

2.3 Impurity scattering at a junction between waveguides

It is implicit in the model by Glazman *et al.* (1988) for a single constriction that an electron wave leaving the constriction is focused. A wave propagating adiabatically along a guide of variable width like that pictured in Figure 1 conserves the mode number N. As the width of the guide narrows, the transverse energy increases while the longitudinal energy decreases continuously. When the wave abruptly encounters the end of the adiabatic taper, the transverse wave-vector is now conserved (at least in the interval $\pm\pi/W_f$), and the wave is focused to a width $\pi k W_f$. The collimation or focusing of the electron wave in the forward direction can produce a vanishing resistance in a second constriction in series or nonadditivity of the resistance of two constrictions in series (Beenakker and van Houten 1989).

Two surprising consequences of the collimation of the electron wave in the forward direction are the negative four-terminal resistance of a bend (Timp *et al.* 1988, Takagaki *et al.* 1989) and the suppression of the Hall effect at low magnetic fields ($H < 1\,\mathrm{T}$) (Roukes *et al.* 1987, Chang *et al.* 1989, Ford *et al.* 1989). If the gate electrodes of the device of Figure 6 are biased similarly, then the potential distribution of the junction ideally resembles a cross. We can use the four constrictions as the current and voltage leads to measure the resistance. Figure 8(a) shows the four-terminal resistances $R_B \equiv R_{14,32}$ and $R_H \equiv R_{12,43}$ measured as a function of the gate voltage applied to all of the

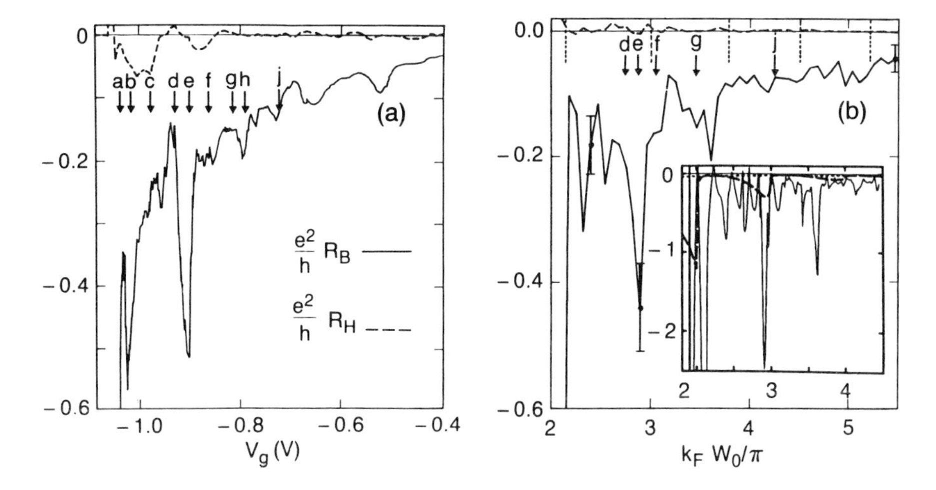

Figure 8: *Two characteristic resistance measurements made in a cross geometry, R_B (solid) and R_H (dashed), as (a) a measured function of gate voltage V_g, and (b) a calculated function of the Fermi wave-vector, k_F. The measured resistances in (a) were obtained from a junction between four submicron constrictions in a* 2DEG *with a mobility of* $115\,\mathrm{m^2\,V^{-1}\,s^{-1}}$ *and a carrier density of* $2.8\times10^{15}\,\mathrm{m^{-2}}$. *The labels (a–h, j) refer to the V_g associated with the corresponding traces in Figure 9(a). The results of a quantum mechanical calculation for a rounded-corner structure with impurities are shown in (b). The labels (d, e, f, g, j) refer to the k_F associated with the corresponding traces of Figure 9(b). The inset shows R_B for three additional calculations: a classical calculation with square, sharp corners (dotted) and quantum mechanical calculations with square corners (dashed) and rounded corners (solid) without impurities. The wire width at the threshold of the third subband, W_0, is used for normalisation of the calculated results.*

electrodes, V_g, in the absence of a magnetic field ($H = 0$) at 280 mK (Behringer *et al.* 1991, Kakuta *et al.* 1991). Figure 9(a) shows R_H at 280 mK as a function of magnetic field.

Figure 8(a) shows that R_B becomes globally more negative as V_g becomes more negative, but there are substantial fluctuations about the average at low temperature ($T < 2$ K). Sharp minima are observed in R_B near $V_g \approx -1.02$ V, -0.90 V and -0.80 V, for example. *Identical* features are also found in $R_{13,42}$ at the same V_g in the same device, and similar, but not identical, features have been observed in other devices in this and other heterostructures. Note that the minimum near $V_g = -0.9$ V accounts for 80% of the resistance!

The Hall resistance, R_H, can be used to determine the number of occupied 1D subbands, N, at a particular V_g since one expects $i = 2N$ quantum Hall plateaus (Büttiker 1989, 1988a,b, 1986). We find that there is a correspondence between: (i) the minima observed in R_B at $H = 0$; (ii) the value of the resistance associated with the first plateau found in R_H with increasing H; and (iii) the number of plateaus found as a function of H. For example, when $V_g = -1.050$ V, only the $N = 1$ subband is occupied in the constrictions ($G_{12,12}$ and $G_{34,34}$ vanish beyond $V_g = -1.050$ V when all the split gates are biased). Also, the Hall resistance for $-1.035\,\mathrm{V} \leq V_g \leq -0.98\,\mathrm{V}$ [traces (a), (b) and (c) in Figure 9(a)] changes dramatically from a value near zero to $(h/e^2)/(2.05 \pm 0.07)$, which indicates the depopulation of the spin-polarised edge state $i = 2$. There is no evidence of the depopulation of a higher-index edge state for $V_g < -1.0\,\mathrm{V}$. There is a second sharp minimum in R_B when $V_g \approx -0.90\,\mathrm{V}$. We identify this feature with the occupation of $N = 2$ or $i = 4$ in all of the constrictions and associated it with the threshold for $N = 2 \to 3$. This identification is supported by traces (d) and (e) of Figure 9(a) which show a plateau in R_H beyond $H = 0.5\,\mathrm{T}$ at $(h/e^2)/(4.1 \pm 0.2)$, corresponding to the magnetic depopulation of the $i = 4$ edge state. Furthermore, we associate the third sharp minimum found near $V_g = -0.8\,\mathrm{V}$ with the threshold for $N = 3 \to 4$ and support that interpretation with traces (g) and (h) of Figure 9(a).

The correspondence between the minima in R_B and the Hall resistance traces shows that there are sharp features in R_B on a V_g scale much smaller than the subband spacing. Furthermore, these features seem to be associated with thresholds of the 1D subbands and are therefore quantum mechanical in origin but are not caused by random impurities. In order to elucidate the possible mechanisms for such quantum mechanical features, we now turn to our theoretical results on the resistance of such junctions.

Theoretically, the four-terminal resistance is related to the transmission probabilities between the leads. Following Büttiker (1986, 1988) we can express the four-terminal resistance in terms of quantum mechanical transmission probabilities,

$$R_{mn,kl} = \left(\frac{h}{2e^2}\right)\frac{T_{km}T_{ln} - T_{kn}T_{lm}}{D}, \tag{3}$$

where T_{jk} is the total transmission probability between leads j and k,

$$T_{jk} = \sum_{m,n} t_{jk,mn}.$$

Here $t_{jk,mn}$ is the probability for a carrier in subband n and lead k to be transmitted to subband m in lead j. The denominator D is always positive and depends on the geometry of the leads, but is independent of permutations in the indices mn, kl. Notice that $R_B \equiv R_{14,32} = (T_{31}T_{24} - T_{34}T_{21})/D$, and that the Hall resistance $R_H \equiv R_{12,43} = (T_{41}T_{32} - T_{42}T_{31})/D$. Thus R_B is a measure of the isolation between two crossed wires because it measures the difference in transmission probabilities for going around a corner and straight through the junction.

The inset to Figure 8(b) presents the results of several calculations of R_B for different structures. In the calculations we used a recursive Green function technique (Baranger *et al.* 1988) applied to a tight-binding Hamiltonian of four perfect leads (width d) attached to a junction. In the square, sharp-corner case, $d = 21$ sites; in the rounded-corner case, the junction potential was $U = ax^4y^4/d^8$ matched at $x = d$ onto a y^4 cross-section in the leads ($d = 31$, $k = 650$); thus, the width of the lead at the junction was twice the asymptotic width. Disorder in the junction region was simulated by using an Anderson model with disorder $W = 0.5$. This produced a mean free path of about 80 sites ($\approx 7\,W_0$), slightly larger than the total distance across the junction from the start of rounding until the straight lead is recovered.

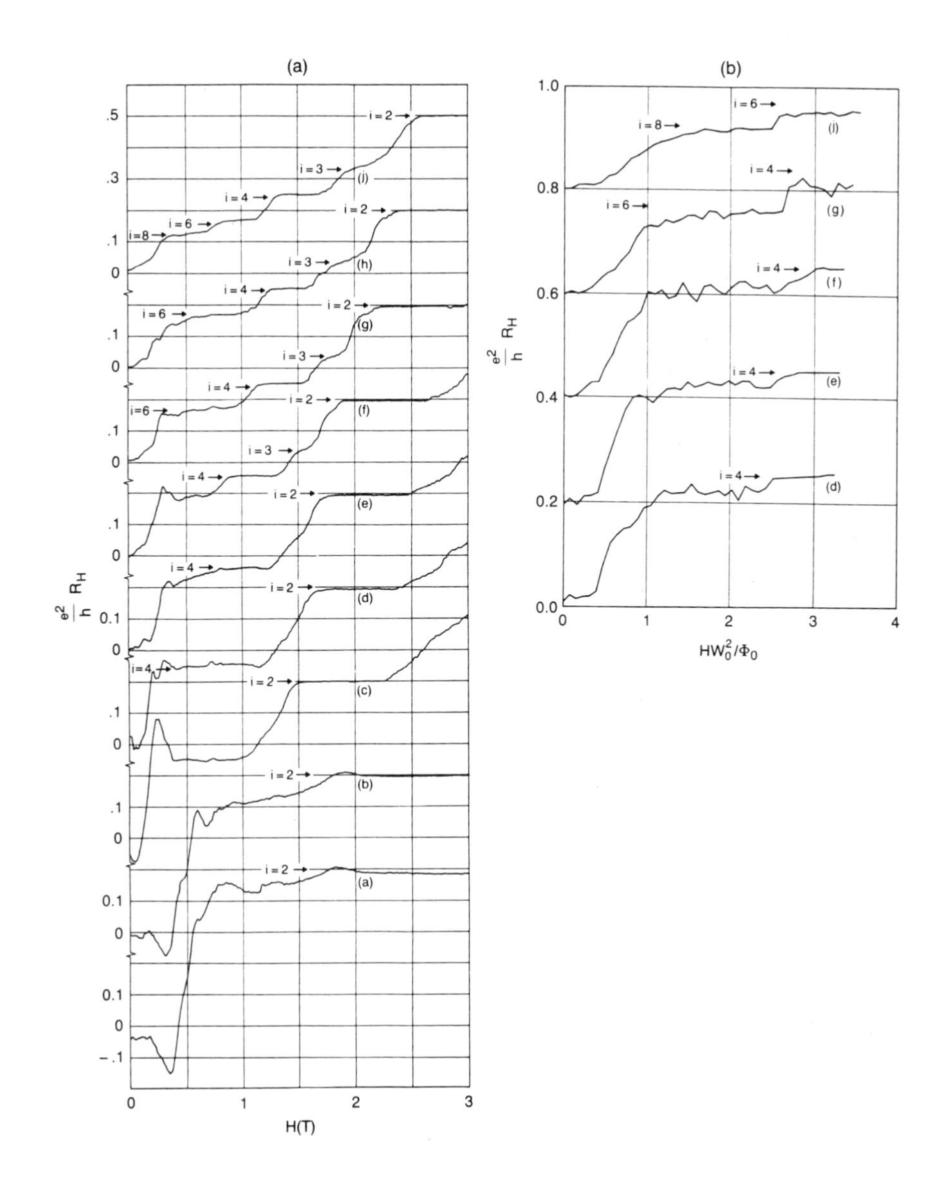

Figure 9: *R_H versus magnetic field, (a) measured for different V_g, and (b) calculated for different wave-vectors (a-h,j), corresponding to Figure 8. The quantum of magnetic flux $\Phi_0 = hc/e$ and W_0 is the width of the wire at the threshold of the third subband; $\Phi_0/W_0^2 \approx 0.3\,\mathrm{T}$for a typical $W_0 \approx 120\,\mathrm{nm}$. For traces (a–h, j) in (a) $V_g = -1.035\,\mathrm{V}$, $-1.025\,\mathrm{V}$, $-0.980\,\mathrm{V}$, $-0.936\,\mathrm{V}$, $-0.900\,\mathrm{V}$, $-0.871\,\mathrm{V}$, $-0.817\,\mathrm{V}$, $-0.795\,\mathrm{V}$ and $-0.712\,\mathrm{V}$. For traces (d), (e), (f), (g) and (j) in (b), $k_F W_0/\pi = 2.7$, 2.9, 3.05, 3.6 and 4.3. These wave-vectors and gate voltages are indicated by the arrows in Figure 8.*

The width of the constriction is kept constant in each of the numerical calculations of Figure 8(b) while the Fermi wave-vector varies. The simplest structure to consider is the intersection between straight wires in a square cross in which the equipotential lines make a right-angle. The inset compares a classical calculation (dotted) with a quantum mechanical calculation (dashed) for such a structure with a hard-wall confinement potential. Both calculations of R_B reveal a negative resistance; however the classical calculation is featureless and smaller than the experimental values, while the quantum mechanical calculation has sharp features at the threshold for occupation of the 1D subbands and a resistance comparable to or larger than the experimental values shown in Figure 8(a). The sharp features in the (dashed) quantum mechanical calculation are due to coherent scattering from the junction near threshold (Avishai 1989, Baranger 1990, Kirczenow 1989). As the energy increases from a value just above the threshold for occupation of a subband to a value just below the threshold for occupation of another subband, the electron wave-vectors become more forward directed. Consequently, the forward transmission T_{21} increases relative to the transmission around the corner T_{31}, and the resistance R_B becomes more negative. Over a larger energy range, R_B becomes less negative as the Fermi energy increases because N increases.

While the sharp features in the quantum mechanical calculation for the square junction are reminiscent of the experimental minima, quantitatively large differences remain, *e.g.* the magnitude of the minima beyond $N = 3$, minima which occur below threshold, and the difference in shape of the minima. In fact, the confinement potential in the actual junction is not square but tapered because of depletion (Baranger and Stone 1989). This raises the question of whether the sharp features evident in the simple theory persist in more realistic junctions. The classical model for the resistance of a ballistic cross with tapered corners yields a smoothly varying R_B as a function of energy or width, and underestimates R_B near the experimentally observed sharp minima. Thus, we are led to consider a quantum mechanical calculation for a junction with rounded or tapered corners [solid line in the inset to Figure 8(b)]. In this case R_B fluctuates rapidly and wildly with an amplitude larger than any of the other calculations, and much larger than the experimental observations. The large amplitude (in some cases larger than $R_B < -100\,h/e^2$) is caused by the partial collimation of electrons in the forward direction by the rounded corners (Beenaker and van Houten 1989), which acts to increase the magnitude of the bend resistance. In sum, none of the three calculations depicted in the inset of Figure 8(b) adequately represents the experimental results of Figure 8(a).

Agreement with the experimental data of Figure 8(a) can only be obtained by including elastic impurity scattering within a quantum mechanical calculation of the resistance of a junction with rounded corners. An average over weak impurity scattering has the dual effect of averaging out the large fluctuations (which arise from interference between long paths trapped in the junction) and degrading the collimation and hence the average value of R_B. Figure 8(b) shows the calculated resistances, R_B and R_H, as a function of the Fermi wave-vector k_F for the rounded-corner geometry in which an average over weak, short-range impurity scattering has been performed. The calculation now shows remarkable agreement with the experiment in terms of both the magnitude and the form of R_B.

Notice that the calculated R_B is plotted as a function of $k_F W_0$, while the measured resistances are plotted versus V_g. While there is not necessarily a direct relationship between the two abscissae of Figure 8, we have used $R_H(H)$ to establish that approximately the same range in subband index is covered in both Figures 8(a) and (b), $1 < N \leq 5$ (Behringer 1991). The close correspondence between the experimental results of Figure 8(a) and the numerical results of Figure 8(b) is therefore not coincidental.

The deepest minima found in R_B correspond closely to the threshold for the occupation of 1D subbands in the constrictions comprising the junction, independent of the impurity configuration. For example, the deepest minimum in Figure 8(b), near $k_F W_0/\pi \approx 2.9$, corresponds to the threshold for $N = 2 \rightarrow 3$ and, according to our interpretation, the deepest minima observed experimentally near $V_g = -0.9\,\mathrm{V}$ correspond to the same threshold (vertical dashed lines mark subband thresholds for $N \geq 2$; the statistical error involved in averaging over different impurity configurations is indicated by the error bars; and W_0 is the effective width of the lead at the threshold for the third subband). We believe that the cause of the minima is essentially the same as in the case of sharp corners, that the electrons become more forward directed as the energy is increased within a subband. While the effect is robust at the threshold for $N = 2 \rightarrow 3$, complicated scattering in the junction may obscure this effect for higher subbands.

The correspondence between Figures 8(a) and (b), and the correspondence between the numerical calculations of the magnetoresistances $R_H(V_g)$ shown in Figure 9, support the contention that the deep minima correspond to the threshold for occupation of a subband (Timp 1990, Behringer *et al.* 1991). The dependence on temperature supports this contention too. Similar features to those depicted in Figure 8(a) are found even after cycling the device to room temperature. Unlike aperiodic fluctuations, which change after cycling the device to room temperature because of minute changes in the impurity configuration, the quantum mechanical features shown in Figure 8(a) are *average* resistance effects which persist after ensemble averaging. Therefore these quantum mechanical features, associated with subband thresholds, are fundamentally different from the aperiodic fluctuations studied earlier in normal metal and semiconducting wires (Washburn and Webb 1986) because they depend on interference between particular ballistic paths rather than phase differences between random paths.

The close correspondence found between the numerical calculation and the transport measurements through submicron wires at low temperature is derived only by including (i) the quantisation of the transverse momenta, (ii) coherent scattering in a junction with rounded corners, and (iii) elastic impurity scattering in the calculation of the resistance. The quantisation of the momenta gives rise to a threshold for occupation of a subband; coherent scattering gives rise to sharp minima in the resistance at threshold; the collimation of the electron wave associated with a tapered lead yields a large negative bend resistance near threshold; and impurity scattering diminishes the effect of the collimation producing a resistance consistent with our observations. While elastic impurity scattering diminishes the effect of collimation, it does not eliminate it, as is evident from the observation that the Hall resistance remains suppressed for low H.

3 Manipulating an atomic beam with light

We have established that coherent impurity scattering has a deleterious effect on transport through an electron waveguide. Because of the sensitivity of a coherent device to the position of a single impurity, positioning of atoms with high precision ($\ll \lambda_F$) is required for reproducibility. Even without coherent transport, crystal growth of high precision is desirable because of the sensitivity to statistical fluctuations.

We propose to use the spatial structure of a light field, and the dipole force between the light and a neutral atom, to deflect atoms in an atomic beam to the appropriate positions on a substrate during deposition. The dipole force is just the Lorentz force which results from the average over time of the interaction between the spatial gradient of the light field and the dipole moment induced in the atom by the light (Ashkin 1980, Letokhov and Minogin 1981, Stenholm 1986, Minogin and Letokhov 1987). Provided that the surface diffusion of the atomic species is low relative to the deposition rate, features associatd with the structure of the light field could be rapidly transferred onto the film deposited on the substrate. Thus we propose to use light as a stencil for atoms *during* deposition, in contrast with conventional photolithography which uses atoms as a stencil for light to pattern a film *after* deposition. This technique represents a fundamentally new type of contactless lithography.

The qualitative features of the dipole force can be understood by examining a classical oscillator in a field. The dipole moment, $\mathbf{P}$, induced by the light on the atom, oscillates at the frequency of the field and can be expressed as $\chi(\mathbf{E})\mathbf{E}$. The average of $\mathbf{P}$ over time depends on the difference, $\Delta\omega = \omega - \omega_0$, between the resonant frequency of the dipole, ω_0, and the frequency of the field, ω. The time-averaged component of the force depends on the scalar product of $\mathbf{P}$ and the slope of the electric field. Thus, if the slope of the field is in phase with $\mathbf{E}$, the force can be expressed as $\frac{1}{2}\chi\nabla|\mathbf{E}|^2$, which is proportional to the gradient of the intensity. This component is referred to as the 'stimulated' force. Alternatively, the component of the induced dipole moment which is in quadrature with the field is referred to as the 'spontaneous' component. Generally, the force due to the pressure of the light on a two-level atom interacting with a monochromatic light field can be represented as

$$\mathbf{F} = \hbar(P_{\text{in}}\boldsymbol{\alpha} + P_{\text{quad}}\boldsymbol{\beta}), \tag{4}$$

(Hoeneisen and Mead 1972). Here $P_{\text{in}}\boldsymbol{\alpha}$ and $P_{\text{quad}}\boldsymbol{\beta}$ represent the stimulated and spontaneous components. The vectors $\boldsymbol{\alpha}$ and $\boldsymbol{\beta}$ are defined by the gradient of the intensity: if $g = (i/\hbar)\boldsymbol{\mu}\cdot\mathbf{E}$, then $\nabla g = (\boldsymbol{\alpha} + \boldsymbol{\beta})g$. P_{in} and P_{quad} are proportional to the real and imaginary components of the induced dipole moment P; $P_{\text{in}} = (\Delta\omega)p/2(1+p)$ and $P_{\text{quad}} = \Gamma p/2(1+p)$, where $p = |g|^2/\gamma^2$, $\gamma^2 = \Gamma^2 + (\Delta\omega)^2$ and Γ is the natural linewidth of the atomic transition driven by $\mathbf{E}$.

While both the stimulated and spontaneous components of the force can be used for lithography, the stimulated force is the superior choice because (i) variations in the film morphology can be obtained by varying the spatial distribution of the intensity, so it is this component that is relevant to submicron as well as centimetre scales; and (ii) the sign and magnitude of the force are directly controlled by $\Delta\omega$, and the force is unbounded. In our experiments we have investigated the force on an atom exerted by a Gaussian standing wave, and how that force changes the morphology of a film. For a standing wave, $\beta \to 0$, so the contribution of the spontaneous component to the force from the light field vanishes.

The strength of the stimulated force depends on the intensity gradients in the light field. For a Gaussian beam there are gradients over several different length scales: the confocal parameter, the beam waist, and the optical wavelength. Specifically, if we consider a Gaussian beam standing wave propagating along the z-axis (Yariv 1985), we have

$$g \approx g_0 \frac{W_{x0}}{W_x(z)} \frac{W_{y0}}{W_y(z)} \cos(kz) \exp\left\{ ik \left[z + \frac{r^2}{2} \left(\frac{1}{iz_{x0}+z} + \frac{1}{iz_{y0}+z} \right) \right] \right\}, \tag{5}$$

where $k = 2\pi/\lambda$, λ is the wavelength of the light, $W_x^2(z) = W_{x0}^2[1 + (z/z_{x0})^2]$, and W_{x0} and z_{x0} are the size of the waist and the confocal parameter in the x-direction, with similar expressions for y. Mechanical instability and the mobility of Na on the quartz substrate in our experiments will obliterate any variations of the film morphology on the scale of λ, but variations on the scale of W_x and z_{x0} can be easily observed. Specifically the spatially averaged force is dominated by the transverse intensity gradient when $z \gg z_{x0}, W_{x0}$, and $F \approx F_x \approx \exp\{-x^2/W_{x0}^2[1 + (z/z_{x0})^2]\}$.

3.1 Manipulating an atomic beam on a mesoscopic scale

Figure 10 shows a schematic representation of the experiment. A neutral atomic beam collimated to within about ± 20 milliradians emanates along the y-axis from a glass ampoule (supplied by Aldrich Chemical) which is filled with 5 g of 99.95 % pure Na and heated to about 300°C using a conformal Mo wire. The beam is collimated by the mouth of the ampoule which forms a glass flue approximately 5 cm long and 4 mm in diameter. The atomic beam impinges on a 5 cm diameter quartz optical flat separated from the top of the flue by 30 cm. In this configuration, using a vacuum of 10^{-7} torr, it is possible to deposit *in the absence of light* a featureless, metallic Na film, about 20–40 nm thick and 3 cm in diameter, in less than 20 minutes.

In our experiments we were able to change the morphology of the Na film by deflecting a portion of the atomic beam with a light field interposed between the source and the substrate (Figure 11). The light beam, which propagated along the z-axis, was cylindrically focussed ($f = 12$ cm). The focussing of the light allowed us to evaluate the effect of variations in the intensity along z on the deflection of the Na along x. The edge of the light field was separated from the flat quartz substrate by about 2 cm. In most cases a resonant light beam, propagating transverse to both the Na beam and to the deflecting light beam, was used to evaluate the effect of the deflecting light on the atomic beam prior to deposition on the quartz substrate. The frequency was varied about the transition between the $F = 2$ and $F = 3$ sublevels of the (589 nm) $3\,^2S_{1/2} \to 3\,^2P_{3/2}$ transition in Na using an acousto-optic modulator. It was impossible to change the morphology of the film without supressing optical pumping. Optical pumping results in the accumulation of a significant atomic population in the $F = 1$ ground state of Na. To suppress optical pumping (Phillips *et al.* 1985) a light beam, copropagating with the deflecting beam, was present which was nearly resonant with the $F = 1 \to 2$ transition. At the origin, the Gaussian beam had a waist only $W_{x0} \approx 4\,\mu$m wide, $W_{y0} \approx 3$ mm

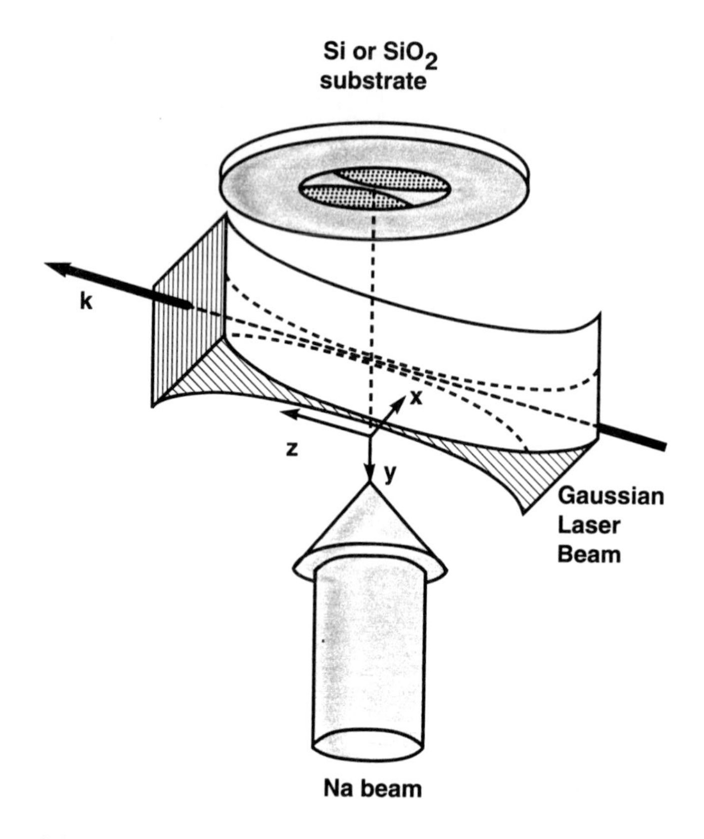

Figure 10: *Schematic representation of the experiment. A Gaussian beam of light, interposed between a collimated atomic beam and an optically flat quartz substrate, is used to deflect the atoms to the desired positions. The coordinate system is shown.*

thick, with an intensity of approximately $10^5\,\mathrm{mW\,cm^{-2}}$ in the $F = 2 \rightarrow 3$ frequency component and approximately $10^4\,\mathrm{mW\,cm^{-2}}$ in the $F = 1 \rightarrow 2$ component.

Figure 11(a) shows a photograph of Na metal films prepared with a standing wave Gaussian beam interposed between the substrate and the oven using a detuning of $\Delta\omega = 80\,\mathrm{MHz}$ above the $F = 2 \rightarrow 3$ transition frequency. The surface profile of the film was measured in air using an Alpha Step 200 (to prevent the deterioration of the Na film in air, a 50 nm thick layer of AuPd was evaporated onto the Na film immediately after deposition while the film was still in vacuum). The profile shown in Figure 11(b) was obtained near the dashed line in Figure 11(a) and is typical of that obtained for $|z| > z_c \approx 1\,\mathrm{mm}$. Corresponding with the position of the light beam, Figure 11(b) shows a gap approximately 20 nm deep, the total thickness of the Na film, and also shows evidence of acumulation on either side of the gap indicative of the material forced out by the stimulated force.

We interpret these observations as evidence of the predominance of the stimulated, velocity-independent force for $|z| > z_c$. Because of the change with detuning, the force which gives rise to the gap in the Na film of Figure 11(a) is associated with the

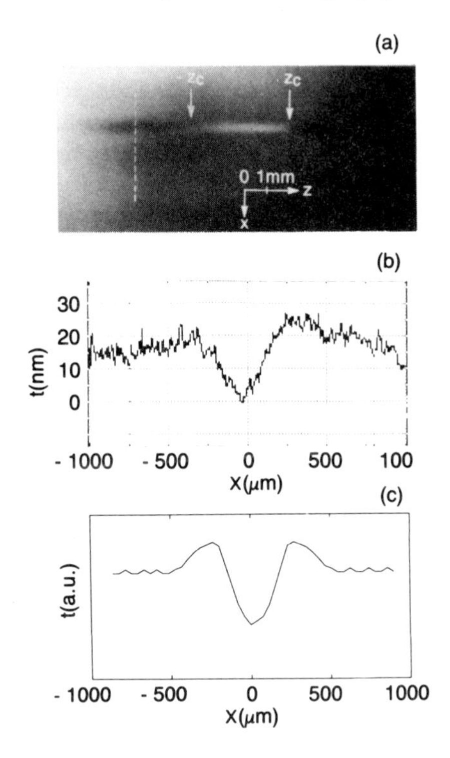

Figure 11: *(a) Photograph of a Na film with a AuPd overlayer obtained using a geometry similar to that represented in Figure 10 with a standing wave, Gaussian light beam interposed between a thermal neutral atomic beam and a quartz substrate, with* $\Delta\omega = 80\,\mathrm{MHz}$ *above resonance. For* $|z| > z_c \approx 1\,\mathrm{mm}$, *(a) shows a gap in the Na film induced by velocity-independent light forces, while there is an accumulation of Na for* $|z| < z_c$, *where velocity-dependent forces are presumed to predominate. (b) Gap found in the surface profile of the Na film obtained for* $|z| > z_c$, *near the dashed line of (a). (c) Results of a numerical simulation with parameters which correspond to the positions where the profile was taken; t is the thickness of the film.*

stimulated force (Prentiss *et al.* 1991). In Figure 11(c), we show the profile determined from a two-dimensional numerical simulation of the effect of the velocity-independent stimulated force given by Equation (4) on a perfectly collimated atomic beam with a thermal velocity along y. In this numerical simulation, 10000 atoms initially at rest in the light field are allowed to propagate through the light field along the x and y-directions subject to the forces given by Equation (4). We assume (i) a mean thermal velocity of $v_y \approx 500\,\mathrm{m\,s^{-1}}$ corresponding to a temperature of about 300°C; (ii) no initial transverse velocity, $v_x = 0$; (iii) a free flight distance, beyond the light field, of about 2 cm; and (iv) $W_{0x} = 4\,\mu\mathrm{m}$ and $W_{0y} = 3\,\mathrm{mm}$ for $z = 5\,\mathrm{mm}$. The simulation uses the measured W_{0x} size and intensity, I, near the position where the profile of Figure 11(b) was measured. The effect of the stimulated force is in close correspondence with the measured profile of the film.

For light of high intensity, additional non-conservative, velocity-dependent forces can predominate, and can drasticaly alter the morphology of the film (Dalibard and Cohen-Tannoudji 1985). Specifically, the non-conservative contribution to the force can change the sign and magnitude of the total force. Thus far, we have not considered the dependence of the light pressure force on velocity because the atomic beam was initially well collimated in the region where it interacts with the light, and because the velocities which result from the interaction with the field are small compared with Γ/k. Moreover, according to our simulations, the dependence of the force on velocity cannot produce both the accumulation and the depletion found for $|z| > z_c$. But the intensity in the Gaussian beam increases algebraically as $z \to 0$ for constant x, p increases, and the dependence of the force on velocity cannot be ignored. Ultimately, the intensity becomes so large that the velocity-dependence of the force predominates and causes it to change sign for a given $\Delta\omega$ (Hoeneisen and Mean 1972). According to our interpretation, the light field has a p large enough for velocity-dependent forces to dominate and even to change the sign of the force when $|z| < z_c$ in Figure 11(a). For $|z| > z_c$ we find a gap in the Na film corresponding to the position of the light when $\Delta\omega = 80\,\mathrm{MHz}$, but for $|z| < z_c$ we find that Na accumulates, which must be associated with a change in sign of the force. We expect that the force due to the standing wave imparts a large velocity to the atom for $|z| < z_c$. Provided that the velocity-dependent component of the force predominates, F_x will change sign when $p^2/(1-p) = \Gamma^2/2|\gamma|^2$ or when $p = 0.075$ (Hoeneisen and Mead 1972). The algebraic decay of the Gaussian profile along z satisfies this condition when $z_c = \pm 1.2\,\mathrm{mm}$ for the parameters used here, in close correspondence with our observations.

3.2 Manipulating an atomic beam on a nanometre scale

We have demonstrated that it is possible to control the morphology of a film during deposition by controlling the lateral position of atoms using the dipole force. The morphology of the film near the position of a Gaussian standing wave beam is in close corespondence with the results of our numerical calculations of the effect of the dipole force on a well-collimated neutral atomic beam. By controlling the longitudinal velocity of the atomic beam and the interaction time between the beam and the light field, it should be possible to produce abrupt nanometre scale forces in a planar standing wave light field. A configuration similar to that shown in Figure 10 could be used to produce features on a nanometre scale in a thin film provided that a planar standing wave replaces a Gaussian beam and that the size of the grains in the film is small enough. The minima in the optical potential which occur every $\lambda/2$ in the standing wave can be positioned precisely over the substrate using a Fabry-Perot interferometer to lock the posibion of the crystal relative to the standing wave and eliminate mechanical vibration. Moreover, if the Fabry-Perot light beam is modulated acousto-optically, it would be possible to vary dynamically the position of the potential minima relative to the crystal by varying the frequency of the light, and so position the minima anywhere in a field defined by $\lambda/2$ with very high precision ($\lambda/50$–$\lambda/10000$).

Figure 12 shows the results of a numerical computation of the atomic distribution on a surface over a distance of $\lambda/2$ showing the effect of a standing wave potentital detuned by 100 MHz on an atomic beam with a longitudinal velocity of $v_y = 500\,\mathrm{m\,s^{-1}}$.

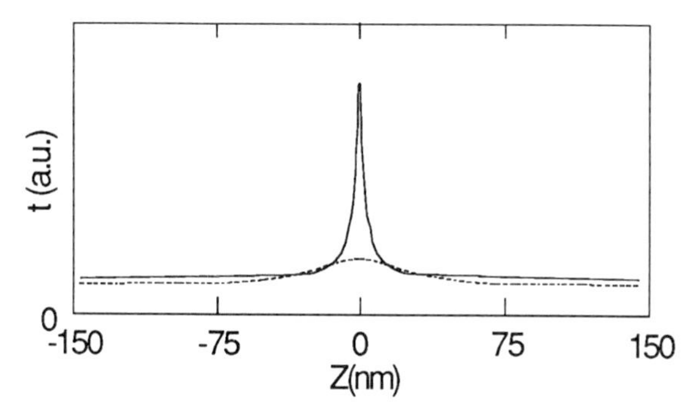

Figure 12: *Numerical calculation of the effect of a standing wave on thin film morphology for two different interaction lengths. It is assumed that the light field clips the surface of the substrate (so there is no free flight time), and that the atomic beam has a velocity of* $500\,\mathrm{m\,s^{-1}}$ *along* y. *The parameter* t *is the thickness of the film. Only a distance of* $\lambda/2$ *is shown in the figure; the data are periodic in* $\lambda/2$. *The solid curve represents an interaction length,* L, *between the atomic beam and the light field of* $50\,\mu\mathrm{m}$; *the dashed curve represents* $L = 25\,\mu\mathrm{m}$.

In the computation we assume that the light beam clips the edge of the substrate (*i.e.* there is no free flight time) and interacts with the atomic beam over a length L. The result of two different interaction lengths between the atomic beam and the light field are shown in the figure. Notice that for L = $50\,\mu$m (solid line), the atomic distribution is very narrow ($\approx \lambda/100$ wide) with a peak-to-valley ratio of about 8, while for $L = 25\,\mu$m (dashed line) the distribution is broad ($\lambda/10$) with a ratio of about 1. The narrow distribution occurs when the interaction time is aproximately equal to $T \approx 1/4\Omega$, where $\Omega = 4\hbar k^2/(\Delta\omega)$. A harmonic potential would yield a δ-function distribution of atoms at $x = 0$ for this interaction time; the finite width of the distribution reflects the anharmonic nature of the optical potential. This calculation does not include the effect of quantum fluctuations or stray spatial variations of the light field on the distribution. The heating associated with quantum fluctuations may be important when the wavelength of the atom becomes comparable to the width of the potential.

The calculation of Figure 12 suggests an opportunity for positioning atoms in a parallel way with high acuracy on a crystal surface. Since the spatial structure of the light field can be registered with respect to the substrate with a precision of $\lambda/50$–$\lambda/10000$, using interferometric techniques to spatially lock the light field to the substrate, an arbitrary feature could be drawn on a substrate surface within a field defined by $\lambda/2$. Furthermore, any atomic species could be manipulated in this way, and multiple species could be manipulated simultaneously.

4 Conclusions

We have shown that the transport characteristics of an electron waveguide are acutely sensitive to defects. This sensitivity develops from both statistical variations in the doping density and from coherent scattering of the electronic wavefunction due to scattering from the dopant configuration. This senstivity is apparent from the deterioration of the quantisation of the two terminals of a waveguide with increasing length, from random telegraph noise observed near the threshold for the occuation of a 1D subband, and from a reduction in the collimation of an electron wave launched into a junction. Using the electron waveguide as a paradigm, we surmise that the utility of very small electronic devices depends upon our ability to control the position of dopants at an atomic level.

We have also given a rudimentary demonstration of a new type of lithography which has the potential for nanometre resolution with atomic precision. This lithography uses the spatial structure in a light field and the dipole force between light and neutral atom to manipulate a neutral atomic beam during deposition on a substrate.

References

Alt'shuler B L, and Spivak B Z, 1986. *JETP Lett.*, **42**, 447.

Alt'shuler B L, 1985. *JETP Lett.*, **41**, 648.

Ashkin A, 1980. *Science*, **210**, 1081.

Avishai Y, and Band Y, 1989. *Phys. Rev. Lett.*, **62**, 2527–2530.

Baranger H U, 1990. *Phys. Rev. B*, **42**, 1479–1495.

Baranger H U, and Stone A D, 1989. *Phys. Rev. Lett.*, **63**, 414.

Baranger H U, Stone A D, and DiVincenzo D P, 1988. *Phys. Rev. B*, **37**, 6521.

Beenakker C W J, and van Houten H, 1989. *Phys. Rev. B*, **39**, 10445.

Behringer R, Timp G, Baranger H U, Cunningham J E, 1991. *Phys. Rev. Lett*, **66**, 930. ,

Berggren K F, Thornton T J, Newson D J, and Pepper M, 1986. *Phys. Rev. Lett.*, **57**, 1769.

Büttiker M, 1989. In *Proc. Int. Symp. on Nanostructure Physics and Fabrication*, Eds. Kirk W P and Reed M A, Boston, Academic Press, p. 319.

Büttiker M, 1988a. *Phys. Rev. B*, **38**, 9375.

Büttiker M, 1988b. *IBM J. Res. Develop.*, **32**, 317.

Büttiker M, 1986. *Phys. Rev. Lett.*, **57**, 1761.

Chang A M, Chang T Y, and Baranger H U, 1989. *Phys. Rev. Lett.*, **63**, 996.

Chu C S, and Sorbello R S, 1989. *Phys. Rev. B*, **40**, 5941.

Dalibard J, and Cohen-Tannoudji C, 1985. *J. Opt. Soc. Am.*, **B2**, 1707.

Das Sarma S, and Stern F, 1985. *Phys. Rev. B*, **32**, 8442.

Davies J H, 1989. Private communication.

Davies J H, and Nixon J A, 1988. *Phys. Rev. B*, **39**, 3423.

Dekker C, Scholten A J, Liefrink F, Eppenga R, Van Houten H, and Foxon C T, 1991. *Phys. Rev. Lett.*, **66**, 2148.

Economou E N, and Soukoulis C M, 1981. *Phys. Rev. Lett.*, **46**, 618.

Fang F, Smith T P, and Wright S L, 1988. *Surf. Sci.*, **196**, 310.

Feng S, Lee P A, and Stone A D, 1986. *Phys. Rev. Lett.*, **56**, 1960.

Fowler A B, 1991. Private communication.

Ford C J B, Washburn S, Büttiker M, Knoedler C M, and Hong J M, 1989. *Phys. Rev. Lett.*, **62**, 2724.

Glazman L I, and Jonson M, 1991. *Phys. Rev. B*, **44**, 0000.

Glazman L I, and Larkin I A, 1991. *Semicond. Sci. Technol.*, **6**, 32.

Glazman L I, Lesovick G B, Khmel'nitskii D E, and Shekhter R I, 1988. *JETP Lett.*, **48**, 238.

Haanappel E G, and van der Marel D, 1989. *Phys. Rev. B*, **39**, 5484.
Hoeneisen B, and Mead C A, 1972. *Solid State Electron.*, **15**, 818.
Imry Y, 1989. In *Proc. Int. Symp. on Nanostructure Physics and Fabrication*, Eds. Kirk W P and Reed M A, Boston, Academic Press, p. 379.
Kakuta T, Takagaki Y, Gamo K, Namba S, Takaoka S, and Murase K, 1991. *Phys. Rev. B*, **43**, 4321.
Kirczenow G, 1989. *Solid State Commun.*, **71**, 469–472.
Kirton M J, and Uren M J, 1989. *Adv. Phys.*, **38**, 367.
Kumar A, Laux S E, and Stern F, 1989. *Appl. Phys. Lett.*, **54**, 1270.
Landauer R, 1985. In *Localization, Interaction, and Transport Phenomena*, Eds. Bergmann G and Bruynseraede Y, New York, Springer-Verlag, pp. 38–50.
Laughton M J, Barker J R, Nixon J A, and Davies J H, 1991. *Phys. Rev. B*, **44**, 1150–1153.
Lee P A, and Stone A D, 1985. *Phys. Rev. Lett.*, **55**, 1622.
Letokhov V S, and Minogin V G, 1981. *Phys. Rep.*, **73**, 1.
Minogin V G, and Letokhov V S, 1987. *Laser Light Pressure on Atoms*, New York, Gordon and Breach.
Nixon J A, Davies J H, and Baranger H U, 1991a. *Phys. Rev. B*, **43**, 2638.
Nixon J A, Davies J H, and Baranger H U, 1991b. *Superlat. Microstruct.*, **9**, 187.
Payne M C, 1989. *J. Phys. Condens. Matter*, **1**, 4939.
Phillips W D, Prodan J V, and Metcalf H J, 1985. *Opt. Soc. Am.*, 1751.
Prentiss M G, and Ezekiel S, 1986. *Phys. Rev. Lett.*, **56**, 46.
Prentiss M, Timp G, Bigelow N, Behringer R E, and Cunningham J E, 1991. Unpublished.
Price P J, 1984. *Surf. Science*, **143**, 145.
Roukes M L, Scherer A, Allen S A Jr, Craighead H G, Ruthen R M, Beebe E D, and Harbison J P, 1987. *Phys. Rev. Lett.*, **59**, 3011.
Sakaki H, 1980. *Jpn. J. Appl. Phys.*, **19**, L735.
Salamon C, Dalibard J, Aspect A, Metcalf H, and Cohen-Tannoudji C, 1987. *Phys. Rev. Lett.*, **59**, 1659.
Stenholm S, 1986. *Rev. Mod. Phys.*, **58**, 699.
Szafer A, and Stone A D, 1989. *Phys. Rev. Lett.*, **62**, 30.
Takagaki Y, Gamo K, Namba S, Takaoka S, Murase K, Ishida S, Ishibashi K, and Aoyagi Y, 1989. *Solid State Commun.*, **71**, 809.
Takagaki Y, Gamo K, Namba S, Takaoka S, Murase K, Ishida S, Ishibashi K, and Aoyagi Y, 1989. *Solid State Commun.*, **69**, 811.
Timp G L, and Howard R E, 1991. *Proc. IEEE*, **79**, 1.
Timp G, 1990. In *Semiconductors and Semimetals*, Vol. ed. Reed M A, New York, Academic Press.
Timp G, Baranger H U, de Vegvar P, Cunningham J E, Howard R E, Behringer R and Mankiewich P M, 1988. *Phys. Rev. Lett.*, **60**, 2081.
Timp G, Behringer R E, and Cunningham J E, 1990. *Phys. Rev. B*, **42**, 9259.
Timp G, Behringer R E, Cunningham J E, Westerwick E W, 1991b. In *Condensed Systems of Low Dimensionality*, Eds. J L Beeby *et al.* , New York, Plenum Press, pp. 347–358.
Timp G, Chang A M, de Vegvar P, Howard R E, Behringer R, Cunningham J E, and Mankiewich P, 1988. *Surf. Sci.*, **196**, 68.
Timp G, Prentiss M, Behringer R E, Bigelow N, and Cunningham J E, 1991a. In *Proc. Int. Conf. on Nanofabrication and Mesoscopic Physics*, Eds. Reed M A and Kirk W P, Boston, Academic Press.
van der Marel D, and Haanappel E G, 1989. *Phys. Rev. B*, **39**, 7811.
van Wees B J, van Houten H, Beenakker C W J, Williamson J G, Kouwenhoven L P, van der Marel D, and Foxon C T, 1988. *Phys. Rev. Lett.*, **60**, 848.

Washburn S, and Webb R A, 1986. *Adv. Phys.*, **35**, 375.
Westbrook C I, Watts R N, Tanner C E, Ralston S L, Phillips W D, Lett P D, and Gould P L, 1990. *Phys. Rev. Lett.*, **56**, 33.
Wharam D A, Thornton T J, Newbury R, Pepper M, Ahmed H, Frost J E F, Hasko D G, Peacock D C, Ritchie D A, and Jones G A C, 1988. *J. Phys. C*, **21**, L209.
Yacoby A, and Imry Y, 1990. *Phys. Rev. B*, **41**, 534.
Yariv A, 1985. *Optical Electronics*, New York, Holt Rinehart and Winston, p. 29.

Semiclassical Motion in Periodic Potentials and Non-local Resistance

P C Main

University of Nottingham
Nottingham, U.K.

1 Semiclassical Motion in Periodic Potentials

1.1 Introduction

Mobilities exceeding $100\,\mathrm{m^2\,V^{-1}\,s^{-1}}$ are now routinely obtained at low temperatures in the two-dimensional electron gas (2DEG) formed in modulation doped heterostructures grown in GaAs-Al$_x$Ga$_{1-x}$As [see, for example, Pfeiffer *et al.* (1989) for a mobility of $1000\,\mathrm{m^2\,V^{-1}\,s^{-1}}$]. These mobilities correspond to electron elastic mean free paths of order tens of microns. It is possible using modern lithography techniques to introduce electrostatic potential variations on a scale much smaller than a mean free path. One possibility, which has been achieved in a variety of ways (Weiss *et al.* 1989, Winkler *et al.* 1989, Alves *et al.* 1987, Heitmann 1989, Ismail *et al.* 1989), is to impose a periodic potential, or lateral superlattice, on the 2DEG, generally with a period between 100 nm and 500 nm. This length scale is much smaller than the elastic mean path but much longer than the periodic crystal potential of the GaAs itself. There are, therefore, two possibilities for the electron motion in the periodic potential: (a) quantum coherent formation of a miniband structure analogous to the band structure obtained from a nearly-free electron model, and (b) the semi-classical ballistic motion of the electrons of the 2DEG in the periodic potential.

Although the observation of quantum behaviour was the principal motivation for these investigations it turns out that semi-classical effects are dominant in most of the experiments performed to date.

Section 1.2 describes briefly the construction of devices with periodic potentials, and Sections 1.3 and 1.4 describe their observed magnetotransport properties at low temperature together with explanations in terms of the semi-classical model of electron transport. Finally the limits to the applicability of this model are discussed, with a view

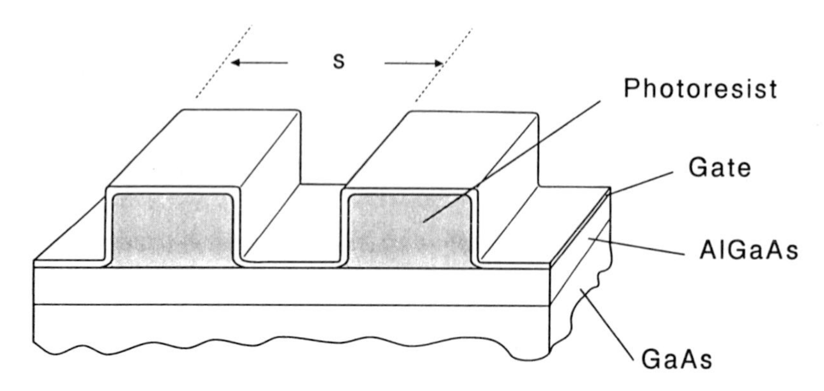

Figure 1: *Schematic diagram of sample fabrication.*

to determining what is necessary to construct a device which exhibits true quantum effects.

1.2 Experimental

A schematic diagram of the device is shown in Figure 1. A high mobility 2DEG is formed in a standard modulation-doped GaAs-$Al_xGa_{1-x}As$ heterostructure. Various heterostructures were used; typically, the mobility of an untreated layer was around $100\,m^2\,V^{-1}\,s^{-1}$. Processing involved laying a layer of resist on the surface of the heterostructure, usually negative resist, which was patterned with electron beam lithography into a series of lines. A Ti-Au layer was then evaporated over the pattern to form a gate, as shown in Figure 1. The gate acts more effectively when it is in contact with the semiconductor surface than when it is raised away by the resist. The application of a negative bias to this Schottky gate then enables us to impose a periodic potential on the 2DEG embedded in the heterostructure. The samples were fabricated into Hall bars, of typical dimensions $5\,\mu m \times 10\,\mu m$, in which the current flows parallel to the longer direction and perpendicular to the lithographed lines. Using this technique we have been able to fabricate and investigate samples with periods down to 100 nm.

Even with no applied gate bias, trapped surface charge gives a periodic potential at the 2DEG which is estimated to be less than 1 meV (see below).

1.3 Oscillations at low magnetic field

The magnetoresistance at 4.2 K of a sample with period 300 nm is shown in Figure 2 for various gate voltages. The bottom curve, for gate voltage $V_g = 0$, shows three essential features. For magnetic field, B, below 0.1 T there is a positive magnetoresistance. This is discussed in detail in Section 1.4. As B is increased, there is a series of oscillations periodic in $1/B$ (Weiss *et al.* 1989, Winkler *et al.* 1989). The peaks in magnetoresistance are given approximately by

$$2R_c = \left(n + \tfrac{1}{4}\right) s, \tag{1}$$

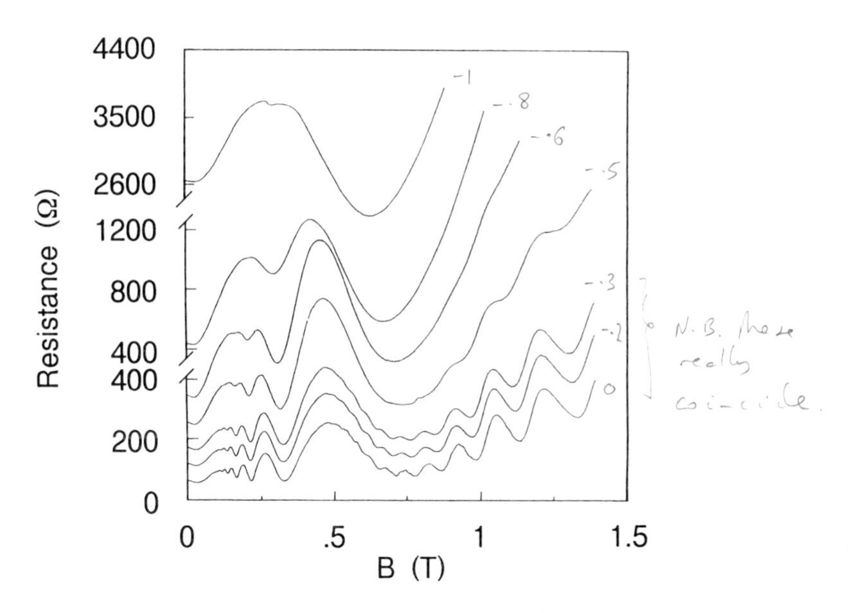

Figure 2: *Magnetoresistance for* $s = 300\,\text{nm}$ *and* $T = 2\,\text{K}$ *for various gate voltages;* $V_g = -1.0\,\text{V}$ *(top),* -0.8, -0.6, -0.5, -0.3 *(displaced by* $100\,\Omega$*),* -0.2 *(displaced by* $50\,\Omega$*) and* 0 *(bottom).*

where R_c is the classical cyclotron radius, s is the period of the potential and n is a positive non-zero integer. The positions of the minima obey the formula

$$2R_c = \left(n - \tfrac{1}{4}\right) s. \tag{2}$$

The third feature of the magnetoresistance is the onset of Shubnikov–de Haas (SDH) oscillations for $B > 0.5\,\text{T}$.

Figure 2 shows what happens when the magnitude of V_g, and hence the amplitude of the periodic potential, is increased. For $V_g > -0.5\,\text{V}$ the electron density ($n = 3.7\times10^{15}\,\text{m}^{-2}$) is only very weakly dependent on V_g. However, the oscillations at lowest field quench as V_g is made more negative while, at the same time, the peak at highest field becomes larger. Similarly, the SDH oscillations are quenched as the amplitude of the potential is increased. The latter observation can be understood readily. When $2R_c < s$, which is the experimental situation for the observation and quenching of the SDH oscillations, there is a broadening of the Landau levels by an energy eV_0, where V_0 is the amplitude of the electrostatic periodic potential (Winkler *et al.* 1989). When this becomes comparable with $\hbar\omega_c$ we expect quenching. Values of V_0 extracted from the minimum fields where SDH oscillations occur are given in Table 1 as V_0^{LL}. These are consistent with the assertion that $V_0 \approx 1\,\text{mV}$ made above.

Various explanations have been put forward to explain the origin of the oscillations at low field (Winkler *et al.* 1989, Gerhardts *et al.* 1989, Beenakker 1989, Vasilopoulos and Peeters 1989). Beenakker (1989) has pointed out that the oscillations are semi-classical. His argument is best explained with reference to the numerical simulations of semi-classical trajectories shown in Figure 3 (a)–(f). All the simulations refer to a

V_g (V)	B_c (T)	V_0 (mV)	V_0^{LL} (mV)
0.0	0.11	1.2	0.73
−0.2	0.11	1.2	0.73
−0.3	0.12	1.3	0.75
−0.5	0.15	1.7	1.06
−0.6	0.18	2.0	1.34
−0.8	0.24	2.6	...
−1.0	0.31	3.4	...

Table 1: *Nominal gate voltages and values of magnetic field for the peak in positive magnetoresistance* B_c. V_0 *is the amplitude of the electrostatic potential derived from* B_c *and* V_0^{LL} *from the quenching of the Landau levels.*

2DEG with Fermi energy $E_F = 10\,\mathrm{meV}$. The straight lines represent equipotentials and the curves are electron trajectories determined by numerical integration of the semi-classical equations of motion. The two left-hand columns refer to a magnetic field such that $2R_c/s = 6.25$, that is a peak in the magnetoresistance, and the two right-hand columns $2R_c/s = 5.75$ which is a condition where a minimum occurs.

In Figure 3(a), the right-hand orbit drifts as a function of time. This can be understood in terms of the drift velocity, $(1/B^2)(\mathbf{E} \times \mathbf{B})$, experienced by a charged particle in crossed electric and magnetic fields. This drift averages out for motion perpendicular to the equipotentials, but there is no cancellation for the extrema parallel to the equipotentials. When the extrema of the orbit experience the same electric field, there is an enhancement of the drift. This will occur whenever the cyclotron diameter is approximately equal to a whole number of periods. This explains the periodicity of the effect given by Equation (1). The phase factor of $\frac{1}{4}$ arises from the integration over the electron orbit and appears naturally in the numerical simulation. This enhanced drift depends on the position of the orbit centre; the left-hand orbit of Figure 3(a) is stationary due to its symmetry, despite the $\mathbf{E} \times \mathbf{B}$ drift, and other orbits drift in the opposite direction.

The important point is that mean square drift velocity, $\langle v_d^2 \rangle$, averaged over all orbit centres is non-zero. Figure 3(b) shows the situation for a magnetic field corresponding to a minimum in the magnetoresistance. Here all orbits are stationary so $\langle v_d^2 \rangle$ is zero. The effect of the non-zero drift is to enhance the diffusion in the direction parallel to the equipotentials (y-direction).

Beenakker (1989) determined $\langle v_d^2 \rangle$ analytically and obtained

$$\frac{\Delta\rho_{xx}}{\rho_0} = \left(\frac{eV_0}{E_F}\right)^2 \left(\frac{l^2}{sR_c}\right) \cos^2\left(\frac{2\pi R_c}{s} - \frac{\pi}{4}\right) \tag{3}$$

for the relative increase in resistivity. Here l is the mean free path, ρ_0 is the resistivity for $B = 0$ and τ is the transport relaxation time. Equation (3) has also been obtained using other approaches (Gerhardts *et al.* 1989).

According to Equation (3), the amplitude of the oscillations is predicted to be

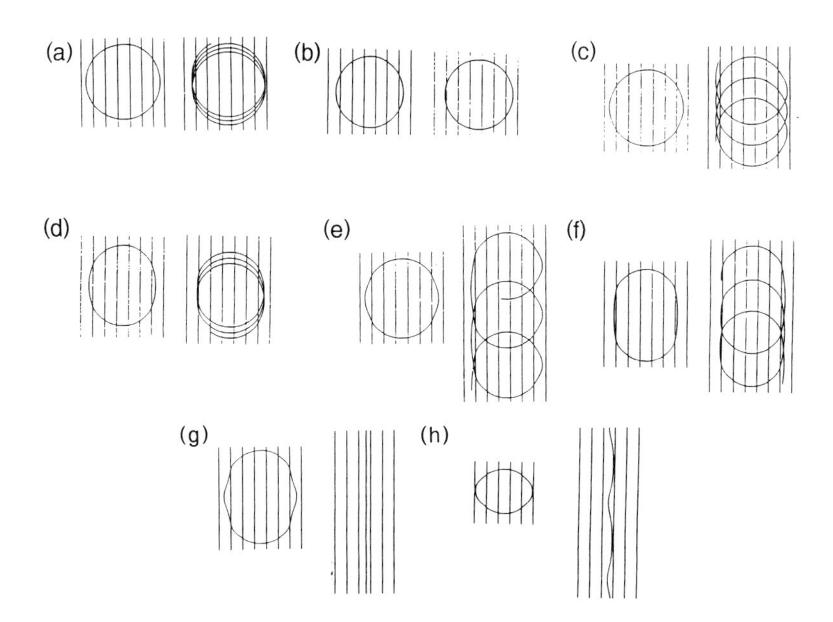

Figure 3: *Numerical simulations of classical particle trajectories; (a), (c), (e) and (g) have* $2R_c/s = 6.25$ *and (b), (d), (f) and (h) have* $2R_c/s = 5.75$. *Values of* eV_0/E_F *are* 0.01 *for (a) and (b),* 0.05 *for (c) and (d)* 0.09 *for (e) and (f), and* 0.15 *for (g) and (h).*

proportional to B. A common feature of all experimental results (Weiss *et al.* 1989, Winkler *et al.* 1989, Alves *et al.* 1987) is that this is not the case. Also, as V_g is made more negative, the amplitude of the highest-field oscillation does increase as predicted qualitatively by Equation (3), but the lower field oscillations actually decrease in size and eventually disappear. Examination of Figure 2 indicates that it is the minima of the oscillations which are quenched first, not the maxima. This indicates that the resonant enhancement is relatively unaffected but the orbits corresponding to the minima are no longer stationary as they are in Figure 3(b). As V_0 is increased relative to E_F at constant B, the resonant orbits drift more rapidly but the previously stationary orbits also drift [Figures 3(d) and (f)]. This arises due to distortions of the electron orbit introduced by the potential. Figures 3(e) and (f) show this distortion clearly and also indicate that the contrast between the 'drifting' and 'stationary' conditions is drastically reduced. An equivalent set of curves can be produced for the variation of B at constant V_0 and similar distortions are produced at low magnetic fields.

Numerical calculations of this effect (Dellow 1991) show that it is a plausible explanation for the quenching of the oscillations in low magnetic field.

1.4 Positive magnetoresistance

A common feature of all our magnetoresistance curves shown above is the positive magnetoresistance which occurs at magnetic fields below those where the oscillations appear. Figure 2 shows clearly that the positive magnetoresistance becomes stronger

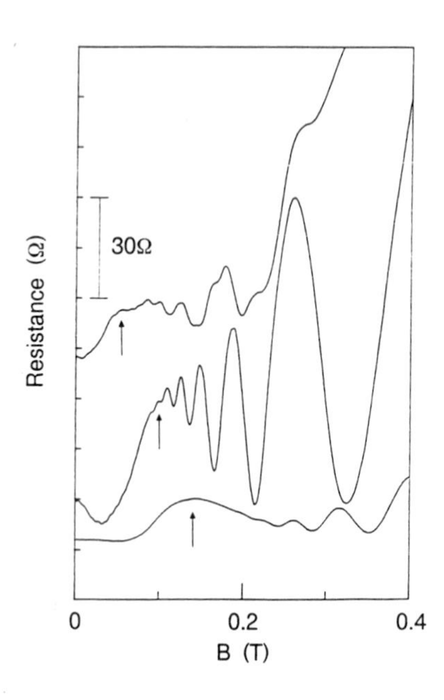

Figure 4: *Magnetoresistance for* $s = 150\,\mathrm{nm}$ *(bottom),* $300\,\mathrm{nm}$ *and* $500\,\mathrm{nm}$ *(top). The curves are offset for clarity and the arrows mark the positions of* B_c.

and persists to higher magnetic fields as the amplitude of the potential is increased. Figure 4 shows the relevant magnetoresistance for three samples with periods of 150 nm (bottom), 300 nm and 500 nm (top) with electron densities of $4.5\times10^{15}\,\mathrm{m}^{-2}$, $3.7\times10^{15}\,\mathrm{m}^{-2}$ and $3.8\times10^{15}\,\mathrm{m}^{-2}$ respectively. All the curves were taken at $V_g = 0$ and at 4.2 K except the middle curve which was at 2.7 K. Since the devices were all constructed in an identical fashion, V_0 should be approximately the same for all of them. The positive magnetoresistance has been ascribed to magnetic breakdown in the miniband structure created by the periodic potential (Weiss 1990, Středa and MacDonald 1990). However, as has been pointed out by Beton *et al.* (1991), this theory predicts that the maximum field for the positive magnetoresistance, B_c, *decreases* as the period increases, in clear contradiction to the experimental data of Figure 4.

Motivated by the success of the semi-classical model in explaining the oscillatory magnetoresistance, the simulations were extended to higher values of V_0 or, alternatively, lower values of B. The results are shown in Figure 3(g)–(h) where $eV_0/E_F = 0.15$. Some of the orbits are still closed but there is now the possibility of classical open orbits. These are analogous to, but distinct from, the open orbits which appear in band conduction. Physically the electrons do not receive sufficient momentum from the magnetic field to overcome the electrostatic force of the periodic potential. The potential causes the electrons to stream parallel to the equipotentials. The streaming velocity is very close to v_F so these trajectories make a very large contribution to the magnetoresistance of the form

$$\frac{\Delta R_{xx}}{R_0} \approx (\omega_c\tau)^2 f(B) \tag{4}$$

where $f(B)$ is the fraction of the electrons in the streaming orbits (Beton *et al.* 1990). Numerical simulations indicate that $\Delta\rho_{xx}/\rho_0 \propto B^2$ for low magnetic fields, as seen in the experimental data. These streaming orbits do not occur at all magnetic fields. For fixed V_0 and s, the critical magnetic field B_c is determined by the condition that the maximum electrostatic force experienced by the electron is equal to the magnetic force at the Fermi velocity. Beton *et al.* (1990) showed that this gives

$$B_c = \frac{2\pi V_0}{s v_F}. \tag{5}$$

This can be compared with the data of Figures 2 and 4, where we identify B_c with the maximum magnetoresistance. Experimentally the agreement with Equation (5) is excellent for both the variation of v_F and V_0, confirming the validity of the semi-classical approach. Furthermore, the values of V_0 determined from the data in Figure 2 are shown in Table 1 and are in reasonable agreement with the V_0 estimated from the SDH oscillations. The lack of precise agreement is of no concern, given the approximate nature of the determination of V_0 from the SDH oscillations. The values determined from B_c should be more reliable.

1.5 Quantum versus classical

The circumstances in which quantum effects might become important has been considered in detail by Beton *et al.* (1990). The Schrödinger equation for an electron in a periodic electrostatic potential and a uniform magnetic field is

$$\left\{\frac{1}{2m}(-i\hbar\nabla + e\mathbf{A})^2 + eV_0\left[\cos\left(\frac{2\pi x}{s}\right) + 1\right]\right\}\Psi = E\Psi, \tag{6}$$

where $\mathbf{A} = (0, xB, 0)$ is the vector potential in the Landau gauge. This has solutions of the form $\Psi = \exp(ik_y y)\Phi(x)$ which gives, after substitution into Equation (6),

$$\left[-\frac{\hbar^2}{2m}\frac{\partial^2}{\partial x^2} + V_{\text{eff}}(x)\right]\Phi(x) = E\Phi(x), \tag{7}$$

where

$$V_{\text{eff}}(x) = \tfrac{1}{2}m\omega_c^2(x - x_0)^2 + eV_0\left[\cos\left(\frac{2\pi x}{s}\right) + 1\right] \tag{8}$$

and

$$x_0 = -\frac{\hbar k_y}{eB}. \tag{9}$$

Solutions to (7) for the form of $\Phi(x)$ are very complicated since V_{eff} depends on B, V_0 and the position of the centre of the orbit. However, we can understand the basic physics by inspecting Figure 5. This shows the effective potential for three magnetic fields for $V_0 = 1.25\,\text{meV}$ and $s = 300\,\text{nm}$. The value of the orbit centre (x_0) for each curve, that is for each value of B, is chosen so that the point of minimum gradient occurs at an energy of 10 meV, which is taken to be the Fermi energy. Using Equation (5) we can calculate B_c for the values of V_0, s and v_F given above. The right-hand curve in

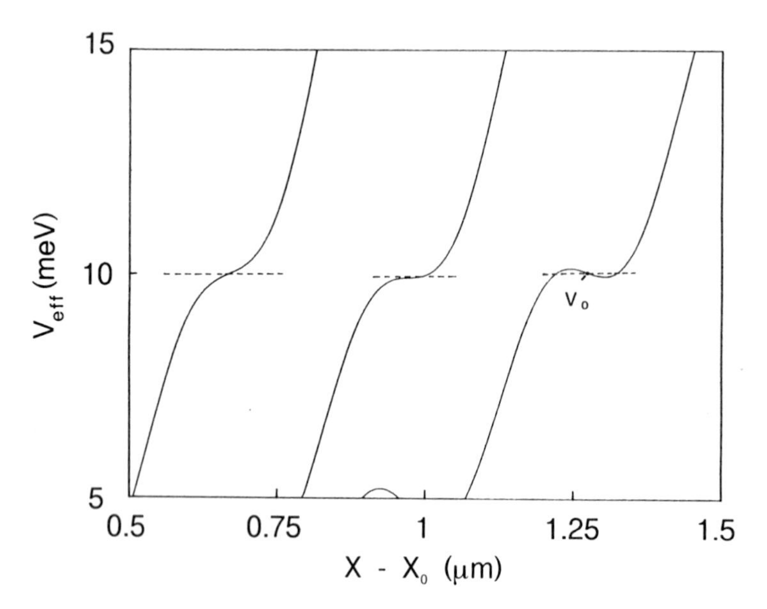

Figure 5: *Effective potential, $V_{\text{eff}}(x)$, for an electron in a periodic potential with* $s =$ 300 nm, $V_0 = 1.25$ meV *and* $B > B_c$ *(left),* $B = B_c$ *(middle) and* $B < B_c$ *(right).*

Figure 5 is for $B < B_c$ so there are quasi-bound states which occur at the Fermi energy. These states are equivalent to the streaming orbits discussed above and have a group velocity (Beton *et al.* 1991) $v_g = \hbar^{-1} dE/dk_y \approx v_F$, consistent with the simple classical argument. The left-hand curve is for $B > B_c$ and there are no quasi-bound states at all (remember that the values of x_0 are chosen for each curve for the situation most likely to give bound states). The periodic potential can be treated as a perturbation and the eigenstates acquire a drift velocity (Winkler *et al.* 1989) $v_d \ll v_F$, which is responsible for the $1/B$ oscillations. At $B = B_c$ (centre curve) the local minimum is just forming.

For electrons in local minima, as shown in the right-hand curve of Figure 5, there is the possibility of tunnelling out of the quasi-bound state through the potential barrier in V_{eff}. This is of course forbidden classically, and the semi-classical picture would not be valid if there were a high probability of tunnelling. For an electron in a magnetic field just below B_c, the transmission coefficient of an electron at the Fermi energy is given by $T = \exp(-\Gamma)$ within the WKB approximation, where

$$\Gamma = 4\left(\frac{B_c - B}{B_c}\right)^{5/4}\left(\frac{eV_0}{E_G}\right)^{1/2}. \tag{10}$$

Here $E_G = h^2/2m^*s^2$ is the energy of an electron with wavelength equal to the period of the applied potential. For the semiclassical model to be valid $T \ll 1$ or $\Gamma > 1$. This requires

$$\left(\frac{B_c - B}{B_c}\right) > \left(\frac{E_G}{16eV_0}\right)^{2/5}. \tag{11}$$

The critical field will remain well defined provided $E_G < 16eV_0$, which is true for typical values of $s = 150$ nm and $V_0 = 1$ meV, giving $E_G \approx 1$ meV $\approx V_0$. If structures could be made with $s \approx 50\,\text{nm}$, then $E_G > 16eV_0$ would be possible and quantum magnetic breakdown should dominate, as has also been pointed out by Středa and van de Konijnenberg (1991). This implies that quantum effects should be observable in 1D periodic potentials, but only if structures can be fabricated with periods smaller than at present.

1.6 Conclusions

The search for quantum coherent effects in lateral periodic potentials in 2DEGs has been complicated by the existence of a range of semi-classical effects. Consideration of these effects, however, enables us to predict conditions where quantum effects will occur. We require structures with a small period (approximately 50 nm) of high homogeneity and $V_0 < 1\,\text{meV}$. The small period will require high quality lithography *and* high mobility 2DEGs very close to the surface (approximately 50 nm), so that the periodicity of the potential is not smeared. Even where miniband structure does occur, observation will require low temperatures, since the band-gaps will be small, and low magnetic fields to avoid magnetic breakdown. The co-existence of quantum and semi-classical effects may prove to be an interesting field of study.

2 Non-local resistance

2.1 Introduction

Advances in fabrication have enabled experiments to be performed which have changed the way in which we think about electrical resistance. At very low temperatures the collisions which determine the classical electrical resistance are associated with crystalline imperfections, impurities and, in very pure materials, the walls of the conductor. In parallel with this is the central idea that resistance is associated with the dissipation of electrical energy. In other words, the appearance of a voltage between two contacts implies that dissipation is occuring in the region between them.

In recent years the classical idea of resistance has been modified in two distinct ways. First, it has been possible to investigate effects which depend on the quantum character of electrons. The classical theory, of course, treats the electrons as semi-classical particles, albeit obeying quantum statistics, and it does not allow the possibility of interference between the electron waves. These interference effects have appeared in a wide range of phenomena including the Aharonov-Bohm effect (Sharvin and Sharvin 1981, Washburn *et al.* 1985), weak localisation (Abrahams *et al.* 1979), universal conductance fluctuations (UCFs) (Lee and Stone 1985) and various other similar effects based on ingenious semiconducting and metallic devices. Secondly, two-dimensional electron gases (2DEGs) in GaAs-$Al_xGa_{1-x}As$ semiconducting heterostructures have been made with such high mobilities that transport mean free paths, l_e, in excess of $10\,\mu\text{m}$ are routinely possible. This enables devices to be fabricated, using optical and electron-beam lithography, on a length scale which is small compared with l_e. Studies of transport in such devices (ballistic transport) has led to the discovery of quantised conduction in quantum point contacts (QPCs) (van Wees *et al.* 1988, Wharam *et al.* 1988), anomalous bend

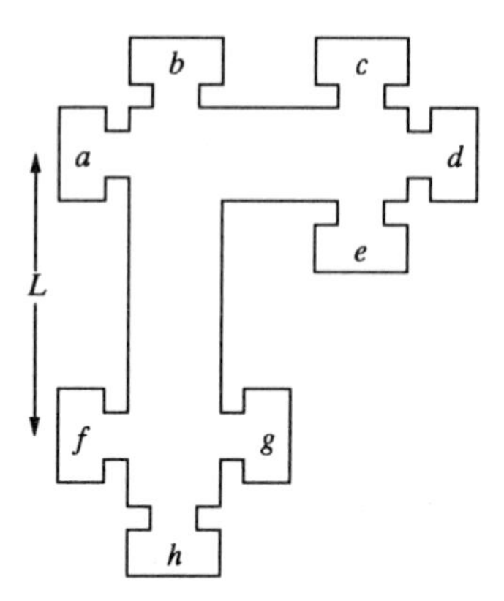

Figure 6: *Typical geometry for the measurement of non-local resistance.*

resistance, electron focussing (van Houten *et al.* 1988), *etc.* Closely related to ballistic transport is adiabatic transport in high magnetic fields. Here the current is carried by edge states which are located near the edges of the sample. The direction of motion of the electrons is fixed by the combination of the applied magnetic field and the confining electrostatic field, so that back-scattering is strongly suppressed. Electrons are able to travel macroscopic distances in these edge states without undergoing any sort of relaxation. To describe phenomena in both these non-classical transport regimes, the idea of non-local resistance has emerged: a four-wire resistance measurement in which the voltage probes are spatially separated from the region in which one would expect the classical dissipation to occur. This concept is described in more detail in Section 2.2. Then in the following sections the non-local resistance is discussed in the quantum, ballistic and adiabatic regimes. In the final section there is a more detailed discussion of an experiment exploring the crossover between classical and quantum adiabatic transport. In this experiment, unlike all the other cases, the non-local resistance does not increase monotonically as the temperature decreases. It provides an excellent illustration of the importance of dissipation in the measurement of resistance.

2.2 Non-local resistance

A typical geometry for the measurement of non-local resistance is shown in Figure 6. The diagram represents a conductor with a series of ohmic contacts a–h which allow the introduction of a current or the measurement of a voltage. The conductor itself may be ballistic, adiabatic or normal but we require the contacts to be sufficiently disordered that the electrons within them can be considered to be in equilibrium. For convenience, the contacts are assumed to be in zero magnetic field even if the conductor is not. It is customary to use the convention for a four-wire resistance measurement, $R_{ij,kl} = V_{kl}/I_{ij}$ where V_{kl} is the voltage difference between contacts k and l due to a current flowing between contacts i and j. Thus, $R_{ad,bc}$ in Figure 6 would correspond to a local (*i.e.* conventional) measurement of resistance leading, in zero magnetic field, to a value for the bulk resistivity of the material. $R_{ad,ce}$ corresponds to a Hall resistance and, classically at least, is zero in zero field. Non-local resistance configurations would be $R_{ad,fg}$, $R_{ad,gh}$

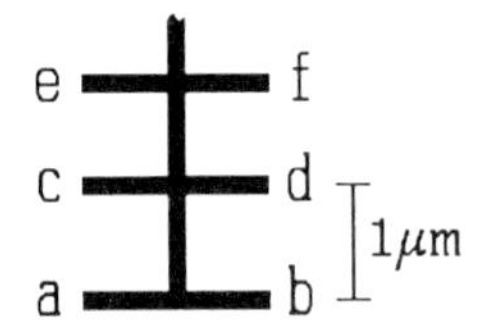

Figure 7: *Schematic diagram of an* n^+-GaAs *wire.*

or $R_{hb,ce}$ for example. A bend resistance might correspond to $R_{hd,ab}$. Before discussing these resistances in the quantum, ballistic and adiabatic regimes it is worth pausing to consider what they are for a classical ohmic conductor. A straightforward solution of Laplace's equation shows that

$$R_{ad,fg} \approx R_\square \exp\left(-\frac{\pi L}{W}\right) \tag{12}$$

where $R_\square$ is the resistance per square of the material, W is the width of the conductor and L is the separation of the voltage and current contacts. For $(L/W) \approx 5$, $(R_{ad,fg}/R_\square) \approx 1.5\times10^{-7}$ and is entirely negligible. Physically the reason that the non-local resistance is so small is that the current is zero in the vicinity of the voltage contacts.

The bend resistance depends more strongly on the geometry of the contacts. However, it is clear that for a current flow from h to d in a classical conductor the potential at a will always be higher than the potential at b.

2.3 Mesoscopic regime

The mesoscopic regime occurs when the distance over which the electrons retain their quantum phase coherence, l_ϕ, is comparable with the relevant dimension of the device. For $R_{ad,fg}$ this would require $l_\phi \approx L$. Note that it is not important for the electronic mean free path to be large provided that the collisions are elastic, *i.e.* the electrons retain their phase during a collision. Indeed, the collisions provide a mechanism for the existence of different interfering electron paths. The quantum interference alters the probability that an electron will be transmitted to another region of the device.

Consider a device as shown in Figure 7. Current is passed between a and b and the voltage measured at c and d. The classical voltage can be made negligibly small but now there is the possibility of coherent quantum transport between the current and the voltage leads (Skocpol *et al.* 1987).

Figure 8 shows the non-local magnetoresistance measured in the sample shown schematically in Figure 7. The device is made from n^+-GaAs ($n = 1.1\times10^{24}\,\mathrm{m^{-3}}$) grown by MBE with a low-temperature transport mean free path $l_e \approx 50\,\mathrm{nm}$ so that $\omega_c\tau \approx 1$ at 6 T. The material is grown on a semi-insulating substrate and the conducting thickness is estimated to be approximately 30 nm with a mobility around $0.16\,\mathrm{m^2\,V^{-1}\,s^{-1}}$ and a resistance of $800\,\Omega\,\square^{-1}$. The wire on which the measurements of Figure 8 were made had a width of 150 nm and leads c and d were 1 μm from a and b, whereas e and f

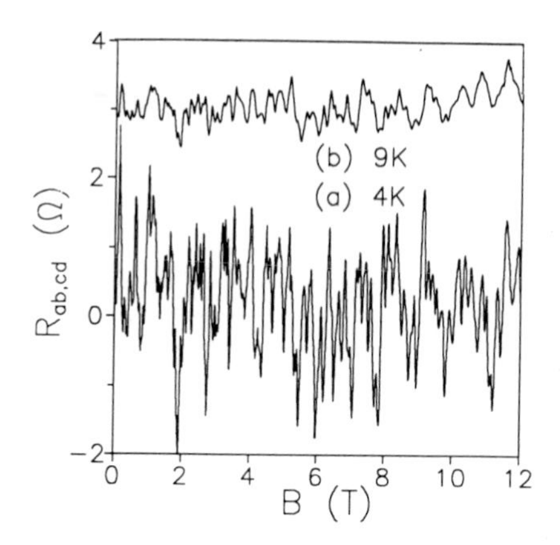

Figure 8: *Non-local magnetoresistance $R_{ab,cd}$ at 4 K (curve a) and 9 K (curve b).*

were 2 μm from a and b. All measurements were made using standard a.c. techniques at 12 Hz. Classically $R_{ab,cd}$ is totally negligible on the scale of Figure 8.

The effect of changing the magnetic field is to change the magnetic flux through the interference loops formed by the scattered electrons. The Aharonov-Bohm effect tells us that changing the flux through a closed loop changes the phase difference of the interfering electron waves. A change in flux of h/e is equivalent to a change in phase of 2π but, since the interference loops have a wide range of areas, the effect of changing B is to introduce a wide range of phase shifts, and the transmission probability for an electron to reach, say, one of the voltage contacts fluctuates randomly as the magnetic field is altered. This is seen in traces (a) and (b) in Figure 8 which are taken at 4 K and 9 K respectively. The fluctuations in magnetoresistance are variously called mesoscopic fluctuations, UCFs and 'magnetofingerprints'. The term 'mesoscopic' reflects the intermediate size regime between the classical macroscopic scale and the truly microscopic region, $l_\phi \approx L$ or W but $l_\phi \gg l_e$. The term 'magnetofingerprint' reflects the observation that, although the general amplitude and periodicity of the fluctuations can be described within a general theory, the detailed positions of the peaks and troughs depend on the actual positions of the scattering centres and the Fermi energy and are unique for a given conductor.

There are several qualitative features apparent about the non-local resistances shown in Figure 8. First, the size of the fluctuations decreases with increasing temperature. This is well understood and is a consequence of l_ϕ decreasing with increasing temperature due to increased scattering and/or thermal spreading of the Fermi surface (Skocpol *et al.* 1987, Lee *et al.* 1987). This feature also occurs in the local resistance measured on the same device. Secondly, the magnitude of the fluctuations decreases with increasing L and, thirdly, the average resistance at each temperature is zero, consistent with the classical prediction. For $T < 10$ K the r.m.s. amplitude of the fluctuations is well described by the expression

$$\Delta R \approx R_{\square}^2 \left(\frac{e^2}{h}\right) \exp\left(-\frac{L}{l_\phi}\right) \tag{13}$$

with a temperature-dependent l_ϕ which is about $0.3\,\mu$m at 4.2 K and decreases as $T^{-1/2}$ at higher temperatures.

The conditions for the observation of non-local resistance are, as stated above, that the current exists in the vicinity of the voltage contacts and that dissipation occurs. In this case, for bulk resistance in such a sample, dissipation always occurs. However, only at low temperatures is the quantum coherence strong enough for a current to flow near the voltage contacts. Note there is nothing significant about the resistance becoming negative. This does not imply negative energy dissipation, it simply means that the quantum current is random in magnitude and direction. Since, in general, l_ϕ increases as T decreases, the non-local resistance due to quantum coherence is a low temperature effect and is a maximum as $T \to 0$.

2.4 Ballistic transport

Whereas the length scale for the quantum transport regime is l_ϕ, the ballistic regime is defined by $L < l_e$, *i.e.* a typical sample dimension is smaller than the elastic mean free path. We require, therefore, long mean free paths such as those found in the high-mobility 2DEGs in GaAs-Al_xGa_{1-x}As heterostructures. In these systems values of l_e in excess of $100\,\mu$m are possible (see, for example, Pfeiffer *et al.* 1989) so $L \ll l_e$ is easily achievable with modern lithography. To discuss the resistance in this regime we employ the formalism devised by Landauer and Büttiker (1986).

Referring to a multi-terminal ballistic conductor such as that shown, for example, in Figure 6, each contact reservoir, α, injects a current I_α into the conductor given by

$$\left(\frac{h}{2e}\right) I_\alpha = \mu_\alpha N_\alpha - \sum_i T_{i\alpha}\mu_i \tag{14}$$

where μ_i are the electrochemical potentials of each contact; $\mu_i = eV_i$ where V_i are voltages measured at the contacts. N_α is the number of channels in the lead from contact α and the summation is over all the leads of the device. The coefficient $T_{i\alpha}$ is the probability that an electron injected from reservoir i emerges into reservoir α and $T_{\alpha\alpha}$ is the probability of reflection. It follows that $\sum_j T_{ij} = N_i$. The origin of Equation (16) is discussed in more detail in the chapter by Stone. Using Equation (16) and the current continuity conditions for each contact (*e.g.* the total current drawn from a voltage contact must be zero), it is possible to generate values of resistance from a knowledge of the coefficients T_{ij}. This is a formidable problem in a conductor as complicated as that shown in Figure 1 but the basic origin of the bend resistance, say, can be simply understood. Ballistic electrons travel in straight lines so many of the electrons injected by h may enter b; but b is a voltage contact so its electrochemical potential, μ_b will adjust so that as many electrons leave it as enter it. Similarly electrons injected by d may enter contact a ballistically. Therefore if current is passing between d and h, so that $V_d > V_h$, then $V_a > V_b$. Remember that classically $V_b > V_a$ so the bend resistance has the opposite sign. This anomalous bend resistance has been seen in a number of ballistic devices (see the chapter by Timp for example) although the magnitude of the effect depends on the detailed geometry.

In this example of non-local resistance (and others in ballistic devices), the current paths are non-classical due to the lack of scattering in the device. The lack of scattering

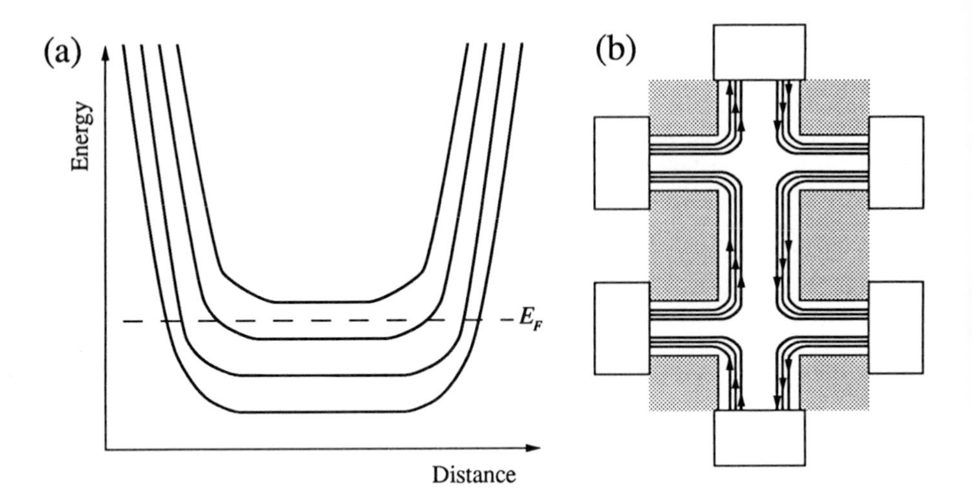

Figure 9: *(a) Schematic representation of edge states. (b) Hall bar with edge states. Arrows represent direction of* $(\mathbf{E} \times \mathbf{B})$.

also means that no dissipation can occur in the bulk of the device. In this case the dissipation occurs in the contacts themselves when the injected electrons come into equilibrium with the reservoir.

2.5 Adiabatic transport

The electron states of a 2DEG in a strong magnetic field ($\omega_c \tau \gg 1$) are Landau levels. Near the boundaries of the conductor these states are affected strongly by the electrostatic confinement potential to form edge states (Beenakker *et al.* 1989) (Figure 9a). Regardless of the relative position of the Fermi energy and the Landau levels in the bulk of the conductor, Figure 9a shows clearly that each occupied edge state contains a boundary between filled and empty states and is therefore able to carry a current. Furthermore, since an electron in an edge state experiences both a magnetic field, $\mathbf{B}$, and an electric field, $\mathbf{E}$, (due to the confinement) it moves with a velocity given by $\mathbf{v} = (1/B^2)(\mathbf{E} \times \mathbf{B})$. Figure 9b shows a typical Hall bar and the direction of electron motion along the edge states is shown by the arrows. Reversing $\mathbf{B}$ also causes $\mathbf{v}$ to reverse. The contacts inject carriers into the edge states and, because the electrons can only move in one direction, there is a strong suppression of back-scattering. In fact, it appears that there is also strong suppression of scattering *between* edge states, so that once the carriers have been injected into an edge state they are able to travel macroscopic distances adiabatically in an quasi-ballistic fashion (Komiyama *et al.* 1989, Alphenaar *et al.* 1990). Assuming ideal contacts it is possible to use Equation (15) to generate resistances in an analogous manner to that for a ballistic device.

Assuming perfect adiabatic transport, contacts b and c in Figure 9b will be at the same voltage. Although there is current in the vicinity of b and c the dissipation occurs only near the current source and drain. Adiabatic transport is not valid if, for example, the Fermi energy lies within a bulk Landau level (see Figure 9a), and dissipation can then occur away from the source and drain and there will be a voltage between b and c. This gives rise to the well-known peaks in resistance between the quantum Hall zeros of resistance. However, even under these circumstances the conductor does not behave classically. For example, in a conductor with a geometry similar to that shown in Figure 6, non-local resistances such as $R_{ad,fg}$ would be non-zero. The origin of this effect is subtle and is discussed by McEuen *et al.* (1990). The basic principle is that current is taken by all edge states to classically inaccessible regions. However, transport in the occupied edge state with the highest energy may now involve back-scattering so that there must always be some redistribution of electrons between edge states, and hence dissipation at the contacts. Our criteria for a non-local resistance are therefore satisfied. The dissipation necessary for non-local resistance to be non-zero occurs at the contacts and is therefore not strongly affected by temperature. The current arrives at the classically inaccessible regions by the enhanced quantum conduction along the edge states. This process is improved by a reduction in temperature so the non-local resistance is essentially quantum in origin and is maximum at the lowest temperatures.

2.6 Crossover between quantum and classical regimes

The crucial importance of the existence of dissipation in measuring non-local resistance has been emphasised in a new experiment by Geim *et al.* (1991) which studies the crossover region between classical and quantum effects. The sample is the same n^+-GaAs wire described in Section 2.3 and shown schematically in Figure 7. At low temperatures ($T < 10\,\mathrm{K}$) we have already seen that there are mesoscopic fluctuations in the non-local magnetoresistance. In Figure 10, the non-local resistance $R_{ab,ef}$ is plotted as a function of magnetic field from 8 T to 12 T. The curves refer to different temperatures and are offset for clarity with the dashed lines representing the respective zeros. In this configuration the voltage probes are separated from the current probes by $2\,\mu\mathrm{m}$ so that the classical non-local resistance is vanishingly small for the effective line width of 150 nm (Equation 13).

At 4.2 K the mesoscopic contribution to $R_{ab,ef}$ is small but dominant. The average resistance is zero, consistent with the classical prediction. As the temperature is increased the resistance fluctuations die away, as expected, but there is an extra contribution to the magnetoresistance which appears to grow as the temperature is increased to approximately 30 K only to disappear again as the temperature is increased further. The new magnetoresistance has an oscillatory form periodic in $1/B$ rather like conventional Shubnikov–de Haas (SDH) oscillations which are seen in the local resistance. However, there are important differences between the non-local oscillations (NLO) and the local SDH oscillations (LSDHO).

1. The LSDHO have maximum amplitude at $T = 0$ whereas the NLO depend on temperature as shown in Figure 11 for two of the peaks. In particular, the NLO disappear at very low temperatures.

2. The LSDHO are only a few percent of the average local resistance whereas the NLO form essentially all of the non-local resistance.

3. The NLO are not, in general, in phase with the LSDHO. The phase difference

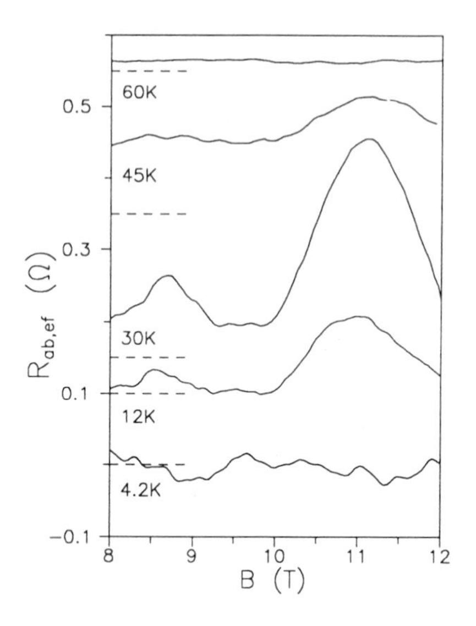

Figure 10: *Non-local resistance $R_{ab,ef}$ as a function of magnetic field at various temperatures. The dashed lines represent the respective resistances at zero field.*

may depend on the direction of magnetic field, although $R_{ab,ef}(B) = R_{ef,ab}(-B)$ always, as required by symmetry (Landauer 1988, Buttiker 1986). Also, the phase difference may alter after thermal cycling to room temperature.

Similar behaviour is seen in $R_{ab,cd}$ and other non-local configurations. This new effect is a direct consequence of the conditions for the observation of non-local resistance. For example, consider $R_{ab,cd}$ shown schematically in Figure 12. The onset of quantum conduction along the edge states will occur at high magnetic fields and low temperatures. The edge states are represented by thick lines in Figure 12. These carry current to the vicinity of the voltage contacts but this is not itself sufficient for a voltage drop between c and d — there must also be dissipation.

We expect the voltage drop to have the form

$$V_{cd} \propto I_{ab} \exp\left(-\frac{L}{\lambda_E}\right) P(T) f(T) \tag{15}$$

where L is the separation between the current leads and the non-local voltage probes. This expression is made up of three parts.

1. The factor $\exp(-L/\lambda_E)$ is the probability that an electron injected at lead a reaches the region between leads c and d, where λ_E is the mean free path of an electron in an edge state which has been discussed elsewhere in the context of adiabatic transport (Haug *et al.* 1989). In a high mobility 2DEG, λ_E may reach 1 mm. In this relatively dirty system, with three or four two-dimensional subbands occupied, we may still expect $\lambda_E \approx 1\,\mu$m if we scale λ_E with the density of scattering centres. The bulk elastic mean free path, l_e, is approximately 50 nm

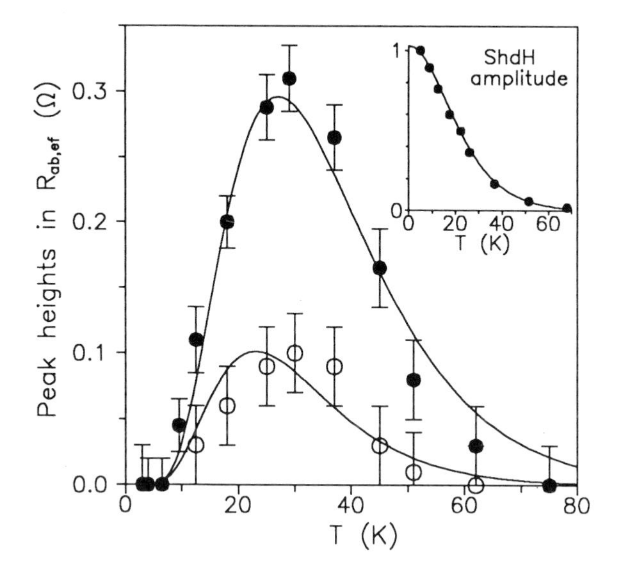

Figure 11: *Dependence on temperature of the peaks in $R_{ab,ef}$ shown in Figure 10. Solid lines are fits to Equation (17). The inset shows the dependence of the amplitude of the* LSDHO *on temperature.*

so the large difference between λ_E and l_e allows a difference to be sustained between the chemical potentials of the edge and bulk states.

2. The factor $f(T) = (k_BT/\hbar\omega_c)\,\mathrm{cosech}(2\pi^2k_BT/\hbar\omega_c)$ is the usual temperature dependence for oscillatory magnetoresistance and arises from thermal smearing of the Fermi level. It is the same factor which determines the temperature dependence of local SDH oscillations (see the inset to Figure 11).

3. The two factors described above give the number of electrons which arrive at the intersection of the vertical section of the wire with probes c and d. $P(T)$ is the fraction of these electrons which then thermalise. This can be estimated in a form resulting from the equation for energy balance,

$$P(T) \propto \left[1 + \exp\left(\frac{\Delta\epsilon}{k_BT}\right)\right]^{-1}, \tag{16}$$

where $\Delta\epsilon$ is expected to be the difference between the Fermi energy and the energy of the maximum of the density of states in the bulk. This is made more clear in Figure 13 which shows schematically the energy of Landau levels across a section of the wire.

Dissipation involves a transition from edge to bulk followed by a relaxation to local equilibrium. When the Fermi energy is in between bulk Landau levels (as in Figure 13), λ_E will be a maximum but at this magnetic field $P(T)$ will be small since $\Delta\epsilon$ will be large. In other words, the electrons have a high probability of reaching the voltage probe region but there is little dissipation. Conversely, there will be strong thermalisation

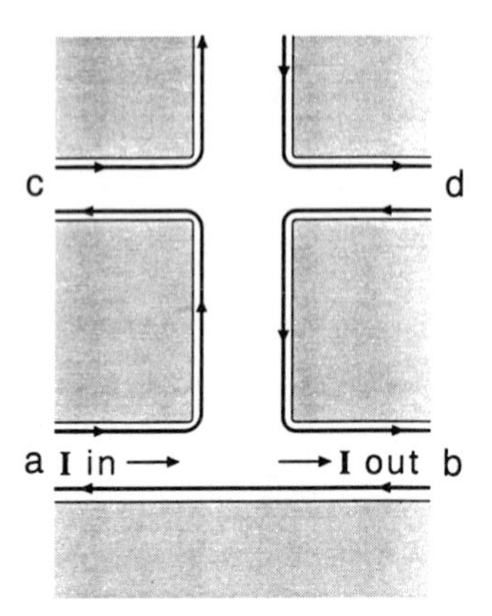

Figure 12: *Schematic diagram showing edge states in a* n^+-GaAs *wire.*

when the Fermi energy lies within a bulk Landau level but λ_E will be correspondingly small. This argument also explains why there is a phase difference between the local and non-local oscillations. Indeed, this phase shift enables a direct determination of $\Delta\epsilon$ since we know that the the period of local SDH oscillations is determined by the Fermi energy being within a bulk Landau level. For the two peaks shown in Figure 11, $\Delta\epsilon = 4$ meV for the upper curve and 3.2 meV for the lower curve. The solid lines in Figure 11 are the best fits to Equation (17) where the overall amplitude is the only adjustable parameter. The agreement is excellent, given the approximate derivation of Equation (17).

Note the unique feature of the non-local resistance in this experiment is that it disappears at $T = 0$. This highlights the importance of dissipation in any measurement of the non-local resistance.

2.7 Summary

All the examples of non-local resistance have two features in common. First, there is some mechanism for carrying the current to non-local regions. This may be semi-classical, as in the ballistic case, or it may be quantum, as for the mesoscopic oscillations and adiabatic transport through edge states. Secondly, there must be dissipation in the vicinity of the voltage probes. Often this dissipation occurs in the contact reservoirs and may not be immediately apparent, but experiments in the crossover region between quantum and classical transport demonstrate clearly its importance.

Acknowledgments

Many people contributed to work on periodic potentials described here including Elmo Alves, Mark Dellow and, particularly, Peter Beton. For the work on non-local resistance, I am very grateful to Andrei Geim, Laurence Eaves and Pavel Středa for their

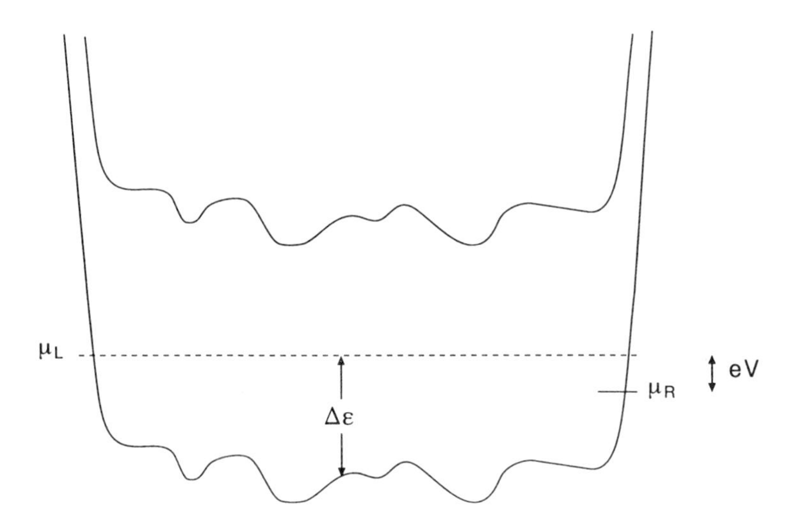

Figure 13: *Landau levels in the disordered wire. The electrochemical potentials on the left and right hand sides of the wire are shown to differ by eV.*

collaboration and their arguments, which were fierce but fun.

References

Abrahams E, Anderson P W, Licciardello D C, and Ramakrishanan, 1979. *Phys. Rev. Lett.*, **42**, 673.

Alphenaar B W, McEuen P L, Wheeler R G, and Sacks R N, 1990. *Phys. Rev. Lett.*, **64**, 677.

Alves E S, Beton P H, Henini M, Eaves L, Main P C, Hughes O H, Toombs G A, Beaumont S P, and Wilkinson C D W, 1987. *J. Phys. Condens. Matter*, **1**, 8257.

Beenakker C W J, 1989. *Phys. Rev. Lett.*, **62**, 2020.

Beenakker C W J, van Houten H, and van Wees B J, 1989. *Superlattices and Microstructures*, **5**, 127.

Beton P H, Alves E S, Main P C, Eaves L, Dellow M W, Henini M, Hughes O H, Beaumont S P and Wilkinson C D W, 1990. *Phys. Rev. B*, **42**, 9229.

Beton P H, Dellow M W, Main P C, Alves E S, Eaves L, Beaumont S P and Wilkinson C D W, 1991. *Phys. Rev. B*, **43**, 9980.

Büttiker M, 1986. *Phys. Rev. Lett.*, **57**, 1761.

Dellow M W, 1991. Thesis, University of Nottingham.

Geim A K, Main P C, Beton P H, Středa P, Eaves L, Beaumont S P, and Wilkinson C D W, 1991. *Phys. Rev. Lett.*, **67**, 3014.

Gerhardts R R, Weiss D, and von Klitzing K, 1989. *Phys. Rev. Lett.*, **62**, 117.

Haug R J, Kucera J, Středa P, and von klitzing K M, 1989. *Phys. Rev. B*, **39**, 10892.

Heitmann D, 1990. In *Electric Properties of Multilayers and Low-Dimensional Semiconductor Structures*, Eds. Chamberlain J M, Eaves L, and Portal J C, Plenum, p. 151.

Ismail K, Smith T P, and Masselink W T, 1989. *Appl. Phys. Lett.*, **55**, 2766.

Komiyama S, Hirai H, Sasa S, and Hiyamizu S, 1989. *Phys. Rev. B*, **40**, 12566.

Landauer R, 1988. *IBM J. Res. Dev. B*, **2**, 306.
Lee P A, Stone A D, and Fukuyama H, 1987. *Phys. Rev. B*, **35**, 1039.
Lee P, and Stone A D, 1985. *Phys. Rev. Lett.*, **55**, 1622.
McEuen P L, Szafer A, Richter C A, Alphenaar B W, Jain J K, Stone A D, Wheeler R G, and Sacks R N, 1990. *Phys. Rev. Lett.*, **64**, 2062.
Pfeiffer L, West K W, Stormer H L, and Baldwin K W, 1989. *Bull. Am. Phys. Soc.*, **34**, 549.
Seeger K, 1973. *Semiconductor Physics*, Springer-Verlag.
Sharvin, D Yu, and Sharvin Yu V, 1981. *JETP Lett.*, **34**, 272.
Skocpol W J, Mankiewich P M, Howard R E, Jackel L D, Tennant D M, and Stone A D, 1987. *Phys. Rev. Lett.*, **58**, 2347.
Středa P, and MacDonald A A, 1990. *Phys. Rev. B*, **41**, 11892.
Středa P, and van de Konijnenberg, 1991. To be published.
van Houten H, van Wees B J, Mooij J E, Beenakker C W J, Williamson J G, and Foxon C T, 1988. *Europhys. Lett.*, **5**, 721.
van Wees B J, van Houten H, Beenakker C W J, Williamson J G, Kouwenhoven L P, van der Marel D, and Foxon C T, 1988. *Phys. Rev. Lett.*, **60**, 848.
Vasilopoulos P, and Peeters F M, 1989. *Phys. Rev. Lett.*, **63**, 2120.
Webb R A, Washburn S, Umbach C P, and Laibowitz, 1985. *Phys. Rev. Lett.*, **54**, 2696.
Weiss D, 1990. *Physica Scripta.*, **T35**, 226.
Weiss D, von Klitzing K, Ploog K, and Weinmann G, 1989. *Europhys. Lett.*, **8**, 179.
Wharam D A, Thornton T J, Newbury R, Pepper M, Ahmed H, Frost J E F, Hasko D G, Peacock D C, Ritchie D A, and Jones G A C, 1988. *J. Phys. C.*, **21**, L209.
Winkler R W, Kotthaus J P, and Ploog K, 1989. *Phys. Rev. Lett.*, **62**, 1177.

Double Barrier Resonant Tunnelling Devices With Lateral Gates

L Eaves

University of Nottingham
Nottingham, U.K.

1 Introduction

The strongly non-linear current-voltage characteristics, $I(V)$, of double-barrier resonant tunnelling diodes (RTDs) are a direct manifestation of the quantum confinement of electrons in a potential well formed between two barriers (Chang *et al.* 1974). This leads to quantisation of motion perpendicular to the barriers. However, in conventional devices of this type, the electrons remain free to move in the plane of the well since the cross-sectional area is relatively large, corresponding to typical diameters between approximately 2 and 200 μm.

By further quantising the carrier motion in the plane of the barriers, it should be possible to form laterally confined *zero-dimensional* states. This aim has been pursued by fabricating RTDs with dimensions small enough ($< 1\,\mu$m) to produce lateral confinement (Reed *et al.* 1988, Tewordt *et al.* 1990, Su *et al.* 1991). New features appearing in the $I(V)$ of such devices have been attributed to lateral confinement and Coulomb blockade (Groshev 1990). However, it is difficult to show experimentally that these features are directly related to the lateral dimension of the device, since two-terminal devices with fixed lateral dimensions have been used.

We have recently fabricated a three-terminal RTD in which the effective lateral dimension may be varied using a gate electrode (Dellow *et al.* 1991). In this paper we describe the basic features of this type of device and give a detailed account of the variation of the $I(V)$ characteristics with gate voltage, temperature and magnetic field. The paper is arranged as follows. Section 2 describes the structure of the device and the overall form of the $I(V)$ characteristics; Section 3 examines the new peaks which are observed at voltages close to the threshold for resonant tunnelling; and Section 4 describes the strong asymmetry of the $I(V)$ characteristics with respect to the sign of the source-drain voltage, when the negative gate voltage gives rise to significant lateral confinement. The results in Section 2 are relevant to recent work on ungated submicron RTDs. We observe peaks below the predicted threshold voltage for resonant tunnelling which have a weak dependence on the lateral dimension of the device, and so cannot be explained simply by either Coulomb blockade or laterally quantised states. We propose that the additional peaks are due to resonant tunnelling through a highly localised region of the device which is probably formed by a shallow donor impurity in the quantum well of the device. Furthermore we are able to use our gate to probe the spatial extent of this region. Our results indicate that the electrical properties of small RTDs close to threshold can be strongly influenced by the presence of a single ionised donor atom.

2 Structure and overall form of $I(V)$

Our device, shown schematically in the inset of Figure 1, is similar to one described by Kinard *et al.* (1990). It is fabricated from a semiconductor layer grown by molecular beam epitaxy (MBE) with $Al_{0.4}Ga_{0.6}As$ barriers of thickness $b = 5.7\,nm$, separated by a GaAs quantum well of thickness $w = 12\,nm$, in which the lowest quasi-bound state energy E_1 is calculated to be 24 meV. The barriers are separated from the doped n-type GaAs contact regions by an undoped GaAs spacer layer 3.4 nm thick, and the doping varies from a low value, $2{\times}10^{16}\,cm^{-3}$, close to the barriers, to a high value, $2{\times}10^{18}\,cm^{-3}$, over a thickness of order $1\,\mu m$. The detailed composition of the layers in the device is shown in Figure 2. By increasing the *negative* voltage, $-V_g$, on the gate, the extent of the depletion region may be increased thereby reducing the effective conducting area of the device.

Figure 1 shows the $I(V)$ curves at low temperature ($T = 4.2\,K$) for various V_g, measured on a typical device (Device 1). The gate voltage is applied with respect to the earthed substrate. The current I is measured from the voltage across a $100\,k\Omega$ series resistor. In the figure, V_{SD} refers to the voltage drop between the top contact and the substrate. Peaks are observed in each polarity due to resonant tunnelling of electrons through the quasi-bound state in the well. As V_g is increased, the peak current, I_p, is reduced. Thus the current, and therefore the effective cross-sectional area of the device, may be varied by applying a voltage to the gate. By comparing the values of peak current in reverse bias for samples with widths $1\,\mu m$, $1.5\,\mu m$ and $2\,\mu m$, we can calculate the reverse bias peak current density, j, and find $j = 3.6\times10^4\,A\,m^{-2}$. To a good approximation $I(V)$ in reverse bias scales with the peak current over a large voltage range; this is not the case in forward bias as discussed below. We may reliably deduce a value for A, the effective cross-sectional area, from the relation $A = I_p/j$, and therefore determine the variation of A with V_g which is shown in Figure 3. Thus A may be decreased down to below $0.1\,\mu m^2$ by raising V_g from 0 to 5 V. Note that $A < 1\,\mu m^2$, the physical area of the device, even for $V_g = 0$.

3 Sub-threshold structure in $I(V)$

We now focus on the $I(V)$ characteristics close to the threshold for resonant conduction. These are plotted for $V_g = 0$ in Figure 4 (for Device 2, a $1\,\mu m^2$ mesa, similar to Device 1)

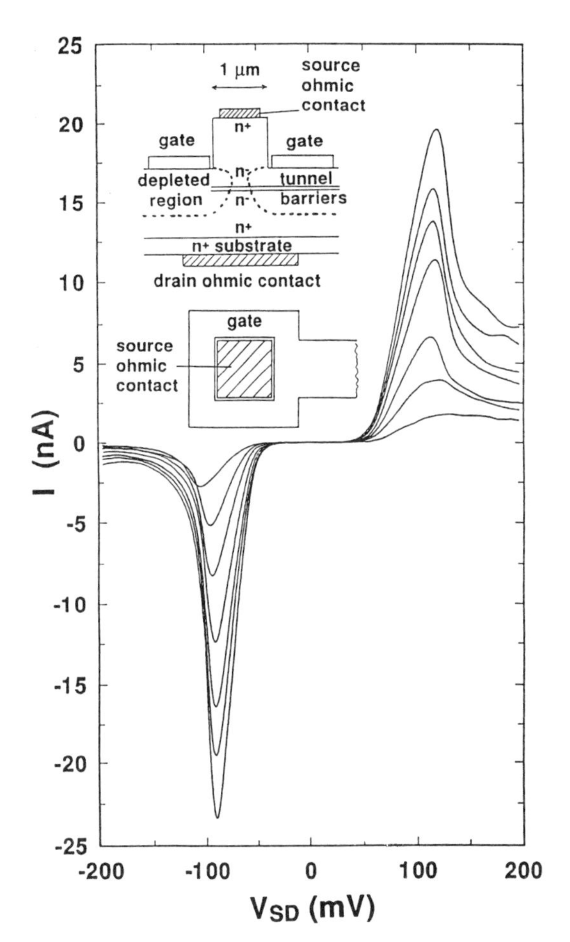

Figure 1: *The source-drain $I(V)$ characteristic measured at 4.2 K for gate voltages V_g = 0, 0.2, 0.5, 1.0, 3.0 and 5.0 V in order of decreasing peak current. Forward bias corresponds to the top contact biased positive for a typical device (Device 1). A schematic diagram of the device is shown in the top inset.*

in which a series of peaks are clearly seen. A peak-to-valley ratio (PVR) of approximately 3 is observed for the first peak in reverse bias. We have observed similar, although not identical, structure in several devices. Figure 4 (inset) shows $I(V)$ at temperatures from 2 K to 20 K at $V_g = 0$. The structure clearly persists up to approximately 10 K. The peaks in reverse bias occur at −26, −36 and −44 mV, and those in forward bias at 31, 40 and 50 mV. The voltage threshold for resonant tunnelling is $V_{\text{th}} \approx 2E_1/e \approx 50$ mV, calculated from the width of the well and doping profile (Leadbeater *et al.* 1989). This is consistent with the position in voltage of the main peak in current in both small and bulk devices. The structure shown in Figure 4, therefore, occurs at sub-threshold values of voltage. To elucidate the origin of these sub-threshold resonances we have studied their variation with V_g.

500 nm	GaAs	$n = 2 \times 10^{18}\,\mathrm{cm}^{-3}$
200 nm	GaAs	$n = 2 \times 10^{17}\,\mathrm{cm}^{-3}$
200 nm	GaAs	$n = 1 \times 10^{17}\,\mathrm{cm}^{-3}$
300 nm	GaAs	$n = 4 \times 10^{16}\,\mathrm{cm}^{-3}$
300 nm	GaAs	$n = 2 \times 10^{16}\,\mathrm{cm}^{-3}$
3.4 nm	GaAs	undoped
5.7 nm	$Al_{0.4}Ga_{0.6}As$	undoped
11.9 nm	GaAs	undoped
5.7 nm	$Al_{0.4}Ga_{0.6}As$	undoped
3.4 nm	GaAs	undoped
300 nm	GaAs	$n = 2 \times 10^{16}\,\mathrm{cm}^{-3}$
300 nm	GaAs	$n = 4 \times 10^{16}\,\mathrm{cm}^{-3}$
200 nm	GaAs	$n = 1 \times 10^{17}\,\mathrm{cm}^{-3}$
200 nm	GaAs	$n = 2 \times 10^{17}\,\mathrm{cm}^{-3}$
500 nm	GaAs	$n = 2 \times 10^{18}\,\mathrm{cm}^{-3}$
	n^+-GaAs	substrate

Figure 2: *Composition of the layers in the wafer grown by* MBE *and used to fabricate the devices.*

In Figure 5 we plot $I(V)$ measured at $T = 39\,\mathrm{mK}$ for various gate voltages in reverse and forward bias. For $2.0\,\mathrm{V} > V_g > 0$ the amplitude and position in voltage of the first maximum in reverse bias is unchanged, while the position of the first peak in forward bias varies by 2 mV. Over the same range of V_g the position in voltage of the structure at higher source-drain voltages varies by roughly 1 mV, and its peak magnitude is only slightly changed. For $V_g > 2.0\,\mathrm{V}$ the sub-threshold structure is more strongly affected by gate voltage. In particular, there is no longer any structure at $-26\,\mathrm{mV}$ and 31 mV for $V_g > 2.5\,\mathrm{V}$.

As we now demonstrate, this body of experimental evidence cannot be explained by a model based either on lateral quantisation or on Coulomb blockade, resulting from the electrostatic confining potential of the gate. Assuming a square well potential, the states laterally confined by the electrostatic potential from the gate are separated in energy by $E_q \approx h^2/2md^2$, where d ($\approx A^{1/2}$) is the width of the conducting region. Coulomb blockade is related to the charging energy, E_c, of the device for one electron (Groshev 1990). The capacitance of the device $C = \epsilon_r \epsilon_0 A/t$, and $E_c = e^2/2C = e^2 t/2\epsilon_r \epsilon_0 A$, where ϵ_r is the dielectric constant and $t \approx 40\,\mathrm{nm}$ is the total thickness of the undoped barrier, the well, and the associated depletion region. Since both E_q and E_c vary as A^{-1}, the positions in voltage of any structure related to either effect should be strongly dependent on A. In contrast to these predictions, we see from our measurements that the position in voltage of the sub-threshold structure is unaffected as A varies by a factor of about 4 (as V_g is varied from 0 to 2.0 V). In addition, the absolute values of the energy shifts $E_q \approx 0.02\,\mathrm{meV}$ and $E_c \approx 0.03\,\mathrm{meV}$ (calculated for $V_g = 0$) are equivalent to temperatures of approximately 200 mK and 300 mK, respectively. In Figure 4 (inset) the sub-threshold structure is seen to persist up to $T \approx 10\,\mathrm{K}$. Furthermore, both Coulomb blockade and lateral quantisation would give rise to an *increase* in the energy

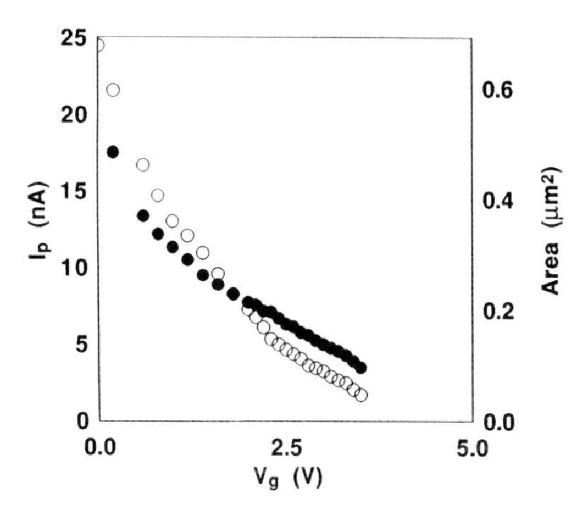

Figure 3: *The variation of peak current with gate voltage in forward (open circles) and reverse (closed circles) bias for Device 2 ($T = 40$ mK).*

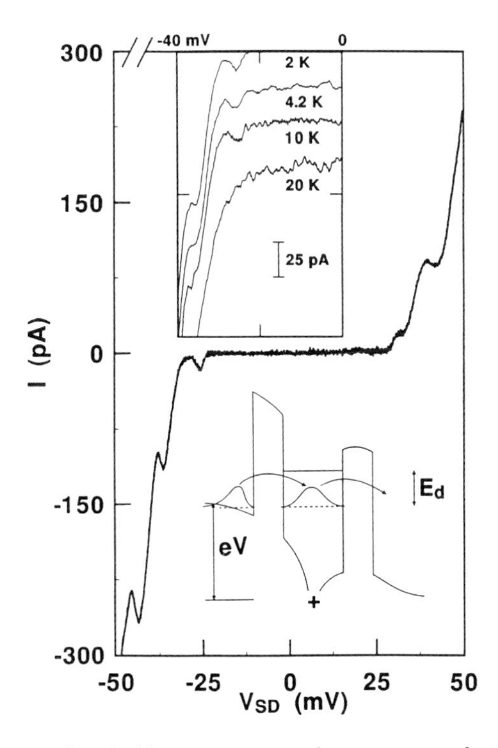

Figure 4: *$I(V)$ close to threshold at zero gate voltage measured at 39 mK for Device 2. The lower inset shows the conduction band profile close to an ionised donor for an applied voltage V. The upper inset shows $I(V)$ in reverse bias close to threshold at zero gate voltage for temperatures $T = 2$, 4.2, 10 and 20 K from top to bottom.*

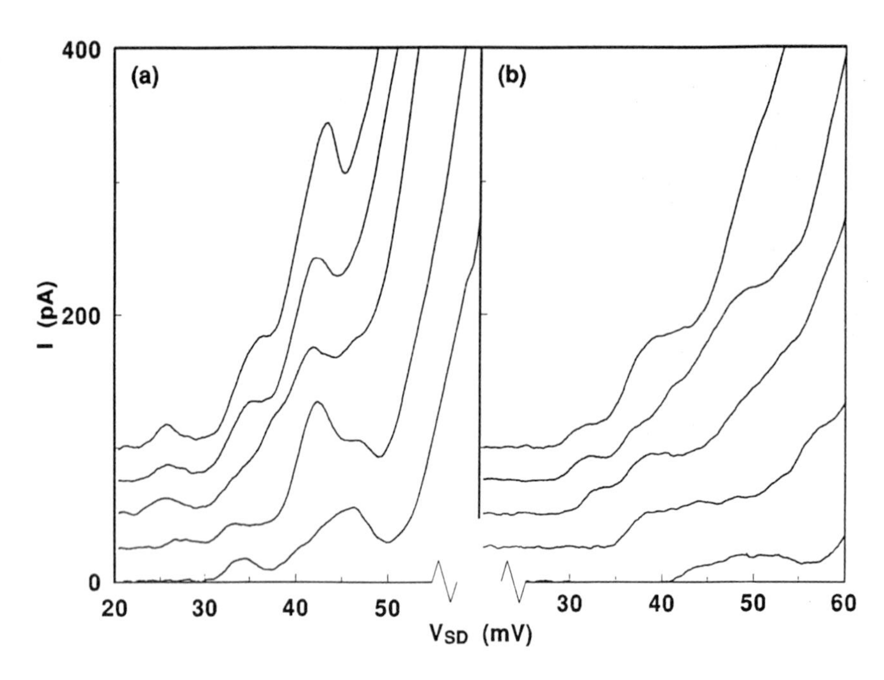

Figure 5: *The dependence of $I(V)$ on gate voltage in (a) reverse and (b) forward bias for gate voltages $V_g = 0$, 1, 1.5, 2.25 and 3 V from top to bottom (Device 2).*

of the bound state, *i.e.* $E_q, E_c > 0$.

Our explanation of these effects is that current flows through a locally favourable path associated with an inhomogeneity in the active region of the device. We propose that a likely microscopic origin of the inhomogeneity is a donor impurity atom unintentionally situated in the quantum well of the RTD. A local minimum in the potential is formed due to this donor with associated bound states in the well. The binding energy, E_d, of the lowest energy bound state depends on the position of the donor in the quantum well. For a well width $w = 12\,\text{nm}$ E_d has been calculated (Greene and Bajaj 1983) to be 12 meV for a donor at the centre of the well and 8 meV for a donor at the well-barrier interface. [See also Jarosik *et al.* (1985) for measured values of the binding energies of donors in quantum wells.] The presence of a single donor also gives rise to weakly bound states at the interface between the undoped GaAs spacer regions and the AlGaAs barriers as shown schematically in the lower part of Figure 2. The binding energy of these states, E_I, is less than 1.5 meV which is the binding energy of a donor at the interface between a semi-infinite GaAs region and an inpenetrable barrier (Bastard 1981). Current flows when sufficient voltage is applied for an electron in the weakly bound interface state to tunnel through the donor-bound well states. Taking $E_d \approx 10\,\text{meV}$ and $E_I = 1\,\text{meV}$, the energy difference between the interface state and the lowest bound state in the well, $E_1 - E_d + E_I$, is approximately 15 meV. The corresponding applied voltage for current to flow is approximately $2(E_1 - E_d + E_I)/e \approx 30\,\text{mV}$, in good agreement with our data. In effect, we have proposed that lateral quantum confinement does occur. However it does not arise from the depletion potential provided by the gate but rather from the presence of a randomly placed single ionised donor which forms bound states through which electrons may resonantly tunnel. These bound states are true zero-dimensional states.

We can use the gate to probe the spatial extent of these states as follows. A peak in $I(V)$ is unaffected until the depletion edge impinges on the region of the corresponding localised state through which electrons tunnel. However, the structure in $I(V)$ will be modified as the depletion encroaches on this region, since the energy of the state will be perturbed. When it extends throughout the region the peak position and amplitude will be strongly affected. This corresponds qualitatively to what is observed, as shown in Figure 5. For example, the lowest peak is unaffected until $V_g \approx 2.0\,\mathrm{V}$. The peak is strongly affected as V_g is further increased by a small amount to a value $V_g = 2.5\,\mathrm{V}$. The change is so rapid that it is not possible to tell if the peak disappears or is shifted rapidly to higher voltage. The effective sample dimension, d, for a given gate voltage may be obtained from Figure 3, and we estimate the spatial extent of the lowest state as $\Delta x = \frac{1}{2}\left[d(V_{g1}) - d(V_{g2})\right]$, where $V_{g1} = 2.0\,\mathrm{V}$ is the highest gate voltage for which the peak is unaffected and $V_{g2} = 2.5\,\mathrm{V}$ is the lowest gate voltage at which the peak is suppressed. This gives a spatial extent of approximately 30 nm for the state corresponding to the lowest peak in each bias direction. The spatial extent of the lowest bound state of a donor in bulk GaAs is approximately $3a_0$, since the expectation value of the radial position of the electron is $3a_0/2$, where a_0, the effective Bohr radius, is approximately 10 nm. In bulk GaAs we therefore expect the lowest state to be localised over a region of width of approximately 30 nm. However, electrons in a quantum well are more tightly bound so this figure represents an upper limit for their spatial extent, but it is unlikely to be in error by more than 20%. Our experimental value for the spatial extent of the state is therefore close to the value expected for the bound state of a single donor.

The slight asymmetry in the positions of the peaks shown in Figure 2 would arise if the donor were not exactly at the centre of the well. This means: (i) the binding energies E_I of the two barrier/spacer layer interface states are different; (ii) the effective barrier heights will be slightly different. Both these effects would lead to asymmetry in $I(V)$. For example, the energies of the interface states could differ by 1–2 meV, resulting in a difference in peak positions in forward and reverse bias in the range 2–4 mV. The observed difference in the lowest peak position in the two polarities is approximately 5 mV, which is slightly above the expected range.

The density of background dopants in nominally undoped material grown in our MBE machine has been measured (Stanaway *et al.* 1989), and is found to be n-type with a concentration in the range 5×10^{19} to $10^{20}\,\mathrm{m}^{-3}$. Therefore it is plausible for just one dopant to be in the active region of the device since the active volume is approximately $d^2(2b+w) \approx 2\times10^{-20}\,\mathrm{m}^3$.

In addition, our model is consistent with the observed values of the peak currents. The current through a RTD is given by $I = Ne/t_D$, where N is the total number of electrons in the well and t_D is the dwell time in the well (Leadbeater *et al.* 1989). We have obtained an approximate value $t_D \approx 3.8\,\mathrm{ns}$ from oscillatory magnetoresistance data for a device of large diameter close to the peak of the main resonance. At low source-drain voltages the collector barrier height encountered by an electron in the state bound to the donor is larger by approximately 20 meV, as compared with electrons tunnelling at the peak of the main resonance, giving rise to an increase in t_D by a factor of approximately 2. We have hypothesised that the lowest peak arises when there is only one available donor bound state through which to tunnel, and so only one electron occupies the well on resonance giving $I = e/t_D \approx 21\,\text{pA}$. In view of the difficulty in estimating t_D this value is in good agreement with the observed peak current of approximately 18 pA for the lowest resonance.

So far we have discussed the peaks at the lowest voltage (−26 and 31 mV). The shoulders on the rapidly rising $I(V)$ curve at higher voltage (Figure 4) could be due either to the presence of other donor atoms embedded at different points in the active region of the RTD, all with slightly different binding energies, or to resonant tunnelling through excited states associated with the ionised impurity, or a combination of these effects. This question can be resolved by studying the effect of a magnetic field on the voltage position of the peaks.

To conclude this section, we have shown that the use of a gate in conjunction with a sub-micron RTD increases our understanding of the additional structure in the $I(V)$ characteristics. In particular we are able to rule out lateral confinement and Coulomb blockade as causes for the sub-threshold structure which we observe. We propose that this structure is instead due to the presence of a locally favourable current path due to an ionised donor in the well region of the RTD. This is qualitatively and, where it is possible to make detailed predictions, quantitatively consistent with the dependence of the sub-threshold structure on gate voltage and temperature.

4 Asymmetry of $I(V)$ at high negative gate bias

In this section, we examine the pronounced asymmetry in the $I(V)$ characteristics of a gated resonant tunnelling device at large negative gate voltages when the cross-sectional area is small. We find that in reverse bias the PVR of the gated devices remains constant at around 20 as the cross-sectional area is reduced down to less than $0.1\,\mu\text{m}^2$. However, we observe a reduction of the PVR of the gated devices in forward bias from approximately 20 to 1 as the cross-sectional area is reduced. We propose that this is due to the lateral non-uniformity of the voltage drop across the barriers in forward bias which leads to a smearing of the resonance, and a consequent reduction of the PVR.

Figure 6a shows $I(V)$ for a device of large diameter in which there is a peak in each direction of bias due to resonant tunnelling of electrons through the well. [For a general discussion of resonant tunnelling in conventional devices of large area, see Leadbeater *et al.* (1989).] $I(V)$ has a high degree of symmetry, as expected for a symmetric RTD. Figure 6b shows $I(V)$ for a gated device (Device 3) with $V_g = 0$, for which $A = 0.45\,\mu\text{m}^2$, and Figure 6c shows $I(V)$ for the same gated device with $V_g = 4\,\text{V}$, for which we deduce $A = 0.06\,\mu\text{m}^2$. The three sets of data in Figure 6 are plotted on different scales in order to emphasise the reduction of the degree of symmetry as the device dimension, $d \approx A^{1/2}$, is reduced. Comparing Figures 6a, 6b and 6c it is clear that all three curves are similar in reverse bias if they are scaled in proportion with their peak current. In contrast, if we compare the curves at forward bias in Figures 6a and 6b we see a reduction in the PVR in the small device as compared with the large device. When V_g is increased to 4 V the PVR is further reduced to near 1 (Figure 6c). A further difference in the forward bias $I(V)$ shown in Figure 6 is the appearance of extra structure as A is decreased.

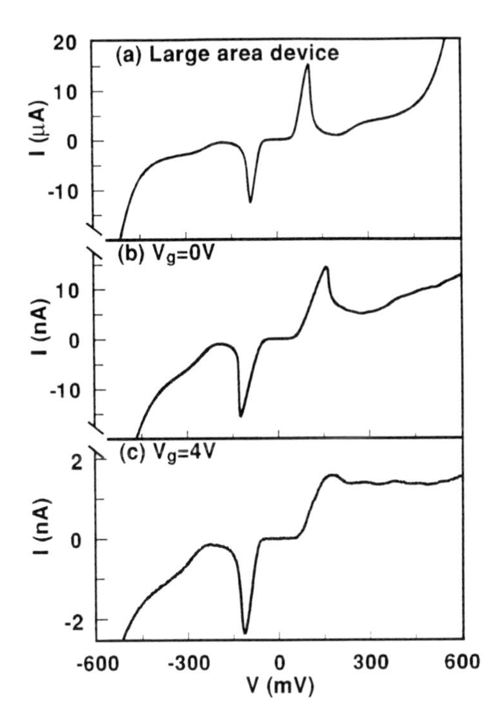

Figure 6: *$I(V)$ characteristics measured at $T = 4.2\,\mathrm{K}$ of (a) a device of large diameter; (b) a gated device (Device 3) at zero gate voltage; (c) a gated device (Device 3) with a voltage $V_g = 4\,\mathrm{V}$ on the gate. The inset shows a schematic diagram of the gated device.*

The structure is absent for the large device, but develops into a series of clear maxima and minima as A is progressively decreased. This behaviour is shown in more detail in Figure 8 below. The data at forward bias are similar to those observed for sub-micron two-terminal devices (Reed *et al.* 1988, Tewordt *et al.* 1990, Su *et al.* 1991, Tarucha *et al.* 1990).

We propose that the origin of the asymmetry and low peak-to-valley ratio in the gated device is related to the shape of the equipotentials close to the tunnel barriers, which gives rise to a voltage drop across the barriers which is not uniform over the device in forward bias. Figure 7 shows a schematic diagram of the potential distribution close to the barriers in zero, forward and reverse bias. The region which is depleted by the gate is hatched in Figure 7a.

In a device of large area, an accumulation layer is formed close to the cathode barrier and a depletion region is formed by the anode barrier when a voltage is applied between the source and drain. The resulting electric field is perpendicular to the barriers and the consequent voltage drop across the barriers is uniform across the device. In contrast, close to the edge of the conducting region (*i.e.* the accumulation layer) of the gated samples in forward bias, the electric field is not perpendicular to the barriers and the potential drop across the barriers is therefore not uniform across the device. To understand this we consider the maximum thickness $t_e(\rho)$ as shown in Figure 7a,

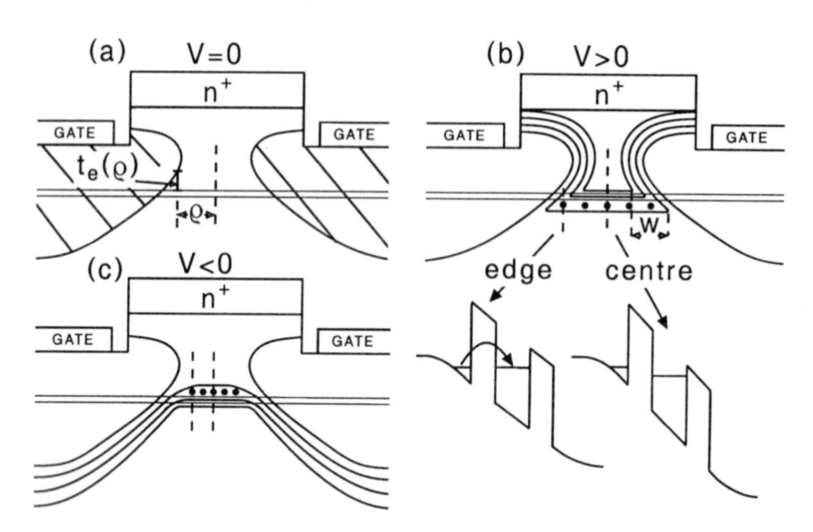

Figure 7: *Schematic diagram of the equipotentials and depleted regions in (a) zero, (b) forward, and (c) reverse bias. The region depleted by the gate is shown hatched in (a). The electrons in the emitter accumulation layer are shown as closed circles. The band diagrams below (b) show the situation with the device on resonance at the edge but off resonance at the centre.*

which may be further depleted at a distance ρ from the centre of the device, when a source-drain voltage V is applied. To a first approximation this is equal to the thickness of the region following the anode barrier which, for $V = 0$, remains undepleted under the application of a gate voltage. For a large device, the depth of the depletion region, t, is related to the voltage drop across the barriers, V_b, by $V_b = Net(b + \frac{1}{2}t)/\epsilon_r\epsilon_0$, where $N = 2 \times 10^{16}\,\mathrm{cm}^{-3}$ is the doping in the GaAs close to the barriers, and $b \approx 30\,\mathrm{nm}$ is the total thickness of the well, barriers and spacer layers. In forward bias there is therefore an 'edge' region of width w for which $t > t_e$, in which the electric field is not perpendicular to the plane of the barriers. The lateral variation of the voltage drop across the device means that the edge of the device goes on and off resonance at a higher voltage than the centre of the device, as shown schematically in the two band diagrams directly below Figure 7b. This results in a smearing of the resonance over a range of source-drain bias with a consequent reduction of the PVR.

We now estimate the voltage over which the resonance is smeared and the relative areas of the 'edge region', where the voltage across the barriers varies, and the centre, where this voltage is constant as it would be in a large device. Assuming the equipotentials cross the tunnel barriers at an angle 45° (Figure 7), we can estimate the effective width, w, of the edge region as $w \approx t$. The ratio, r, of the area of the edge and the central area which remains unaffected by the edge is

$$r \approx \left(\frac{d}{d - 2w}\right)^2 - 1.$$

For $V = 0.15\,\mathrm{V}$, where we would expect to see a minimum in current in a large device, $w \approx t \approx 70\,\mathrm{nm}$. For $V_g = 0$, $d \approx 0.65\,\mu\mathrm{m}$ and $r \approx 0.6$, and when $V_g = 4\,\mathrm{V}$, $d = 0.24\,\mu\mathrm{m}$

and r is large (≈ 5). The shift in voltage of the peak at the edge of the device will be a complicated function of gate voltage, lateral position and source-drain voltage. In particular it is expected to vary over the edge region from 0, close to the centre, up to V_b at the extreme edge of the device. On the basis of this simple model it is clear that edge effects will dominate $I(V)$ in forward bias for $V_g = 4\,\mathrm{V}$ reducing the PVR drastically. Edge effects are also significant even for $V_g = 0$ leading to a reduction in PVR as compared with large devices.

No such complications arise in reverse bias (Figure 7c), since there is no practical limit to the thickness of the depletion layer which may be formed directly below the electron accumulation layer. The voltage across the barriers at the edge of the device is therefore the same as it is at the centre of the device. Accordingly the reverse bias $I(V)$, if scaled by the peak current, closely resembles that of a large device.

Our simple model represents a first-order approximation since the application of a source-drain voltage will interplay with the gate voltage so that the equipotentials shown in Figure 7 become slightly distorted. This combination will increase the convergence of the field lines in forward bias, further exaggerating the asymmetry, whereas in reverse bias it will decrease the convergence of the equipotentials. Our model should be valid if the point of maximum depletion remains on the source side of the barriers, so that electrons always tunnel from a region where the equipotentials diverge to a region where they converge in forward bias, and *vice versa* for reverse bias.

Figure 8 shows the development of the structure in $I(V)$ at forward bias as the gate voltage is increased. We stress that no such structure is observed in $I(V)$ at reverse bias over the same range of gate voltage. Several peaks, each of which moves to higher source-drain voltage as the gate voltage is increased, may be clearly seen in Figure 8. The variation of the peak positions with voltage is shown in Figure 9. Although these data are similar to those observed by Reed *et al.* (1988), Su *et al.* (1991) and Tarucha *et al.* (1990), it is clear that the peaks are due neither to Coulomb blockade, nor to lateral quantisation. This is because (i) they appear in only one direction of bias; (ii) they are observed at a temperature $T = 4\,\mathrm{K}$, whereas such effects would be quenched for $T > 1\,\mathrm{K}$ for $d = 0.5\,\mu\mathrm{m}$; and (iii) the structure is observed in magnetic fields up to 12 T applied perpendicular to the barriers. A possible origin of the peaks is the formation of a hot-electron bottleneck in the region between the barriers and the top contact which leads to a build-up of space charge. We will discuss this model together with the magnetic field data in more detail in a future publication.

To conclude this section, we have shown that the effective voltage drop across the barriers of a sub-micron resonant tunnelling diode is not necessarily uniform across the lateral extent of the device. This can lead to a smearing of the resonance and a consequent reduction of the PVR of the device.

5 Conclusion

The possibility of resonant tunnelling into a 'zero-dimensional quantum box' has attracted great interest since the first work on this subject by Reed *et al.* (1988). Coulomb blockade effects in devices of small area are also of great interest (Groshev 1990). We believe that these effects can only be unambiguously identified in devices in which the

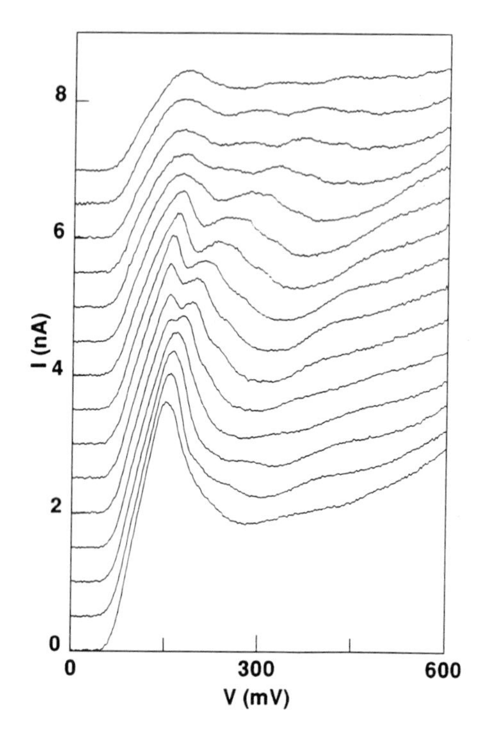

Figure 8: *$I(V)$ characteristics of a gated resonant tunnel device (Device 3) for gate voltages from $V_g = 2.6\,\mathrm{V}$ (bottom) to $V_g = 4\,\mathrm{V}$ (top) in steps of $0.1\,\mathrm{V}$.*

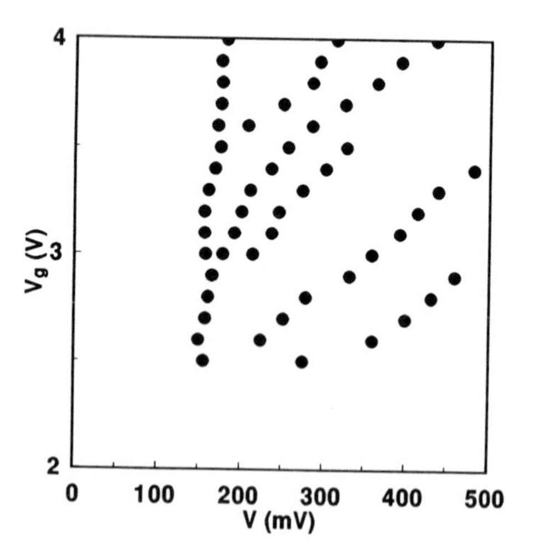

Figure 9: *Fan chart showing the variation of peak position (Figure 8) as a function of gate voltage.*

cross-sectional area can be controlled by a third terminal. We have described how such a device can be fabricated and have discussed its properties. Although we observe additional structure in $I(V)$ which is similar to that reported by Reed *et al.* (1988), Groshev (1990), we can eliminate the possibility that it is due either to Coulomb blockade or to lateral quantisation caused by the small area of the gate confinement potential.

Acknowledgments

This work was performed with my collaborators P H Beton (who thanks the Royal Society for financial support), M W Dellow, T J Foster, M Henini, C J G M Langerak, P C Main, S P Beaumont and C D W Wilkinson, and was funded by the U.K. Science and Engineering Research Council. I am also most grateful to P H Beton for a critical reading of the manuscript.

References

Bastard G, 1981. *Phys. Rev. B*, **24**, 4714.

Chang L L, Esaki L, and Tsu R, 1974. *Appl. Phys. Lett.*, **24**, 593.

Dellow M W, Beton P H, Henini M, Main P C, Eaves L, Beaumont S P, and Wilkinson C D W, 1991. *Electron. Lett.*, **27**, 134.

Greene R L, and Bajaj K K, 1985. *Phys. Rev. B*, **31**, 913.

Greene R L, and Bajaj K K, 1983. *Solid State Commun.*, **45**, 825.

Groshev A, 1990. *Phys. Rev. B*, **42**, 5895.

Jarosik N, McCombe B D, Shanabrook B V, Comas J, Ralston J, and Wicks G, 1985. *Phys. Rev. Lett.*, **54**, 1283.

Kinard W B, Weichold M H, Spencer G F, and Kirk W P, 1990. *J. Vac. Sci. Technol. B*, **8**, 393.

Leadbeater M L, Alves E S, Sheard F W, Eaves L, Hughes O H and Toombs G A, 1989. *J. Phys.: Condens. Matter*, **1**, 10605.

Reed M A, Randall J N, Aggarwal R J, Matyi R J, Moore T M, and Westel A E, 1988. *Phys. Rev. Lett.*, **60**, 535.

Stanaway M B, Grimes R T, Halliday D P, Chamberlain J M, Henini M, Hughes O H, Davies M, and Hill G, 1989. *Inst. Phys. Conf. Ser.*, **95**, 295.

Su B, Goldman V J, Santos M and Cunningham J E, 1991. *Appl. Phys. Lett.*, **58**, 747.

Tarucha S, Hirayama Y, Saku T and Kimura T, 1990. *Phys. Rev. B*, **41**, 5459.

Tewordt M, Law V J, Kelly M J, Newbury R, Pepper M, Peacock D C, Frost J E F, Ritchie D A and Jones G A C, 1990. *J. Phys.: Condens. Matter*, **2**, 8969.

Mesoscopic Localised Transport

A B Fowler

IBM Thomas J Watson Research Center
Yorktown Heights, U.S.A.

1 Introduction

Historically, the first experimental evidence of sample-specific conductance fluctuations or 'footprints' was not observed in metallic systems but rather in systems where conduction was by hopping. The reason is that in hopping systems the conductance can range over orders of magnitude; in metallic systems the fluctuations are of order e^2/h so that the variations are usually in order of parts per thousand or less. Below we will discuss some of the experiments in strongly localised systems.

All such experiments have been performed on samples that were fabricated from metal-oxide-semiconductor (MOS) transistors. See Stiles (1986) for an introduction to these systems. It had been demonstrated in the mid-seventies [see Ando *et al.* (1982) for references] that conduction in such devices near the current threshold, or when the Fermi level lay in the tail of the conduction band, was by hopping and that the equation for Mott hopping in two-dimensions obtained. (The band tail and localised states result from charges and possibly dipoles in the oxides.) MOSFETs allow the experimenter to sweep the Fermi level continuously through the band tail and to observe the variations of conductance. This principle can be applied either in experiments studying hopping or in those studying resonant tunnelling. The first part of this paper will discuss fluctuations in hopping conduction; the latter part will discuss resonant tunnelling—the process that must dominate at low enough temperature. In this format the discussion will be brief but readers will be referred to the original papers and relevant reviews.

2 Fluctuations in hopping conduction

Unexplained fluctuations in the current were seen almost from the earliest experiments on conduction in MOSFETs, even in relatively large samples (Ando *et al.* 1982). However the fluctuations became a dominant feature when small samples were studied. These

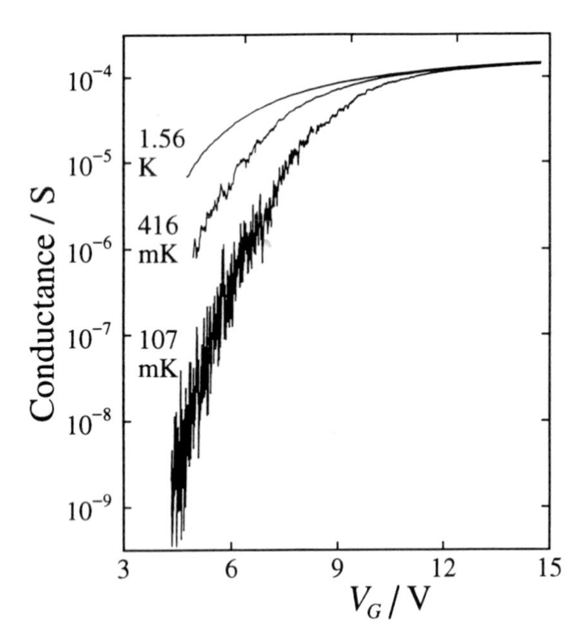

Figure 1: *The conductance of a pseudo-one-dimensional strongly localised* MOSFET *as a function of gate voltage at three temperatures.*

measurements have been made by several groups (Fowler *et al.* 1988, 1991 and references therein) on similarly constructed samples, all based on the principle of using depletion charge to reduce the dimensionality first used by Pepper (1978). In these cases n-channel MOSFETs were made with p-type diffused regions on the sides that could be used to constrict the electron flow to a narrow region.

The conductance of such samples as a function of gate voltage and of temperature is shown in Figure 1. As may be seen, the conductance decreases rapidly as temperature decreases and the relative size of the fluctuations grows. The conductance increases rapidly up to 9 V in gate voltage and the relative fluctuations decrease. It was found that that the temperature dependence was described by $\ln G/G_0 = -(T_0/T)^{1/p}$ where p was 2, up to almost $V_g \approx 6\,\mathrm{V}$. Above 6 V the fit was best for $p = 3$. Because it was expected that the width of the channel would widen as more electrons were induced by the gate, a reasonable interpretation was that the width had increased until the sample was wide compared to the most probable hopping length, so that at low voltages the hopping conductance appeared one-dimensional, whereas at higher voltages (and at higher temperatures) it looked two-dimensional.

These results of course apply to the average $\langle \ln G \rangle$. If a *particular* peak were examined, it increased in height as $\exp(-E/k_BT)$ and broadened. Other peaks with greater E were not evident at low temperature but grew faster. These results are shown in Figures 2 and 3. The positions of the peaks moved with temperature. At high enough temperature (around 300 mK) the peaks overlapped and many occurred so that an average of '$T^{1/2}$' behaviour was seen.

The most straightforward explanation for this behaviour is that which is derived

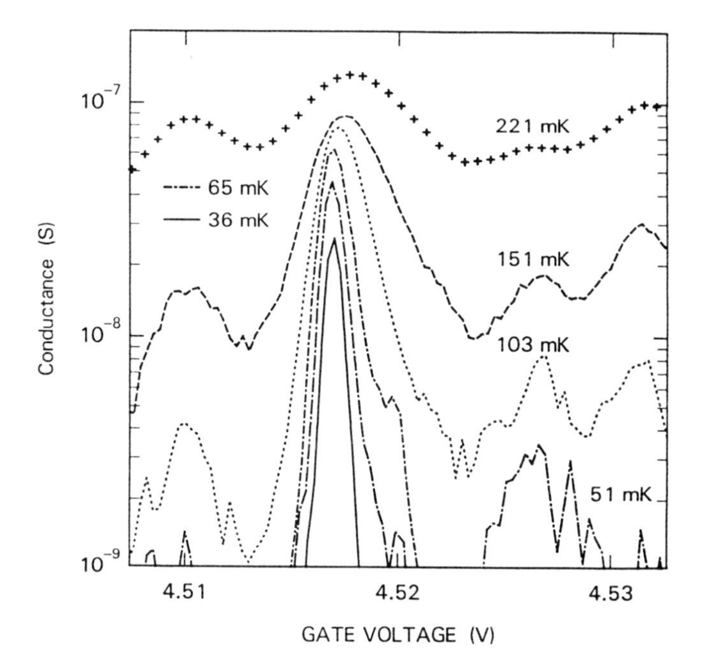

Figure 2: *Conductance as a function of gate voltage for* $T = 36, 51, 65, 103, 151,$ *and* 221 mK. *At* 36 mK *there is only one peak. The peaks lost in the noise grow more rapidly than this peak, and all peaks broaden. As a result,* $\log G$ *is fairly smooth at* 221 mK.

from Kurkijarvi's early argument (Kurkijarvi 1973), that the resistance of some hop will always be exponentially larger that all others for hopping in one dimension. The critical hop changes rapidly as the chemical potential changes. When a single hop controls the conductance, it is given by

$$G = G_{ij} \exp\left(-\frac{2x_{ij}}{\xi} - \frac{|E_i - \mu| + |E_j - \mu| + |E_i - E_j|}{k_B T}\right), \tag{1}$$

where x_{ij} is the distance between the localised states i and j, E_i and E_j are their energies, ξ is the localisation length, and μ is the chemical potential, controlled by the gate voltage. This leads to flat-topped peaks, which are not seen. Simulations (Lee 1984, Serota *et al.* 1986, McInnes and Butcher 1985) find sharp peaks like those actually observed, in addition to the flat-topped ones; these occur when four sites and two hops are involved and show the expected properties. In general the simulations for one dimension mimic the experiments in most particulars.

Figure 4 shows the effect of a magnetic field for the field transverse to and parallel to the current direction. The results are similar but not identical. The average conductance is reduced, the peaks shift position both to higher and to lower gate voltage, and new peaks appear. These results seem to be consistent with the major effect being due to the Zeeman effect. The Zeeman effect is comparable to splittings between states so that states from the apparently overlapping upper and lower Hubbard bands can move relative to each other in a magnetic field. This alters the critical path and the

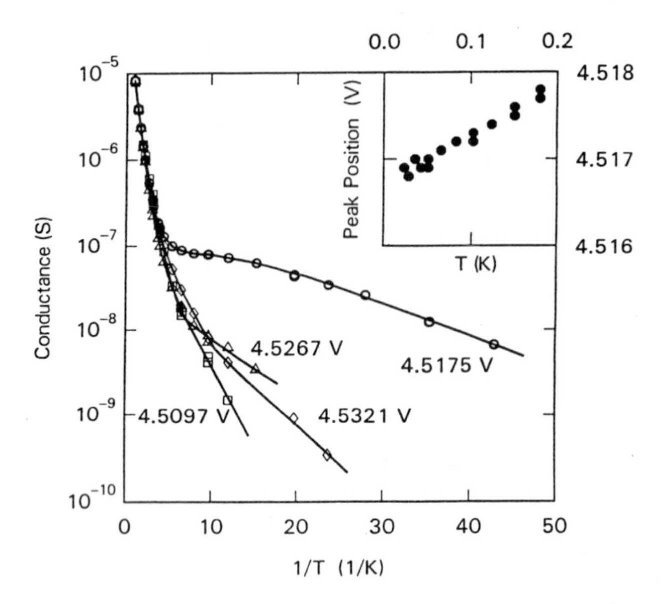

Figure 3: *Dependence on temperature of some of the peaks shown in Figure 2; the highest peak has the lowest activation energy. All peaks show* $\log G \propto T^{1/2}$ *above approximately* 200 mK. *The inset shows the dependence of the position of one peak on temperature.*

filling. Again this system was simulated (Kalia *et al.* 1986). The hops occur from singly occupied to unoccupied sites or from doubly to singly occupied. Again the models and experiments are similar. There was no clear evidence of orbital effects or interference that have been suggested to result in negative magnetoresistance. Most of the above has been extensively reviewed (Fowler *et al.* 1991).

While the pseudo-one-dimensional sample offers the strongest evidence of mesoscopic fluctuations, large fluctuations were seen in much larger two-dimensional samples. Raikh and Ruzin (1987) argued that mesoscopic effects should be seen in large samples in hopping of sizes up to 100 mm. The work above has been summarised in an unpublished review.

Popovic (1990a,b) studied hopping in silicon MOSFETs that were 0.4 mm long and 8.0 mm wide—not of a size where one would expect to see mesoscopic effects. While the effects were relatively small as compared to smaller samples, they were still quite evident with reproducible and sharp conductance variations up to factors of two. One result of this work was that $\langle \ln G \rangle$ was found to vary not simply as $-(T_0/T)^{1/p}$ where p was either 2 or 3 corresponding to one- or two-dimensionality, but rather continuously from low gate voltages where p was close to 2 and changing to 3, as V_g increased. Furthermore the structure shows some degree of correlation; it approaches being periodic. One model is that there is large structure corresponding to the formation of one-dimensional paths in a two-dimensional matrix. These paths change on one scale; the other fluctuations occur for changes in the one-dimensional conductance. One reason these experiments were done on such large samples was that it was easier to study the magnetoconductance

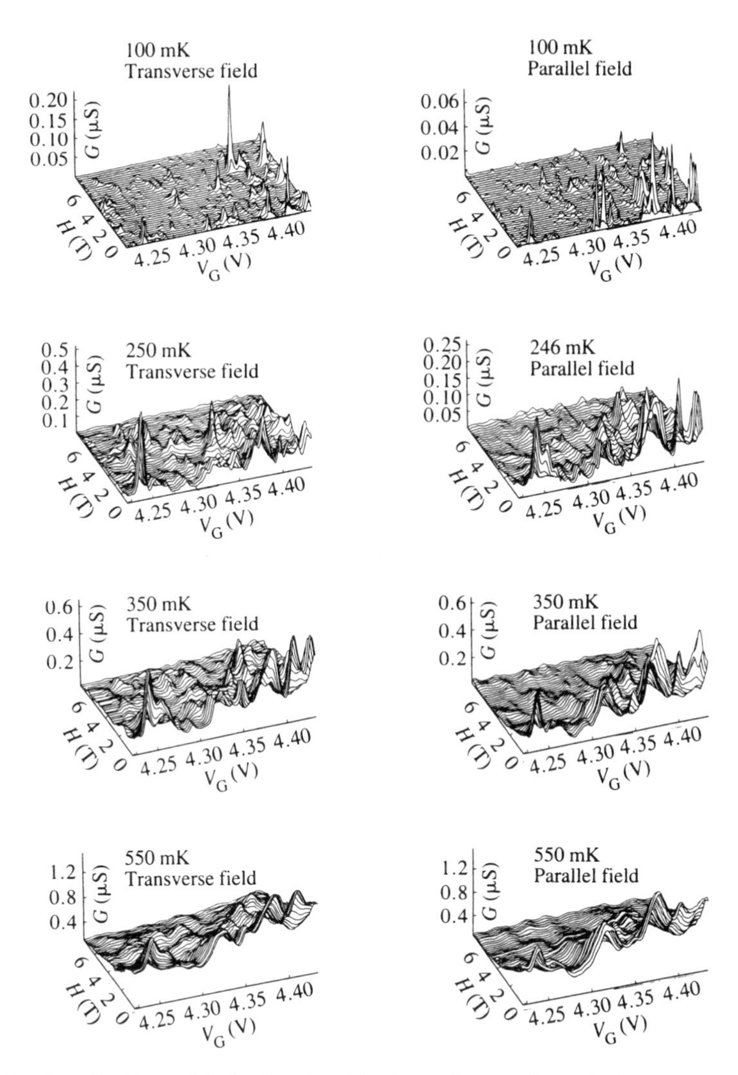

Figure 4: *A collection of data showing the dependence of conductance on gate voltage for different temperatures, magnetic fields, and direction of* **B**. *Figures on the left are for* **B** *perpendicular to the sample surface; those on the right are for* **B** *parallel to the direction of current.*

than in small samples. As in any other studies done in hopping in such samples, there is no evidence for negative magnetoresistance. Why these samples always exhibit positive magnetoresistance and other materials exhibit negative magnetoresistance has not been explained but rather ignored by the theorists who have predicted that interference effects are important.

One result of small samples is that hopping can occur by hops through a small number of sites. Glazman and Matveev (1988) have predicted that a weak power law in the temperature should be observed, increasing with the number of steps (4/3 for 2 steps, 2.5 for 3, 3.6 for 4). Observations that can be fit with such a behaviour have recently been seen (Popovic *et al.* 1991) where at lowest temperatures there is no temperature dependence. There, one supposes the conductance is by resonant tunnelling through one or more sites, which we shall discuss in the next section.

3 Resonant tunnelling

Resonant or direct tunnelling is the only process of interest at low enough temperatures. It was first reported (Fowler *et al.* 1986) in small MOSFET samples where the samples were 0.5–1 μm long and the width was 1–2 μm. Use of the gate in such structures allows a unique way of studying resonant tunnelling because the resonant centre can be tuned into resonance using the gate, with only a small voltage between the source and drain contacts. Most results below were made applying 0.5–1.0 μV a.c. across the sample. Typical data for such measurements are shown in Figure 5. Because the samples are narrow the resonances are fairly well separated so that lineshapes can be studied.

It is generally expected that the line should be described by a Lorentzian,

$$G(\mu) = \frac{2e^2}{h} \frac{\Gamma^L \Gamma^R}{(\mu - \varepsilon_0)^2 + \left[\frac{1}{2}(\Gamma^L + \Gamma^R)\right]^2}, \tag{2}$$

where Γ^L and Γ^R are the escape rates to left and right, and $(\mu - \varepsilon_0)$ is the difference in energy between the chemical potential of the contacts, μ, and the centre of the resonant state due to the impurity, ε_0. The escape rates $\Gamma^{L,R}$ depend exponentially on the distance of the resonant state from the contact. The conductance has a maximum of $2e^2/h$ when the two escape rates are equal, which therefore occurs when the impurity is placed at the centre of the sample (Kalmeyer and Laughlin 1987). The maximum conductance decreases exponentially as the impurity is moved away from the centre and the peak broadens exponentially. At temperatures greater than the linewidth, the lines are thermally broadened; the peak height decreases as $1/T$ and the slopes $dG/d(\mu - \varepsilon_0) = \pm 1/k_BT$. Such behaviour was observed.

These measurements allowed an estimate of the density of states and of linewidth to be made. It was estimated that an overlap of two lines was unlikely. However many lines appeared to be complex with strong central peaks and smaller satellites. A reason for this was that there are Coulomb interactions with nearby sites which shift the energy of the resonant state. These sites can be occupied or not if within a few k_BT. Because the experiment is a time average, it is a measure of the excitations of the system. Simulations by Greene (1990) support this conjecture. So too does the observation of fluctuations in time for a constant gate voltage, although the time-dependent data look richer than what might be expected from the gate-dependent structure.

Recently, tunnelling experiments (Popovic *et al.* 1991) have been carried out in wide samples so that many overlapping tunnelling processes occur. This allows statistical averaging of the tunnelling processes, which is not possible in narrow samples. The samples studied had lengths varying from 1 to 9 μm with $3 \le W/L \le 81$. Data from these samples was discussed earlier because they showed a transition from resonant

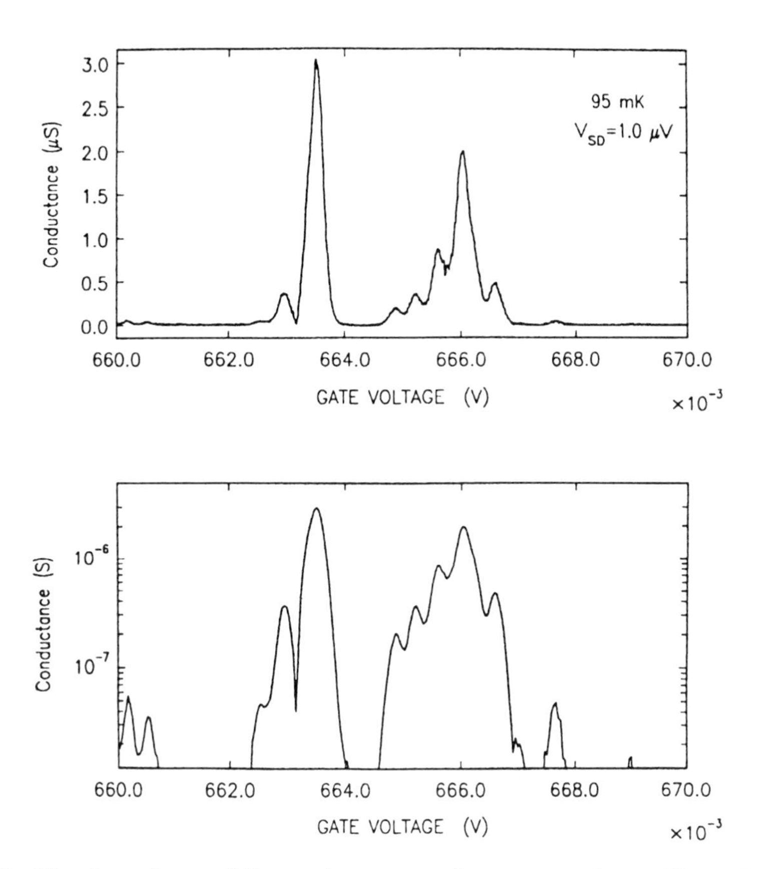

Figure 5: *The dependence of the conductance peaks on gate voltage. These data were taken on a sample with a length of* $0.5\,\mu$m *and a width of* $1.0\,\mu$m.

tunnelling to what might be interpreted as a two- or three-site hopping process. In the tunnelling regime the data were rich enough that $\langle \ln G \rangle$ is easily obtained. The most startling result of these experiments is that $\langle \ln G \rangle$ *increases* as the length of the sample increases. It has been suggested that this results from a statistically increased probability of the numbers of tunnelling processes through more than one resonant site as the length of the sample increased. Different peaks in the autocorrelation function show a different dependence on length so that there is some evidence for this view.

Conclusions

The study of transport by tunnelling and hopping through the localised states below the two-dimensional conduction band edge has demonstrated a wealth of phenomena that can be studied in unique ways by using the field effect. The mesoscopic effects have been found to extend to very large samples indeed. There remain however, many unexplained and intriguing questions.

References

Ando T, Fowler A B, and Stern F, 1982. *Rev. Mod. Phys.*, **54**, 437.

Fowler A B, Wainer J J, and Webb R A, 1988. *IBM J. Res. Develop.*, **32**, 372.

Fowler A B, Wainer J J, and Webb R A, 1991. In *Hopping Processes in Solids*, Eds. Pollak M and Shklovskii B I, North-Holland.

Glazman L I, and Matveev K A, 1988. *Sov. Phys. JETP*, **67**, 1276.

Green M C, 1990. Thesis, University of California, Riverside.

Kalia R K, Xue W, and Lee P A, 1986. *Phys. Rev. Lett.*, **57**, 1615.

Kurkijarvi J, 1973. *Phys. Rev. B*, **8**, 922.

Lee P A, 1984. *Phys. Rev. Lett.*, **53**, 2042.

McInnes J A, and Butcher P N, 1985. *J. Phys. C: Solid State Phys.*, **18**, L921.

Pepper M, 1978. *Phil. Mag. B*, **37**, 187.

Popovic D, Fowler A B, Washburn S, and Stiles P J, 1990. *Phys. Rev. B*, **42**, 1759.

Popovic D, 1990. *Hopping and Related Phenomena*, Eds. Fritzsche H and Pollak M, World Scientific.

Popovic D, Fowler A B, and Washburn S, 1991. To be published.

Raikh M E, and Ruzin I M, 1987. *Sov. Phys. JETP*, **65**, 1273.

Serota R A, Kalia R K, and Lee P A, 1986. *Phys. Rev. B*, **32**, 8441.

Stiles P J, 1986. *Electronic properties in two-dimensional systems*; in *Localisation and interaction in disordered metals and doped semiconductors*, Proc. 31st Scottish Universities Summer School in Physics, Ed. Finlayson D M, Edinburgh, SUSSP, pp. 71–116.

Charge Quantisation Effects in Small Tunnel Junctions

L J Geerligs

Delft University of Technology
Delft, The Netherlands

1 Introduction

1.1 Overview

Quantisation of charge in a metallic island or a semiconductor electron gas, separated from leads by tunnelling barriers ('Coulomb islands') is usually not directly noticeable. However, when the total capacitance of such an isolated Coulomb island becomes small compared to e^2/k_BT, the capacitive energy change associated with the addition or subtraction of one electron from the island becomes significant. With nanofabrication it is possible to create systems with island capacitances C of the order of 10^{-16} F or even smaller. Hence, at temperatures below 1 K the charging energy $E_C = e^2/2C$ is larger than the energy of thermal fluctuations. The charge on the island is a well-defined quantity and adding or removing a single electron is possible only under certain bias conditions. Tunnelling is forbidden (at zero temperature) if the total energy of the system would increase. This phenomenon, now known as Coulomb blockade of (electron) tunnelling, was appreciated and known experimentally as early as 1951 (Gorter 1951; see also Giaever and Zeller 1968, 1969; Lambe and Jaklevic 1969).

The basic theory of Coulomb blockade will be presented in Section 2 of this paper, illustrated by experiments on metal tunnel junctions. Section 3 will discuss the importance of the leads to a system of small capacitance. If the leads to a tunnel junction have a finite impedance, the electron will create electromagnetic excitations during tunnelling, and feel the back-action of those excitations on its own tunnelling process. This effect can drastically change the tunnelling rate. In Section 4 we discuss applications of small tunnel junction circuits. The control of charge transfer at the single electron level seems to be the most useful application of charging effects yet, and we discuss several options which accomplish this. In Section 5 we discuss charging effects in the superconducting state. Cooper pairs are subject to Coulomb blockade similar to that for electrons. The main difference is the inability to convert the energy gained by a Cooper pair during tunnelling into kinetic energy (resembling electron tunnelling in low-dimensional systems, where the kinetic energy is quantised). The study of Coulomb blockade effects in semiconductor devices is catching up rapidly with that of normal metal junctions. In Section 6 we will make a few comments about the similarities and differences between single electron tunnelling in both systems.

This paper will try to give a reasonably up-to-date introduction to charging effects. Recent reviews have been written by Averin and Likharev (1991) and by Schön and Zaikin (1991). For the most up to date information on charging effects the reader is referred to the proceedings of the NATO ASI in Les Houches on single charge tunnelling (Devoret, Martinis and Grabert 1991).

1.2 Fabrication

A planar metal-oxide-metal tunnel junction is like a leaky parallel plate capacitor. It is an almost ideal building block for charging effect devices. A junction area of $(100\,\mathrm{nm})^2$ yields a capacitance of about 10^{-15} F, depending on the barrier thickness. The smallest planar junctions that have been produced so far (*e.g.* Dolan and Dunsmuir 1988, Fulton and Dolan 1987) were all fabricated from aluminium. Useful tunnel resistances (of around $100\,\mathrm{k}\Omega$) are obtained if the aluminium is thermally oxidised at room temperature in oxygen at a pressure of about 1 mbar to create the tunnel barrier. Together with the requirement of high purity metal electrodes this low oxidation pressure means that it is preferable to fabricate the whole junction in one vacuum cycle.

This is conventionally done by shadow evaporation (Figure 1). The two electrodes of a junction are evaporated from two angles. A mask patterned with a small channel interrupted by a bridge results in a junction because of the interruption of the aluminium strips by the bridge shadow. On both sides of the junction the leads are actually also composed of a double aluminium layer with oxide barrier in between, *i.e.* the leads are large junctions. This creation of large junctions in series with the small ones can be partly avoided by using a slightly different geometry (see, for example, Fulton and Dolan 1987 or Kuzmin *et al.* 1989).

In Figure 1 a three-layer resist is used. An alternative used by several groups is the use of a two-layer resist where the high-resolution top layer also functions (after development) as evaporation mask. A three-layer resist is somewhat more flexible due to the separate control of patterning and undercut steps. Any combination of resists for top and bottom layer can be used. A photograph of a Coulomb island (double junction) fabricated by shadow evaporation is given in Figure 2. The fabrication procedure described above has proven to be sufficient for creating junctions with area down to $(30\,\mathrm{nm})^2$.

The four types of devices most widely used for experiments on charging effects are shown schematically in Figure 3. In metal tunnel junction circuits, the charging energies are dominated by the junction capacitances, *i.e.* the capacitances between leads and Coulomb island (or the nearest-neighbour capacitance between Coulomb islands for an array of more than two junctions). In turn, the junction capacitances are well

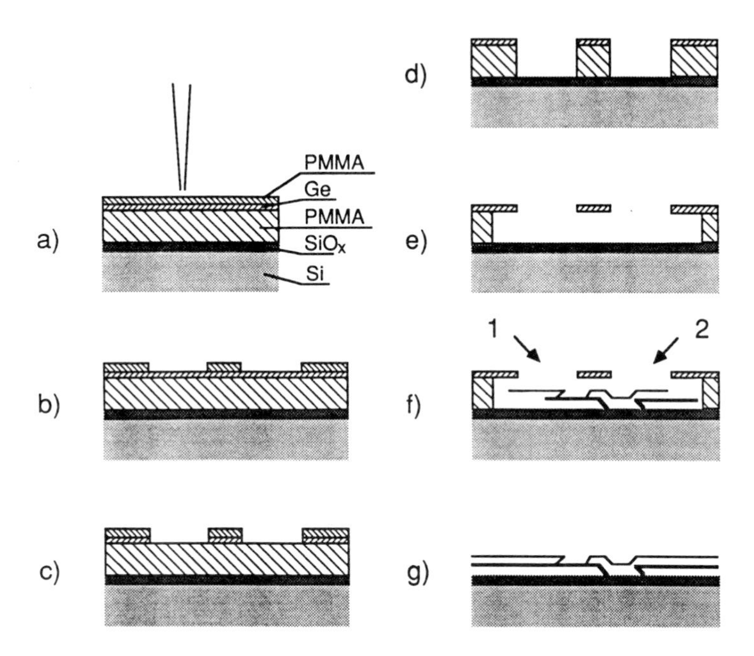

Figure 1: *Side view of the fabrication of a small metal tunnel junction. (a) The pattern is written in the top PMMA layer of a 3-layer resist with an e-beam, (b) developed, (c) etched into the middle Ge layer, and (d) etched into the bottom resist layer. (e) Isotropic etching creates a free-hanging bridge. (f) Oblique angle evaporation of two metal layers results after lift-off (g) in a planar junction.*

approximated by their parallel plate capacitance. Hence, the Coulomb island has a capacitance close to twice the junction capacitance. For 60 nm junctions this is about 0.6 fF, and it could be considerably reduced by reducing the area of the junctions. Eventually, however, the junction capacitance will be given by island-to-lead capacitance, or the self-capacitance of the island (or the capacitance to a ground plane or gate electrode) will become dominant. This is the case for experiments in STM-grain tunnelling

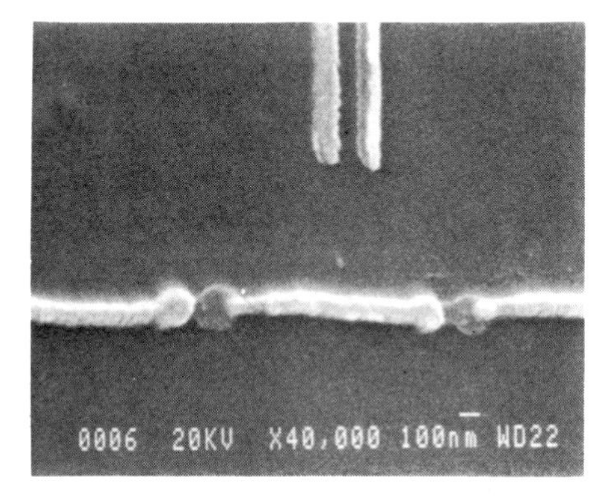

Figure 2: *Scanning electron micrograph of a double metal tunnel junction.*

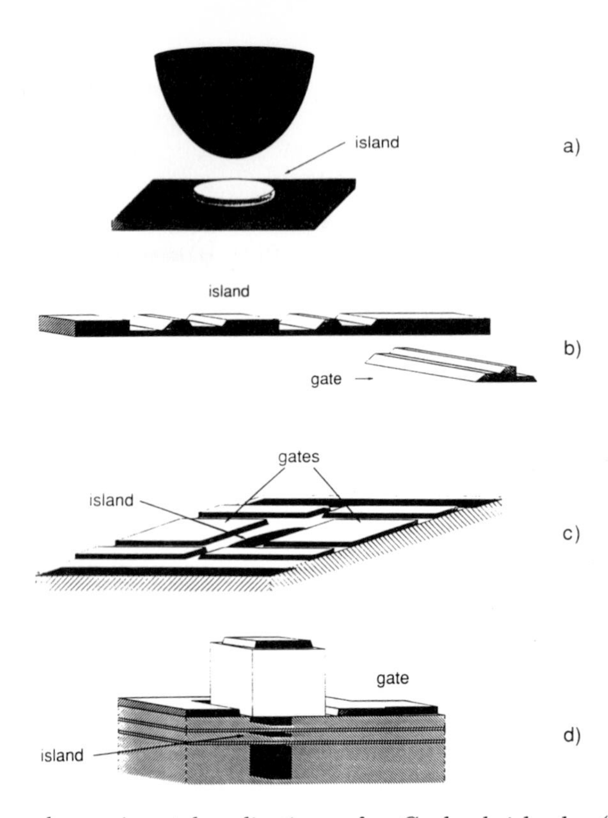

Figure 3: *Several experimental realisations of a Coulomb island. (a) STM-grain-substrate tunnelling device. (b) double planar tunnel junction (with gate electrode). (c) split-gate confined 2-D electron gas in a semiconductor heterostructure. (d) gated resonant tunnelling device. In (c) and (d) the electron gas is indicated by the black areas.*

configurations where the experimentally observed capacitance is of order 10^{-18} F. In Coulomb islands defined by split gates, the gate capacitance is also much higher than the junction capacitance, and therefore contributes most of the island capacitance.

2 Basic theory

2.1 Coulomb blockade of electron tunnelling

In this section we consider tunnel junctions with high tunnel resistance, $R_t \gg h/e^2$. Under the additional constraint that the impedance of the leads is very low, the charge transport can then be calculated by treating the charge Q on the islands as a classical variable.

The kinetic energy transferred to an electron when it crosses a tunnel barrier over which a voltage drop V exists is less than eV. The difference is $e^2/2C_\Sigma$ where C_Σ is some relevant capacitance in the system (we will specify this below). Usually we can neglect this difference, because C_Σ is very large and/or the temperature is relatively high. However, at low temperature and for small capacitance an excess voltage drop is necessary before tunnelling occurs. This is the Coulomb blockade of electron tunnelling.

The energy transferred to the tunnelling electron is (minus) the change of electrostatic energy $\Delta E = E_f - E_i$ of the complete circuit during tunnelling. The electrostatic energy is the sum of the capacitive energies in the system and the work performed by the voltage sources (Likharev 1988, Bakhvalov *et al.* 1989):

$$E = \sum_i \frac{Q_i}{2C_i} - \sum_j Q_{tj} V_j \,. \tag{1}$$

The index i denotes summation over tunnel junctions as well as true capacitors, the summation in j is over all voltages sources in the system, and Q_{tj} denotes the charge transferred through voltage source V_j. Often a large stray capacitor on a chip close to a circuit changes an experimentally applied current bias for high frequencies into a voltage bias.

At $T = 0$ an electron can tunnel into an unoccupied state only if it gains kinetic energy from the decrease of the electrostatic energy (Equation 1) of the system. For arbitrary temperature the tunnelling rate is derived from the golden rule in the usual way, taking into account the Fermi distribution over a constant density of states:

$$\Gamma = \frac{\Delta E}{e^2 R_t} \left[\exp\left(\frac{\Delta E}{k_B T}\right) - 1 \right]^{-1} , \tag{2}$$

where R_t is the tunnel resistance of the junction (the resistance in the absence of charging effects). In the limit $k_B T \ll |\Delta E|$ the tunnelling rate reduces to

$$\Gamma = \begin{cases} \dfrac{|\Delta E|}{e^2 R_t}, & \text{if } \Delta E < 0 \\ 0, & \text{if } \Delta E > 0. \end{cases} \tag{3}$$

Most measurements can be simulated very accurately with this equation. The main deviations from the simulations occur when R_t is not much larger than the quantum resistance $R_K = h/e^2$, so that higher order—co-tunnelling—events become noticeable (discussed later in this section), or when the excitation of electromagnetic modes in the environment during tunnelling becomes possible (Section 3). In the case of semiconductors, the low barrier height may make the assumption of a constant R_t inaccurate. Finite size effects and resonant transmission also make Equation (2) invalid.

As mentioned above, ΔE is of the form $\Delta E = -eV + e^2/2C_\Sigma$ for the tunnelling of a (positive) charge e in the direction of the voltage drop (Esteve 1989). This can be seen from Figure 4. Neglecting dynamics (inductances and resistances in the circuit), any tunnel junction (with capacitance C) is effectively biased by a series circuit consisting of a voltage source and an equivalent capacitor C_e, the capacitance of the circuit shunting the junction. The bias will result in a voltage drop V over the junction. However, when an electron tunnels it will charge the island between C and C_e (with capacitance $C_\Sigma = C + C_e$) by either e or $-e$. Hence its energy gain will be decreased by the charging energy of C_Σ, namely $-\Delta E = eV - e^2/2C_\Sigma$. [This can be checked by calculating ΔE

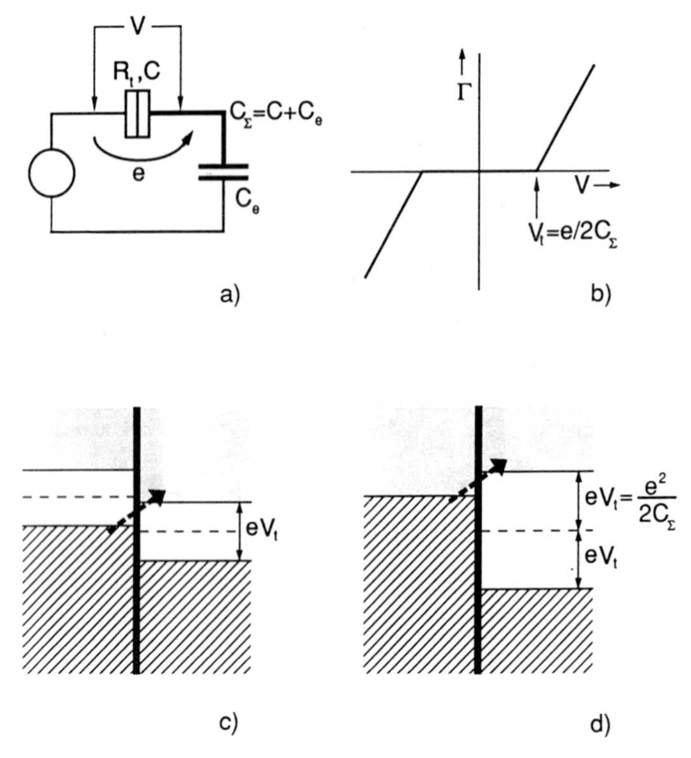

Figure 4: *(a) Equivalent circuit to describe tunnelling through any junction. (b) The resulting tunnelling rate versus voltage drop over the junction. (c) and (d) Two ways to present this information in a potential landscape. Dashed line: electrochemical potential neglecting charging effects. Cross-hatched: occupied states. Shaded: unoccupied states. Electron tunnelling (arrow) is impossible at this bias.*

for this system directly from Equation (1), explicitly including a net charge on the island if one likes.]

For complicated multi-junction systems, the reduction to the above case is very useful. When considering a tunnel event across a given junction k, the parallel circuit of this junction can be replaced by an equivalent capacitor and voltage source. The equivalent capacitor C_{ek} gives the threshold voltage $V_t = e/2(C + C_{ek})$ or junction charge CV_t which has to be exceeded for tunnelling to be possible (Figure 4b).

For double junctions it is common to describe tunnelling under the influence of Coulomb blockade with a potential landscape. All occupied states on the island are drawn a distance eV_t below their conventional position without charging effects, while empty states start a distance eV_t above the potential (Figure 4d). The extension of such a picture to larger arrays is usually problematic. For instance, if two junctions, defining a Coulomb island, have different V_t, where should one draw the above levels with respect to the electrostatic potential? Also, in contrast to the double juction, adding an electron generally does not raise the Fermi level of an island by exactly $2eV_t = e/C_\Sigma$, which makes the landscapes less intuitive to use.

A special case is an array of Coulomb islands with equal capacitances, coupled by tunnel junctions with much smaller capacitances. Here the use of a potential landscape is again helpful.

2.2 The double junction

Consider two tunnel junctions in series, where the island between the junctions is capacitively coupled to a gate voltage. The use of capacitive gates in tunnel junction circuits, introduced by Fulton and Dolan (1987), is an essential feature in experiments on charging effects. In particular, a gate voltage can be used with the same effect as a net island charge Q_0 would be used. However, Q_0 can only change discretely due to tunnelling events, whereas a gate voltage can be swept continuously.

The electrostatic energy of this device is

$$E = \frac{Q_0^2}{2C_\Sigma} - q_l V_l + q_r V_r + \frac{Q_0}{C_\Sigma}\left(C_g V_g + C_l V_l + C_r V_r\right), \tag{4}$$

where V_l is the potential of the left lead, C_l is the capacitance of the left junction and q_l is the charge passed through this junction; the index r denotes similar parameters for the right junction and g for the gate; and $C_\Sigma = C_l + C_r + C_g$ is the capacitance of the island. It is clear that, neglecting an energy offset, the gate voltage can be accounted for by replacing Q_0 by $Q_0' = Q_0 + C_g V_g$. The energy change of all four possible tunnelling events can be written as $\Delta E = e\Delta V$, where $\Delta V = \pm(V_{l,r} - \varphi)$. The effective island potential is either $\varphi = Q_0'/C_\Sigma + (C_l V_l + C_r V_r)/C_\Sigma - e/2C_\Sigma$ for tunnelling of an electron into the island, or $\varphi = Q_0'/C_\Sigma + (C_l V_l + C_r V_r)/C_\Sigma + e/2C_\Sigma$ for tunnelling out of the island. The first two terms yield the conventional electrochemical potential. The last term is the contribution due to charging effects. This is of course in agreement with the remarks of the previous subsection. One specific island charge can be stable at small bias voltage due to this 'double' island potential, *i.e.* no tunnelling occurs. Figure 5 shows the stable island charge as a function of bias and gate voltage for a symmetric system. Outside the stable regions the two tunnelling events (in both junctions) necessary for current flow can occur. For $Q_0' = 0$ a minimum bias voltage $2V_t$ is necessary to obtain tunnelling. The result is a Coulomb gap in the current–voltage ($I(V)$) characteristic. Figure 5 shows that for $Q_0' \approx e/2$ a small voltage is sufficient for current flow, *i.e.* the Coulomb gap is suppressed.

Figure 6 gives the measured $I(V)$ characteristic of a double junction, for the two gate voltages where the Coulomb gap is maximum and minimum. In the case of the maximum Coulomb gap (solid curve) conduction below the threshold voltage is very low, although not completely zero (see next subsection). By applying a gate voltage, the Coulomb gap can be completely suppressed to an almost ohmic curve (dashed curve). At high voltages the same voltage offset $e/2C$ is recovered. As a function of gate voltage the $I(V)$ curve evolves continuously between the two extremes shown. With the average current through the device fixed at a low level, the voltage versus gate voltage can be recorded. An example is given for a similar double junction in Figure 7. The curve is periodic because gate charges V_g and V_g' are equivalent if $C_g(V_g - V_g') = e$. The net island charge changes by e when going from V_g to V_g', after which the potential landscape is

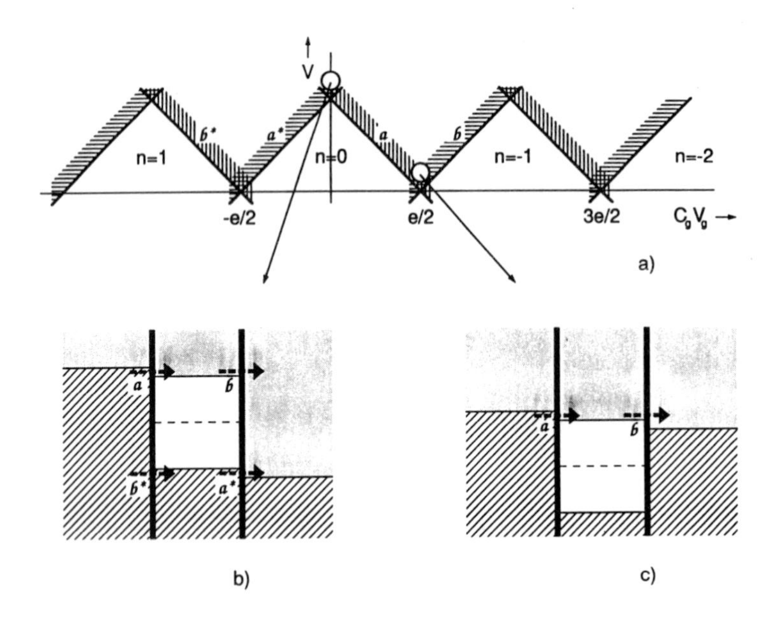

Figure 5: *(a) Stable charge for a symmetric double junction, $V_l = -V/2$, $V_r = V/2$, $C_l = C_r$. Specific tunnelling events can take place on the hatched side of the thick threshold lines. Inside the triangle (diamonds, when considering also $V < 0$) the charge ne is stable. Outside, the sequence of tunnelling events necessary for conduction can take place. (b) and (c) are potential landscapes in the conducting state, for large bias voltage and for gate voltage near $(2k+1)e/2C_g$, respectively.*

the same. The stability of $I(V)$ curves in time shows the absence of noticeable ohmic leakage in aluminium tunnel junctions.

The $I(V)$ curve shown in Figure 6 is typical for symmetrical metal double junctions. It does not show the Coulomb staircase which is well known from, for example, STM-grain tunnelling experiments (Barner and Ruggiero 1987; Van Bentum *et al.* 1988a,b; Wilkins *et al.* 1989). On increasing the voltage bias of a double junction, additional charge states of the island (with net charge e, $2e$, $3e$, *etc.*) become possible. The Coulomb staircase arises when the RC times of the two junctions are very different. In that case, the $I(V)$ curve shows a strong increase in current at the voltage where an additional charge state of the island starts to contribute to the conduction. The slope of the initial current rise is determined by the junction with the highest tunnelling rate. On further increase of the voltage, the current only increases slowly since it is limited by the junction with the slowest tunnelling rate.

2.3 Co-tunnelling

In this section we will discuss events of coincident tunnelling of electrons. This is a process of higher-order tunnelling in which only the energy from initial to final (after all the tunnelling events) state needs to be conserved. The recommended name for these events is co-tunnelling, to emphasise the difference from the correlated but incoherent

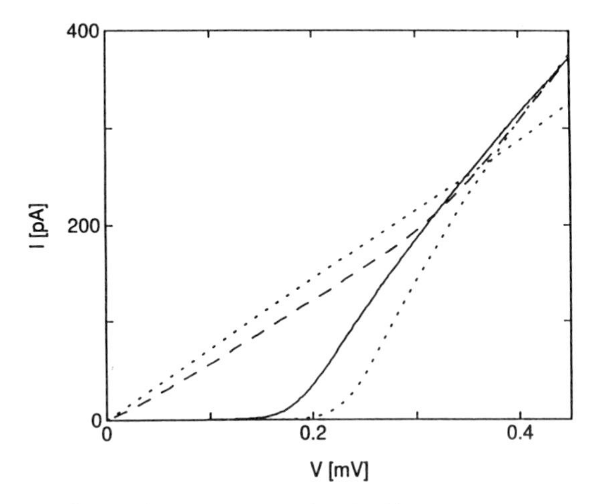

Figure 6: *Current-voltage characteristic of a double junction at* 20 mK, *for gate voltages of* 0 *(solid line) and* $e/2C_g$ *(dashed line). The dotted lines give the theoretical curves for a symmetrical system of the same total capacitance and resistance.*

tunnelling of two electrons which occurs in the first-order description of tunnelling in junction arrays. It has also been called quantum tunnelling of the electric charge (q-MQT), since it is the charge state of the system which tunnels through the Coulomb energy barrier in this process. In arrays of metal tunnel junctions it is an inelastic event where an electron-hole pair is created on the Coulomb island, and two different electrons tunnel(Averin and Odintsov 1989). In semiconductor structures elastic co-tunnelling [the electron has to diffuse virtually to the next junction (Averin and Nazarov 1990)] or resonant tunnelling can occur.

In an array of metal junctions under low bias voltage, extra electrons residing on the central metal islands increase the energy. This produces a barrier against electron transport across the system. Thermal fluctuations can provide the energy for this intermediate state, and thus cause passage of this Coulomb energy barrier. At low temperatures electron transport should be exponentially suppressed in E_C/k_BT. However, tunnelling through one or more virtual intermediate states can cause the system to change the charge distribution to a state where one electron charge has passed through the complete array. Figure 8 shows the inelastic co-tunnelling process for a double junction where an electron-hole excitation is created on the central island, and for three junctions.

For high tunnel resistances the rate is proportional to the product of the junction conductances and proportional to V^{2N-1}, where N is the number of junctions,

$$\Gamma \propto \frac{V^{2N-1}}{h} \prod_i \frac{\alpha_i}{4\pi^2}, \tag{5}$$

where $\alpha_i = R_K/R_t$ for junction i, and the product is over all junctions. In Figure 9 we show experimental $I(V)$ curves of the Coulomb gap of 2, 3 and 5-junction arrays. It is clear that the Coulomb gap sharpens considerably for the longer arrays, in agreement with this equation.

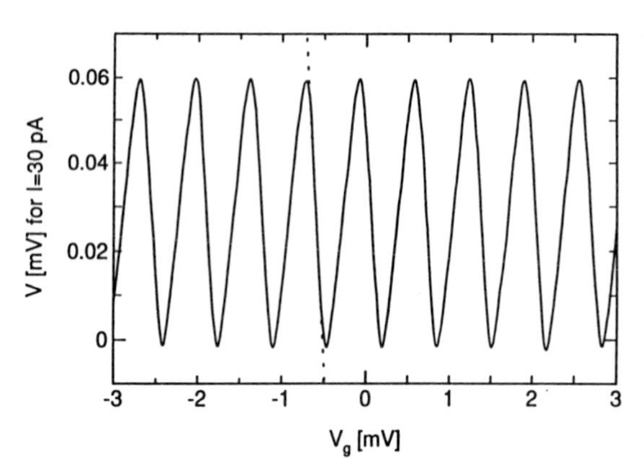

Figure 7: *Voltage across a double junction as a function of gate voltage at a fixed current level, showing highly periodic behaviour.*

For study of the effect of R_t on co-tunnelling, the 2nd-order events are most favorable, since they have the highest rate with the weakest voltage dependence. The rate for voltages inside the Coulomb gap is

$$\Gamma = \frac{h}{(2\pi)^2 e^4 R_{t1} R_{t2}} \left\{ \left(1 + \frac{2}{U}\frac{E_1 E_2}{E_1 + E_2 + U}\right) \left[\sum_i \ln\left(1 + \frac{U}{E_i}\right)\right] - 2 \right\} U, \tag{6}$$

where U is the energy gain in the co-tunnelling event, and E_1 and E_2 are the energies of the (virtual) intermediate state if the first tunnel event occurs in the one junction or the other junction. For a double junction:

$$\begin{aligned} E_1 &= \frac{e}{C_\Sigma}\left[\frac{e}{2} + Q_0 - \left(C_r + \frac{C_g}{2}\right)V\right]; \\ E_2 &= \frac{e}{C_\Sigma}\left[\frac{e}{2} - Q_0 - \left(C_l + \frac{C_g}{2}\right)V\right]. \end{aligned} \tag{7}$$

This rate is in agreement with measurements on double junctions (Geerligs *et al.* 1990b). Figure 10 compares measurements of the $I(V)$ curves of four double junctions [with maximum Coulomb gap, so $Q_0 \approx 0$) with the theoretical prediction of Equations (5)–(6)]. The measurements have been scaled to the dimensionless voltage VC_Σ/e and current IR_tC_Σ/e to allow for easy comparison (the curves should be the same to first order in R_t^{-1}). C_Σ is used as a fit parameter, but corresponds well to the value derived from the Coulomb gap. The slope of $\log I$ versus $\log V$ is about three, as predicted by theory, and the absolute value of the current scales approximately with R_t^{-2}.

2.4 Low junction resistances

For very low R_t a perturbative approach fails. Theory based on a microscopic approach (Ambegoakar *et al.* 1982, Ben-Jacob *et al.* 1983, Eckern *et al.* 1984, Brown and Simánek 1986) predicts a gradual decrease of charging effects, with perhaps an abrupt disappearance around $R_t = 1\,\mathrm{k\Omega}$ (Zwerger 1991). Measurements (Geerligs *et al.* 1989) show at least this gradual decrease of charging effects.

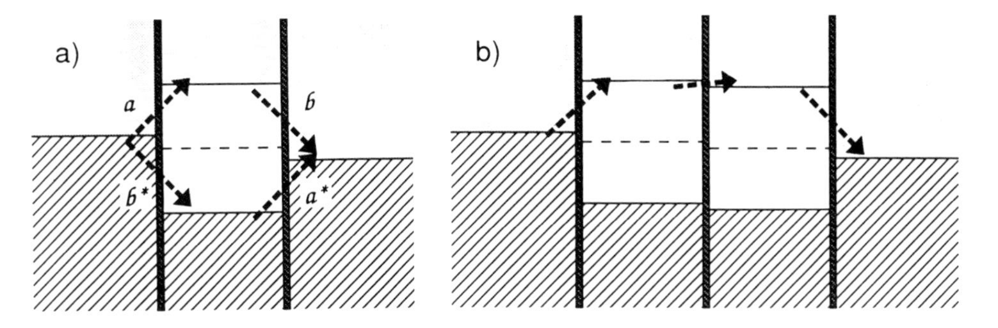

Figure 8: *Potential landscape of co-tunnelling. (a) For a double junction (a, b first tunnel event occurs in the left junction; a^*, b^* first event in the right junction). (b) For three junctions (with island capacitance much larger that junction capacitance, to simplify the Figure). Note that two or three tunnel events, respectively, make one co-tunnel event.*

In a Coulomb island defined in a high-mobility 2DEG, the charge is no longer well defined if the resistance is smaller than $h/2e^2$. In fact, the constrictions are no longer even tunnelling barriers. The suppression of charging effects is therefore much more abrupt and at a higher resistance than for metal tunnel junctions (Kouwenhoven *et al.* 1991b).

3 Effects of the electromagnetic environment

3.1 Single junction — Modified tunnelling rates

Generally, the capacitance between leads to a single (solitary) junction is very large compared to the local junction capacitance. This results in the absence of a Coulomb gap in a single junction unless special precautions are taken (Delsing *et al.* 1989a, Geerligs *et al.* 1989). By using high impedance leads (Cleland *et al.* 1990) the effect of parasitic lead capacitance can be effectively avoided and a Coulomb gap is observed. Apparently the tunnelling rates in the junction are modified by its environment.

This problem has been treated theoretically by various authors (Nazarov 1989a and 1989b, Devoret *et al.* 1990, Averin and Schön 1990). A review which also treats multi-junction systems is given by Grabert *et al.* (1991). The models, which are to first order in R_t^{-1}, predict that the electromagnetic field in the shunt geometry is not only influenced by the tunnelling process, but also has a back-influence on it. A Coulomb gap in a single junction can be interpreted as being due to inelastic tunnelling, *i.e.* it arises if low frequency modes can be excited in the environment during tunnelling. No Coulomb gap arises if only elastic tunnelling is possible (due to small coupling to environmental modes, *e.g.* in a low impedance environment), because the electron always acquires the energy eV from the voltage source. If there is a strong coupling to modes at frequency E_C/h, tunnelling will occur if ΔE is sufficient (*i.e.* $V > e/2C$) to excite these modes. Tunnelling without excitation is much less likely. For lower bias

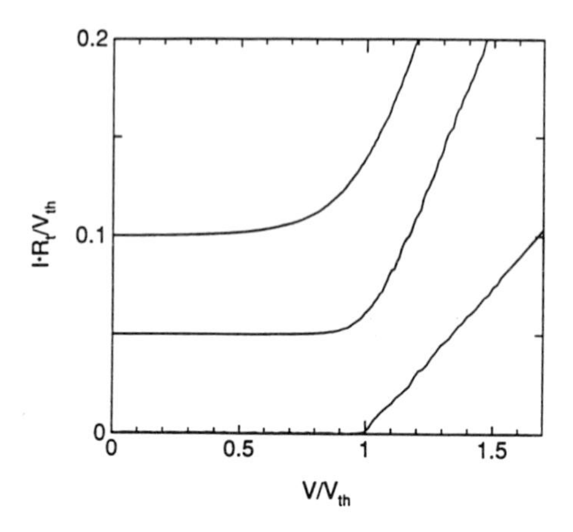

Figure 9: *Scaled current-voltage characteristics for arrays of two, three and five junctions (top to bottom) with approximately equal barrier transmissions.*

voltage the electron will probe the existence of electromagnetic modes around $E_C/\hbar$ through their back-action during a virtual tunnelling event. As an example, Figure 11 shows a junction with purely inductive leads (Devoret *et al.* 1990) where the excitation spectrum is quantised. As a result the differential resistance is also quantised. The slope of the $I(V)$ curve changes with increasing voltage each time a higher energy mode can be excited.

The environment can be taken into account in the form of a probability function for excitation of environmental modes. A function $P(E)$ describes the probability that a tunnelling electron loses the energy E to the environment; it depends on the frequency-dependent impedance $Z(\omega)$ shunting the junction and the junction capacitance. Also, $P(E)$ is normalised and satisfies the sum rule

$$\int_{-\infty}^{\infty} dE\, E\, P(E) = E_C\,, \tag{8}$$

so that at high bias voltage a single junction eventually develops a Coulomb gap, even for a low impedance environment.

The probability function $P(E)$ replaces the δ-function for elastic tunnelling in the conventional 'golden rule' expression,

$$\Gamma(V) = \frac{1}{e^2 R_t} \int_{-\infty}^{\infty} dE \int_{-\infty}^{\infty} dE'\, f(E)\,[1 - f(E')]\, P(E + eV - E') \tag{9}$$

in the direction of the voltage, and a similar expression for the other direction; $f(E)$ is the Fermi function.

If the impedance of the environment is low, $P(E)$ is sharply peaked at low (positive) energies. If there is no interaction with the environment, $P(E) = \delta(E)$ and the above equation reduces to an ohmic $I(V)$ characteristic. If the environment is of high impedance (of order of R_K or larger) at frequencies of order $E_C/\hbar$, $P(E)$ shifts to a curve which is exponentially peaked at E_C and results in a clear Coulomb gap in the $I(V)$ characteristic.

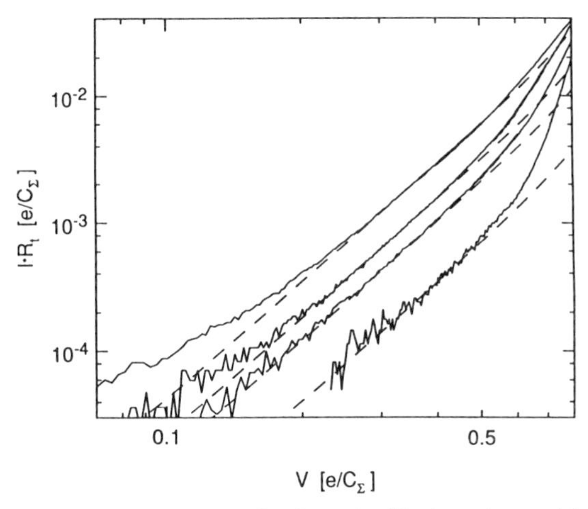

Figure 10: $\log I$-$\log V$ *measurements for four double junctions with* R_t *varying between* $42\,\mathrm{k\Omega}$ *(top) and* $340\,\mathrm{k\Omega}$ *(bottom).*

This case of $P(E) = \delta(E)$ is usually named the 'global rule' to signify the fact that the simple electrostatic energy change determines the tunnelling rates. In contrast, the situation with very high impedance leads, where the tunnelling rate is determined by $\Delta E = (Q - e)^2/2C - Q^2/2C$, is called the 'local rule'.

3.2 Multi-junction circuits

The difference between multi-junction circuits and single junctions is that even the electrostatic energy change for elastic tunnelling (*i.e.* in a low-impedance environment) contains charging energy terms. Therefore these circuits show charging effects without high impedance leads. In Section 2 the tunnelling rates were given neglecting the interaction with the environment, *i.e.* with $P(E) = \delta(E)$.

Tunnelling rates in multi-junction circuits are closely related to the expressions for a single junction (Grabert *et al.* 1991). On tunnelling, only a fraction of an electron charge is transferred through the external circuit, resulting in a modification of $P(E)$. If κ is the ratio of equivalent circuit capacitance to junction capacitance, the effect of the real part of the lead impedance, which is the relevant quantity, is weakened by a factor κ^2. For larger arrays, the tunnelling rates therefore tend more towards the simple case described in Section 2.

3.3 Time correlation of tunnelling events — SET oscillations

The current-biased single junction is a special case for charging effects (Averin and Likharev 1986). It can be realised (see, for example, Büttiker 1986, 1987) by a series circuit of the junction with a very small capacitor C_e. Applying a linearly increasing voltage bias $V = \alpha t$, the junction is subject to a current $I_x = \alpha C_e$, which induces an increase of charge at a constant rate. For $C_e \ll C$ the threshold charge is close to $e/2$.

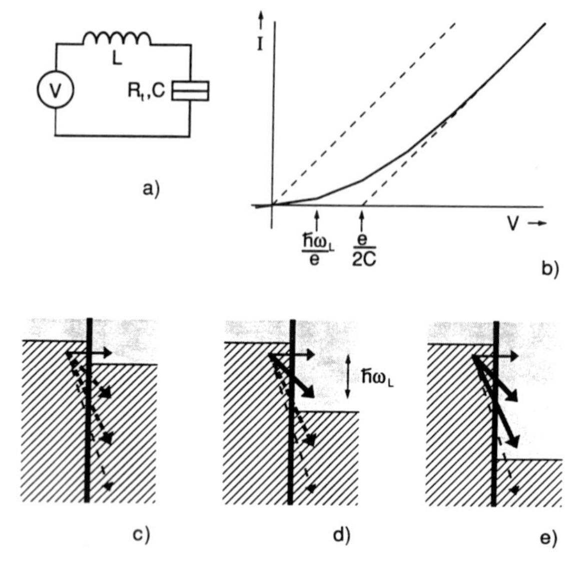

Figure 11: *(a) A single junction with inductive leads. (b) The $I(V)$ characteristic will show a Coulomb gap with quantised differential conductance if the resonant frequency of the circuit is of order E_C/h. (c)–(e) Potential landscapes. The excitation of modes of increasing energy becomes possible with increasing voltage, which increases the tunnelling rate. The creation of excited modes of the environment is expected to be a quantum coherent process. Dashed arrows represent virtual tunnelling.*

If the current is small compared to e/R_tC a tunnelling event will occur at a charge only slightly larger than $e/2$, changing the charge on the junction to about $-e/2$. Then it takes a time period e/I_x to recharge the junction for a new tunnelling event. At $T = 0$ and small current, the resulting d.c. $I(V)$ curve has a parabolic shape. At larger currents the $I(V)$ curve approaches a linear form with voltage offset $e/2C$ and slope $1/R_t$.

The important feature is that the tunnelling events are correlated in time at low currents. The voltage noise spectrum will peak at the Single Electron Tunnelling frequency $f_{\mathrm{SET}} = I_x/e$ and harmonics.

By applying a high-frequency alternating current (frequency f) in addition to the d.c. current, resonances should occur in the $I(V)$ curve at currents $I = (n/m)ef$. This has not yet been observed, but a similar phenomenon has been observed in long one-dimensional arrays of tunnel junctions (Delsing *et al.* 1989b).

In a long array of junctions the current is carried by mutually repelling charge solitons (Likharev *et al.* 1989, Bakhvalov *et al.* 1989). A charge soliton consists of a charged metal island between two junctions, together with the associated polarisation of the neighboring junctions. Due to the repulsion the charge is transferred in a train of regularly spaced solitons. The ratio of the junction capacitance to the self-capacitance C_0 of the islands between the junctions determines the size of a soliton. For small C_0 the decay length of a soliton is about $(C/C_0)^{1/2}$. On a given junction, a tunnelling event occurs each time a soliton passes. Therefore the tunnelling events are again correlated in time. Delsing *et al.* (1989b, 1991) have observed that the $I(V)$ curve of such an SET array shows resonances in the differential resistance and current plateaus at $I = nef$ under high-frequency irradiation.

This soliton character of charges moving in a long array not only causes internal time correlation but can also be used to create a good approximation of a current source. On a given junction the charge increases in small steps if the soliton approaches, decreases by e if the soliton passes, and after tunnelling again increases smoothly in time if one soliton moves away and a new one approaches.

4 Devices

4.1 Introduction

A double junction configured as a 3-terminal device is both a very sensitive charge detector and a high impedance voltmeter. In both functions, it can be configured very close to the source that is to be measured. These characteristics of the double metal junction (also called the SET-transistor or Fulton and Dolan electrometer) have recently been applied in two experiments, discussed in the next subsection.

The controlled transfer of electrons one by one, clocked by a radio frequency (r.f.) gate voltage, is another application of small tunnel junction circuits. A clock frequency f results in a current $I = ef$ or a multiple. It is of obvious importance as a possible high accuracy source of current or charge, and perhaps as an example of a device technology based on single electrons as information carriers.

When a clocked device needs an applied bias voltage to transfer a current it is usually named a turnstile. When it operates with zero bias voltage, or even transfers electrons against the direction of the driving bias voltage, it is called a pump. We will consider the operation of a turnstile device and a pump device constructed of metal junctions. Recently, a Coulomb island in a two-dimensional electron gas has been operated in both turnstile and pump mode (Kouwenhoven *et al.* 1991a).

The low temperature necessary to work with junctions with the presently attainable capacitance forms a limitation for practical use. All experiments presented here have been performed in a dilution refrigerator, with the devices at temperatures down to 10 mK. Low pass filtering of the leads to the devices is important. The filters need to be cooled to low temperatures in order to suppress their own thermal noise.

Several experiments have shown that the islands can have a random offset charge, probably caused by trapped charges near the junctions. This might limit the usefulness of these junctions in large scale applications. However, it has been suggested that with some care or extra precautions, the background charge might relax to a low value. Slow cooling of devices, or heating with light inside the cryostat (Wilkins *et al.* 1989) have been proposed.

Comparison of, for example, the turnstile experiments with theory gives an apparent device temperature of about 60 mK, which is higher than the temperature of the mixing chamber. This temperature increases with increasing r.f. amplitude. It is probably the power input from the voltage modulation which is responsible for the increased temperature. As will be explained later, the operation of the turnstile is intrinsically dissipative. It is unavoidable that hot electrons are created. At the low temperatures of these experiments, the inelastic electron scattering times are very long. The typical electron-electron relaxation time is around 100 ns at 1 K, scaling approximately as T^{-1}; the electron-phonon relaxation time, which is of the same order of magnitude at 1 K, scales approximately as T^{-3}. However, this does not necessarily mean that electron-electron scattering is the dominant relaxation mechanism and accomplishes an equilibrium distribution of the electron gas at an elevated temperature. We expect that hot electrons will also have an increased scattering rate with phonons. Electron and phonon scattering are probably of comparable magnitude for present device parameters, since the tunnelling electron gains of order 1 K in energy. In this case phonons make an important contribution to the diffusion of thermal energy out of the device. This contribution could be increased by increasing the energy gain of the electrons, *i.e.* decreasing C.

In the worst case, where there is strong electron-electron scattering so that the electron gas has a well-defined temperature, this electron temperature should increase as $(P/\Omega)^{1/5}$ where P is the power input, and Ω is the volume of the metal electrode (Wellstood 1988). For the present turnstile devices this temperature increase is close to the values of about 100 mK suggested by the comparison with calculations. Fortunately, on decreasing the junction capacitance, the allowed temperature increases proportional to C^{-1}. In contrast, decreasing C would hardly increase the temperature of the electron gas (only as $C^{-1/5}$).

Co-tunnelling of electrons is usually harmful to the operation of single-electron devices. Recently, however, an interesting application of co-tunnelling has been suggested. It can be used to duplicate tunnelling events from one device to another, so that a current mirror is created. Averin, Korotkov and Nazarov (1991) and Geigenmüller and Nazarov (1991) have proposed two specific circuits. For practical applications the fact that the upper frequency limit scales as R_t^{-2} instead of R_t^{-1} is disadvantageous.

4.2 Electrometer

Obviously, a double junction is a sensitive detector for charge on the gate electrode. It can be used to count electrons, like a DC SQUID is used to count flux quanta. Like the SQUID the sensitivity is higher than the electron charge. In preliminary measurements the gate charge fluctuations, corresponding to the measured voltage noise in curves like Figure 7, is about $10^{-4}\, e/\sqrt{\mathrm{Hz}}$ between 10 and 200 Hz. The ultimate performance could be even better (Korotkov *et al.* 1991). A severe problem compared with the SQUID is the application of the charge to the gate. The input line needs to have a small capacitance compared to C_g, otherwise much of the charge that should polarise the gate capacitor is lost to the parasitic lead capacitance.

One experiment which has already used this high charge sensitivity in the form of a voltage measurement has recently been described by the Saclay group (Lafarge *et al.* 1991). As described in Section 3.3, a single junction biased through a capacitor with a time-dependent voltage should show SET oscillations just like a current biased junction. The problem of how to observe these oscillations was solved by connecting the island between the junction and the capacitor to the gate of a double junction electrometer.

Another experiment that has been proposed by NIST in collaboration with the same group (Williams *et al.* 1991) intends to use the electrometer in a metrological experiment where the electron-pump device is the charge source for a calibrated capacitor. The electrometer is the sensor in a feedback loop to keep the voltage at one electrode of the capacitor at zero. The fact that the electrometer can be placed very close to the capacitor (on the same chip) and operates at mK temperatures is essential in this experiment.

4.3 Turnstile — Dissipative electron tunnelling

A double junction is for given bias and gate voltage either insulating or conducting. This is due to the fact that either precisely one charge configuration is stable, or two charge configurations with island charges Q_0 and $(Q_0 + e)$ are unstable. The result is that the double junction cannot be used (in the normal state) for controlled charge transfer. In the turnstile device (Figure 12, Geerligs *et al.* 1990a, Urbina *et al.* 1991) this problem is solved by adding one junction on each side of the gate capacitor. The charging energy of excess charge on the side islands increases the thresholds for electron tunnelling. Compared to the double junction, the threshold lines for entering of an electron are shifted to more positive gate voltage, whereas the thresholds for leaving of an electron are shifted to more negative gate voltage (Figure 12b). Due to the overlap of the stable charge regions, the bias voltage can alternate between the two without entering a region of uncontrolled charge flow.

The principle can also be explained in terms of threshold voltages. If the gate capacitor C_g is chosen to equal $C/2$, all junctions have the same threshold voltage for tunnelling, $V_t = e/3C$. If a finite voltage V is applied, the gate voltage V_g can be chosen such that the threshold voltage is exceeded for the junctions in the left arm, but not for the junctions in the right arm. An elementary charge will then enter the central island. Once it has entered, it will compensate the gate voltage and the voltage on all junctions will be lower than the threshold. Therefore the elementary charge is trapped on the central island until bias conditions are changed. The charging energy also prohibits another charge from moving to the central island. Of course, making the gate voltage more positive (by another $2e/C$ in this example) will trap a second electron on the central island. The gate voltage is decreased to make the charge leave. The junctions on the right arm will first exceed the threshold voltage, so that the electron leaves through the right arm. Cyclically changing the bias conditions by applying an alternating voltage in addition to a d.c. voltage to the gate capacitor moves one electron per cycle through the array. Increasing the amplitude will move more electrons per cycle.

The first tunnelling event in the traversal of an arm may occur quasi-adiabatically. However, the next event takes place with energy gain. This means that the turnstile is intrinsically dissipative: tunnelling into or out of the central electrode will always create a hot electron. It also causes this turnstile device with superconducting electrodes to be useless for clocking Cooper pairs, because single Cooper pairs have no kinetic degrees of freedom. This is one of the most significant differences between the turnstile and the electron pump of the next subsection.

The physical layout can be very close to the circuit shown in Figure 12, and a SEM

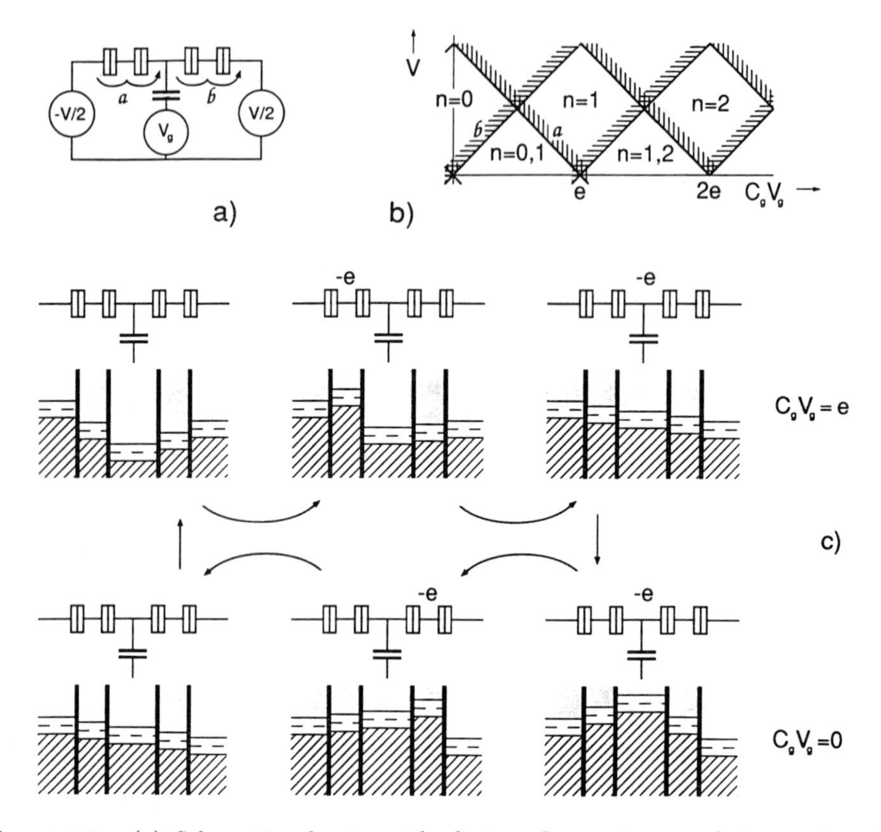

Figure 12: *(a) Schematic of a turnstile device. In event a an electron enters the central island, and in b it leaves through the other arm. (b) Charge stability diagram. The threshold lines for events a and b are indicated. (c) Operation for a block wave modulation;* $V = 0.4\,e/C$, $C_g = C/2$. *Top line, left to right:* $C_gV_g = e$, *an electron enters from the left. Bottom, right to left:* $C_gV_g = 0$, *an electron leaves through the right arm.*

photograph is given in Figure 13. The only refinement over the circuit of Figure 12 is the use of two small auxiliary gate capacitances (0.06 fF) to tune out offset polarisation due to background charges on the remaining two islands. We will present typical results for a device with tunnel junction capacitance about 0.38 fF and tunnel resistance 340 kΩ $[(R_tC)^{-1} \approx 8\,\text{GHz}]$. The gate capacitance C_g is 0.3 fF. The values of R_t and C were determined from the large scale $I(V)$ curve, and C_g was determined from the period of the current modulation by the gate voltage. This period ΔV_g yields the gate capacitance, just as for a double junction, as $C_g = e/\Delta V_g$.

Figure 14 shows $I(V)$ curves of the device, without a.c. gate voltage applied (dotted curve) and with a.c. gate voltage of frequencies 5 and 10 MHz. Without a.c. gate voltage, a large zero-current Coulomb gap is present. With a.c. gate voltage of frequency f, current plateaus develop at a current level $I = ef$ for bias voltage $V > 0$, and at $I = -ef$ for $V < 0$. The plateaus even extend to voltages outside the gap.

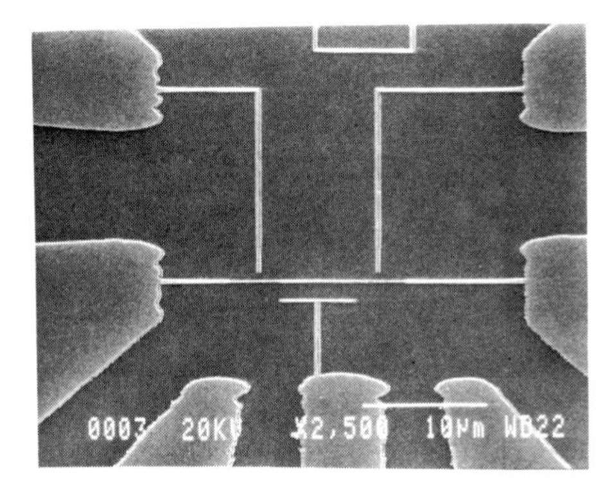

Figure 13: *Scanning electron micrograph of a four-junction turnstile. Figure 2 gives a magnification of one of the arms.*

Figure 15 shows the current versus d.c. gate voltage for different r.f. amplitudes. More than one electron per cycle can be transferred. The curves show plateaus at multiples of ef. Some curves may oscillate between nef and $(n+1)ef$; at other amplitudes the curves might be almost perfectly flat at a multiple of ef. Whether a given amplitude is sufficient for clocking of n or of $n+1$ electrons is dependent on the d.c. gate voltage, as shown in Figure 15b.

Figure 16 shows the dependence of the $I(V)$ curve on a.c. amplitude at a frequency of 5 MHz. Clearly, the height of the plateaus is not dependent on the a.c. amplitude, although the width is. The dependence of the $I(V)$ curves on the a.c. amplitude is very well simulated by numerical calculations based on Equations 1 and 2. A higher temperature (60–115 mK) than the thermometer indicated during the experiments (10–50 mK) was used to account roughly for noise or heating of the device.

Several processes will affect the accuracy of current quantisation in a turnstile device. Rewriting Equation (1) as

$$\Gamma = \frac{1}{R_t C} \frac{\Delta E / 2E_C}{[\exp(\Delta E / k_B T) - 1]}, \tag{10}$$

we see two main prerequisites for accurate electron transfer. The a.c. cycle should last long enough to let tunnelling to and from the central island happen with high probability, *i.e.* f must be much smaller than $(R_t C)^{-1}$ to avoid cycles being lost. On the other hand an electron trapped on the central electrode should have a negligible probability of escape by a thermally assisted transfer. At finite temperature there is a trade-off between the two requirements: a thermally assisted escape will be more probable for lower frequencies. Fortunately, both processes can be suppressed exponentially. The original paper (Geerligs *et al.* 1990a) gives a more quantitative analysis.

Looking at the potential landscapes in Figure 13, in the top row a co-tunnelling process could bring an electron to the central electrode through the right arm. In the bottom row, an electron could leave again through the left arm by a co-tunnelling event. Thus co-tunnelling events would decrease the current from the value $I = ef$. For larger bias voltages, co-tunnelling events actually increase the current. Due to these co-tunnelling events, we expect the accuracy of a four-junction turnstile with parameters

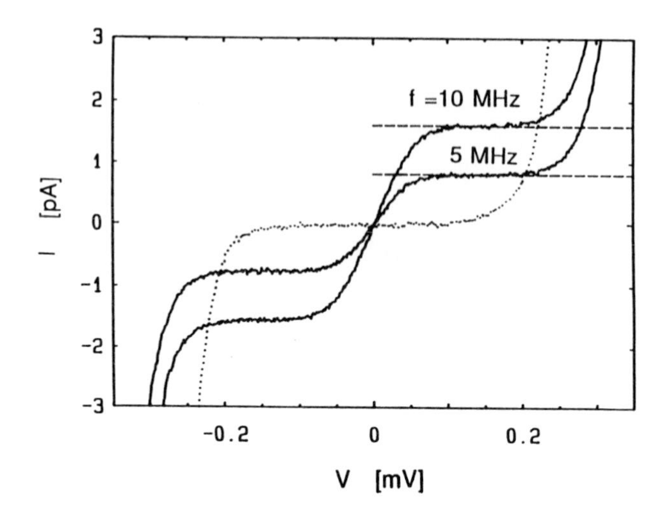

Figure 14: *Current-voltage characteristics of a four-junction turnstile without r.f. (dotted) and with gate voltage modulation at two different frequencies.*

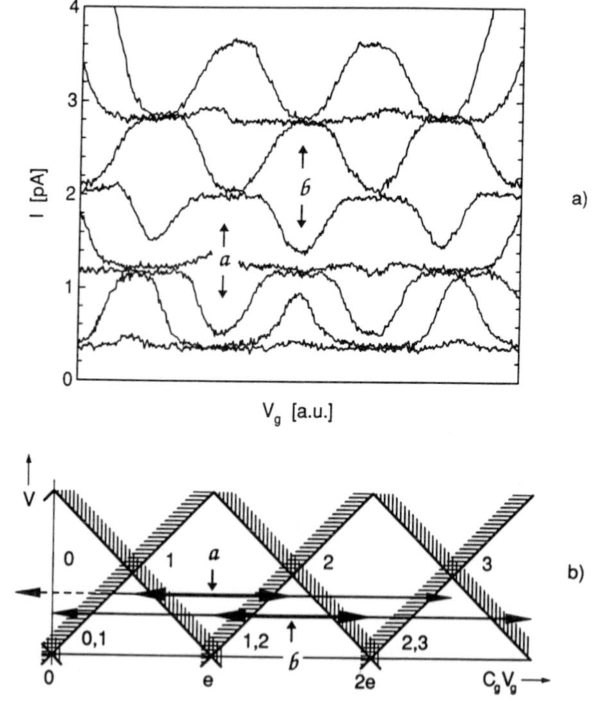

Figure 15: *The dependence of the current at* 0.13 mV *on d.c. gate voltage, for different r.f. gate voltage amplitudes, increasing from bottom (zero amplitude) to top. (b) shows how a small amplitude (thick horizontal arrows) can result in $I = 0$ or $I = ef$, and a larger amplitude (thin arrows) in $I = 2ef$ or $I = 3ef$, depending on the d.c. gate voltage.*

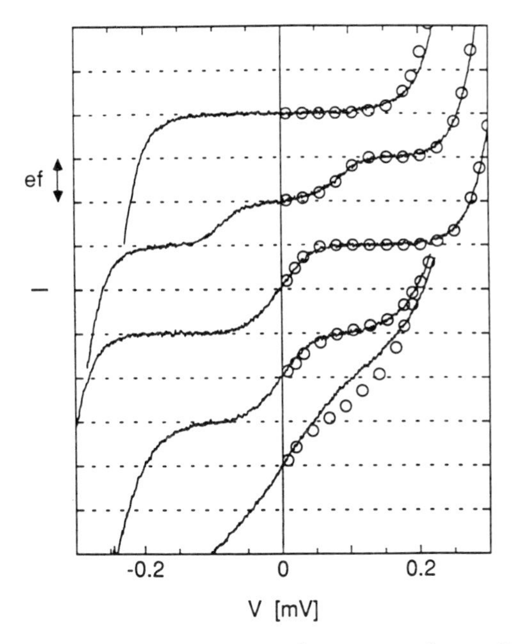

Figure 16: *The dependence of the current plateau on r.f. amplitude at* 5 MHz. *The open circles are numerical calculations with temperature as the fit parameter.*

as in this experiment to be not much better than 10^{-3}. However, a quantitative analysis has yet to be performed. It is worth noting that increasing the frequency will decrease the probability of a co-tunnelling event. If the number of missed regular tunnelling events does not increase significantly at the same time, the relative accuracy of the current quantisation will improve with frequency.

The present current level of turnstile devices is rather low, limited to the pA range by the accuracy requirements described above. For higher current levels one could think of placing more devices in parallel. With a turnstile device as described above, this suffers from the problem that every individual device needs tuned d.c. gate voltages for correct operation. However, Delsing *et al.* (1991) have operated a device which is a hybrid between the SET array and a turnstile device. In a long voltage biased array, the central ten or so junctions are modulated by one gate voltage. This device, which produces current plateaus of comparable quality, does not suffer as much from the need to tune the d.c. gate voltage. Since the gate voltage is applied to many islands, offset charges may be expected to average out.

4.4 Pump — Quasi-adiabatic charge transport

An alternative current source based on charging effects is the charge pump shown schematically in Figure 17 (Urbina *et al.* 1991, Pothier *et al.* 1991). It is a linear array of three tunnel junctions. The two islands between the three junctions are capacitively coupled to gate electrodes. If the system relaxes to the lowest possible energy state by one or more tunnelling events for given values of the bias voltage V and the two gate

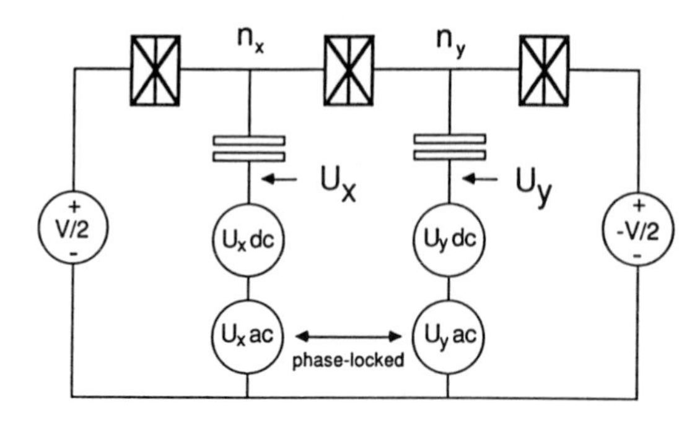

Figure 17: *Circuit of a pump of single charges.*

voltages U_x and U_y, one can induce an arbitrary number of excess charge carriers on either of the islands. This is in contrast to the turnstile. By adjusting the voltages one can cause a single electron to tunnel through any of the three junctions.

One possible cycle which pumps an electron through the device starts by raising the voltage on the left gate. When the voltage across the left-most junction exceeds the threshold voltage one electron can tunnel from the lead to the left island. If subsequently the voltage on the left gate is decreased while increasing the voltage on the right gate, an electron will tunnel to the right island when there is sufficient voltage drop across the middle junction. Finally the voltage on the right gate is decreased and an electron tunnels through the rightmost junction into the lead on the right side. If the amplitudes of the alternating gate voltages are within the proper range, the Coulomb blockade ensures that only one electron is transferred.

If a.c. voltages are applied to the gates, the cycle above described is repeated continuously and charge carriers are pumped one by one through the device. The result is a current proportional to the frequency and the unit of charge: $I = ef$ for electrons. This can be obtained within a certain bias voltage range around zero. The voltage at each gate should have the same frequency but be $\pi/2$ out of phase. Changing the phase difference by π reverses the direction of the current. A description of the operation in terms of a cycle through a stable charge configuration diagram has been given in the original papers.

Again the layout of actual devices is close to the schematic. However, more than for the turnstile, one has to take care to guard the gate electrodes well to avoid cross-capacitances.

Typical $I(V)$ characteristics in the normal state are presented in Figure 18 for frequencies of 10 and 15 MHz. Plateaus at positive and negative current are found for the appropriate choice of the phase difference. The dashed lines are at the expected current values, $I = \pm ef$. The plateaus are much more rounded than for the turnstile of the previous subsection. This is at least partly due to co-tunnelling (double and triple jumps). Because of the intended operation in the superconducting state (Section 5), the junction resistance was chosen lower (closer to the quantum resistance h/e^2) than is usual for single-electron devices.

An important difference between the pump and the turnstile is that, as far as co-

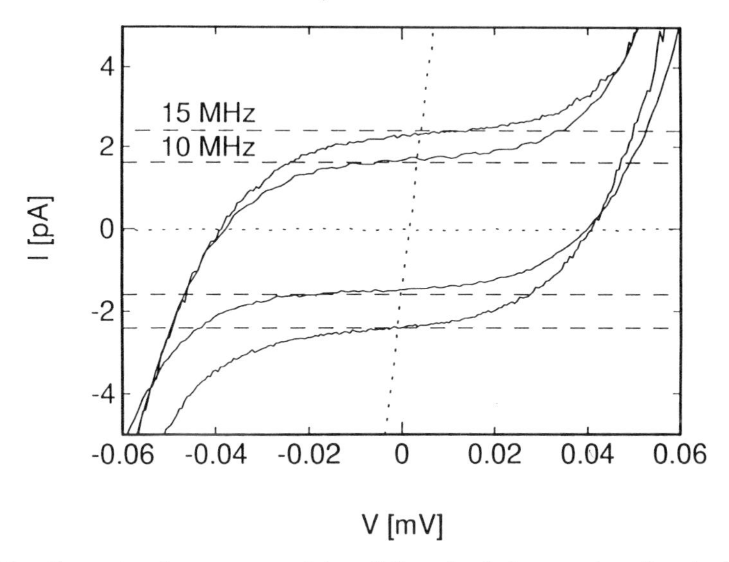

Figure 18: *Current-voltage curves at two different r.f. frequencies of a single-electron pump. Plateaus at $+ef$ or $-ef$ are obtained, depending on the phase difference between the two gate voltages.*

tunnelling is regarded, the pump becomes more accurate at lower frequencies, the turnstile at higher frequencies. During a cycle the pump remains quasi-adiabatically in the lowest energy charge configuration, whereas the turnstile is often in a state which is stable against single tunnelling but not against co-tunnelling.

5 Superconducting junctions

5.1 Cooper pair charging effects

For a superconducting tunnel junction, a tunnelling matrix element of magnitude $E_J/2$ couples charge states differing by one Cooper pair (Josephson 1962). The Josephson coupling energy E_J is of order 1 K for a 10 kΩ aluminium junction, and is proportional to $1/R_t$. The Cooper pairs all occupy the same quantum state and, unlike electrons, can not occupy states over a wide range in kinetic energy. This has some interesting consequences for the tunnelling of single charges in superconductors.

For a small superconducting junction incorporated in an arbitrary circuit, the energy change resulting from transfer of a Cooper pair can be calculated as for an electron, but with a double charge. If this energy change is small compared to E_J, the degenerate charge states differing by tunnelling of one Cooper pair are mixed, yielding new energy eigenvalues which are E_J apart. We will call this situation a resonance for Cooper pair tunnelling (CPT). *If the junction is swept slowly through the energy resonance, one Cooper pair will be transferred.* If $\Delta E < 0$ or if $\Delta E > 0$ no Cooper pair tunnelling occurs because the Cooper pair has no kinetic energy degrees of freedom.

In a double junction the requirement of energy resonance would prohibit d.c. charge transfer in almost all bias situations. For zero bias voltage d.c. current can flow because of co-tunnelling. The Coulomb energy of an excess Cooper pair on the central island, E_i, if large compared to E_J, decreases the coupling for CPT by a factor $E_J/2E_i$. Thus the maximum supercurrent is decreased by a factor of order E_J/E_C (Averin and Likharev 1991). Only if the charging energy E_i of the central state vanishes in the presence of an island charge e does the coupling become of order E_J. Thus, the supercurrent can be modulated by the gate voltage in a double junction.

Experimentally (Geerligs *et al.* 1990c), the supercurrent peaks regularly at all integer island charges, *i.e.* the gate voltage period e/C_g is only half that expected. Theoretically, only for odd island charges, $\pm e$, $\pm 3e$, *etc.* does the change in charging energy during Cooper pair tunnelling vanish, yielding a peak in supercurrent. However, all island charges larger than $e/2$ are energetically unstable against quasiparticle tunnelling (quasiparticles are the single electron excitations in a superconducting metal). The observation of a gate voltage period e/C_g, which is typical for experiments in the superconducting state, demonstrates that a few quasiparticles are still present even at the lowest temperatures and cause transitions between odd and even island charges. On *experimental* timescales quasiparticle tunnelling may therefore be important. In theory the resistance for quasiparticles R_{qp} should increase exponentially with decreasing temperature, approximately as

$$R_{\mathrm{qp}} = R_n \exp\left(\frac{-\Delta}{k_B T} + \frac{\Delta}{k_B T_c}\right), \tag{11}$$

where Δ is the BCS gap, T_c is the critical temperature of the superconducting electrodes ($\Delta \approx 1.7\, k_B T_c$) and R_n is the normal state resistance of the junction (Bol *et al.* 1985). The tunnelling rate is of the order of

$$\Gamma_{\mathrm{qp}} = (R_{\mathrm{qp}} C)^{-1} \tag{12}$$

At typical experimental temperatures (below 50 mK), R_{qp} should be so high according to Equation (10) that we would not expect quasiparticle tunnelling to occur on the experimental time scales. The reason for the survival of quasiparticle excitations at low temperature is not yet clear to us.

As in the normal state, the single current-biased junctions is a rather special case. Time-correlated tunnelling of single Cooper pairs could be interpreted as Bloch oscillations. Recently Kuzmin *et al.* (1991) found evidence for this time-correlated tunnelling. The reviews mentioned in Section 1.1 give ample attention to this subject.

5.2 Coexistence of SET and CPT

In an array of junctions, for non-zero bias voltage, it is generally impossible to realise Cooper pair tunnelling through all junctions, since in some of the events kinetic energy would be transferred to the tunnelling charge. There may still be tunnelling current due to a combination of Cooper pair tunnelling with quasiparticle tunnelling. This was pointed out by Fulton *et al.* (1989) and Averin and Aleshkin (1989).

The quasiparticle tunnelling occurs at the usual single electron tunnelling rate, Equation (2), corrected for the nonlinear junction resistance. If $-\Delta E$ is smaller than the

BCS sum-gap 2Δ, the rate is proportional to the sub-gap conductance R_{qp}^{-1}. If $-\Delta E$ is larger than the BCS sum-gap, quasiparticles can be excited and the rate is proportional to the normal state conductance; at zero temperature $\Gamma_{\text{SET}} = (-\Delta E + 2\Delta)/e^2 R_n$.

A combination of quasiparticle tunnelling and Cooper pair tunnelling was described by Fulton *et al.* (1989) for a double junction as the reason for a peak in current at the BCS sum-gap. Conduction starts by the tunnelling of a quasiparticle to the central metal island, which is followed by Cooper pair tunnelling across one junction. If the the bias voltage $V > 2\Delta/e + e/4C$, this is followed by two quasiparticle tunnelling events across the other junction with energy change so large that the rate is determined by the normal state resistance. Thus a significant current can develop. Although the excess electron on the central island is in an unstable situation, the tunnelling rate out of the island will be governed by the subgap resistance, and hence be very low, under the condition that $V < 2\Delta/e + 3e/4C$. As a result a current peak of width about $e/2C$ arises somewhat beyond the sum-gap.

For a bias between $2\Delta/e + e/4C$ and $2\Delta/e - e/4C$, this catalyzing action of a permanent average charge on the central electrode is less efficient (only one of the quasiparticles will tunnel at a high rate); for still lower voltages it is completely absent. In the latter case only a combination of Cooper pair tunnelling with slow quasiparticle tunnelling ($\Gamma \propto R_{\text{qp}}^{-1}$) would be possible, for which the rate is extremely low. The next subsection discusses mechanisms which increase tunnelling rates at these low voltages.

5.3 Co-tunnelling

For low but non-zero voltage there are two mechanisms which cause d.c. current flow. The energy provided by the voltage source can be dissipated by quasi-particle tunnelling or by dissipative leads. Cooper pair co-tunnelling in combination with a single quasiparticle tunnelling event can cause significant conduction for low voltages (Maassen van den Brink, Schön and Geerligs 1991). This Cooper pair co-tunnelling will yield appreciable current if it results in an intermediate state with energy high enough to excite a quasiparticle. For decreasing voltage, this requires events of increasingly high order. The result is a hierarchy of current peaks in the $I(V)$ curve. Note that peaks arise because co-tunnelling events require a resonance of inital and final energy, like single CPT. Combination of co-tunnelling events with excitations of the environment should likewise show a fine structure, reflecting the excitation spectrum. The combined $I(V)$ curve is very complicated for typical system parameters (Maassen van de Brink *et al.* 1991).

5.4 Pump for single Cooper pairs

The single electron pump as described in Section 4.4 transferred one electron per cycle of two a.c. voltages with a phase difference. The same device has been operated in the superconducting state to investigate whether such control is possible for the superconducting charge carriers (Geerligs *et al.* 1991). Since the pump has been made of aluminium tunnel junctions it is in the superconducting state at low temperature, unless a magnetic field is used to drive it into the normal state.

In the normal state pump, the tunnel junctions exceeded the threshold voltage for tunnelling one after the other, in a controlled way. An electron would tunnel if the threshold voltage was exceeded. In the superconducting state pump, a Cooper pair will tunnel precisely as the junction threshold voltage is passed, keeping the system adiabatically in the lowest energy charge configuration.

In the experiment presented here, the resistance of the junctions is $85\,\mathrm{k}\Omega$ in the normal state, yielding a Josephson coupling energy E_J of about $90\,\mathrm{mK}$. Consequently, the ratio $E_J/E_C = 0.03$ (with $E_J = h\Delta/8e^2R_n$, justified by results on larger aluminium junctions). For successful control of single Cooper pair tunnelling with charging effects, E_J should be much smaller than E_C. Also, one likes to have $E_J > k_BT$ in the experiment to minimise the influence of thermal fluctuations on Cooper pair tunnelling. The quoted values seem a good compromise.

$I(V)$ characteristics in the superconducting state are given in Figure 19 for frequencies between 2 and 20 MHz. Just as in the normal state, positive and negative current plateaus are obtained by appropriate choice of the phase difference between the two r.f. gate voltages. In the superconducting case there is a step in the plateau at $V \approx 0$, in contrast to the normal state in which the plateau goes smoothly through the I-axis. The dotted lines in Figure 19 indicate the expected values $I = \pm 2ef$ (for 2 to 10 MHz). For the lower frequencies the higher of the two levels in a plateau (in absolute value) is close to $2ef$. At high frequencies even the high level is significantly lower.

The step in the current plateau is basically the same effect as the rounding of the plateaus in the normal state due to co-tunnelling. The lower level occurs when the sign of the current is opposite to the bias voltage, *i.e.* when the modulation tries to pump Cooper pairs against a driving voltage. For the 'wrong' voltage sign, the source term in Equation (1) favors transport in the 'wrong' direction. Hence, the resonance for a co-tunnelling event, in which tunnelling in two junctions gives the same effect as a single tunnelling in the third junction, occurs first in the cycle. Since the co-tunnelling events transfer charge in opposite direction, this effect decreases the average current. At $V = 0$ the conditions for resonance coincide. This Cooper pair co-tunnelling probably occurs as an inelastic transition induced by the coupling to the electromagnetic environment (Geerligs *et al.* 1991). For the experimental parameters these are most likely thermally induced transitions.

Around the resonance conditions for Cooper pair tunnelling, the energy eigenstates of the pump are perturbed from the electrostatic value by the Josephson coupling. When driving the system through these regions, a Zener transition can occur with a probability dependent on the drive speed. In the event of a Zener transition, a Cooper pair tunnelling process is missed, and unless an inelastic transition occurs later in the cycle, the cycle will not contribute to the current.

The probability of Zener tunnelling of an undamped system with coupling E_J, driven by a rate of change of the unperturbed (*i.e.* electrostatic) energy dE/dt, is approximately

$$P_z = \exp\left[-\frac{E_J^2}{8\hbar\,(dE/dt)}\right]. \tag{13}$$

In the charge pump dE/dt depends on the r.f. frequency and amplitude, and generally will be different for the three transitions in a cycle. Since it appears exponentially in

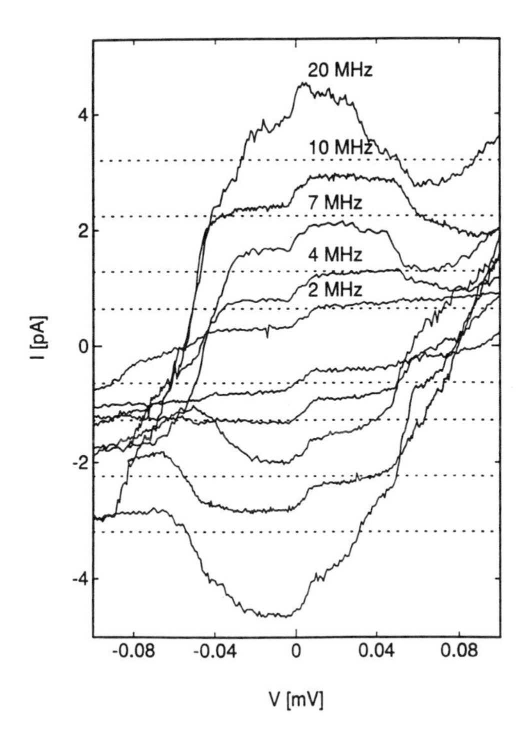

Figure 19: *Curent-voltage curves at several different r.f. frequencies of a single Cooper pair pump. Plateaus at about* $+2ef$ *or* $-2ef$ *are obtained, depending on the phase difference between the two gate voltages.*

Equation (12), one transition will likely dominate the total effect of Zener tunnelling. This should yield a current

$$I = 2ef(1 - P_z), \tag{14}$$

which is in good agreement with the observed frequency dependence of the higher plateau level.

During part of the gate voltage modulation cycle, the charge state of the system is unstable against quasiparticle tunnelling. Quasiparticle tunnelling events may change the system charge configuration in such a way that current is pumped in the wrong direction. The rates and effects of quasiparticle tunnelling will depend strongly on the precise bias parameters. The fact that in the experiment the pumped current was close to $2ef$, and the deviation seemed to be explained by Zener tunnelling, is somewhat puzzling.

5.5 Cooper pair turnstile

Finally, the analogue of a turnstile does exist for the superconducting state and will be discussed in this subsection. In a double junction in the normal state, an electron cannot be trapped on the central island if it can tunnel out with energy gain. In contrast, in a superconducting double junction, such an excited charged state is metastable if excess

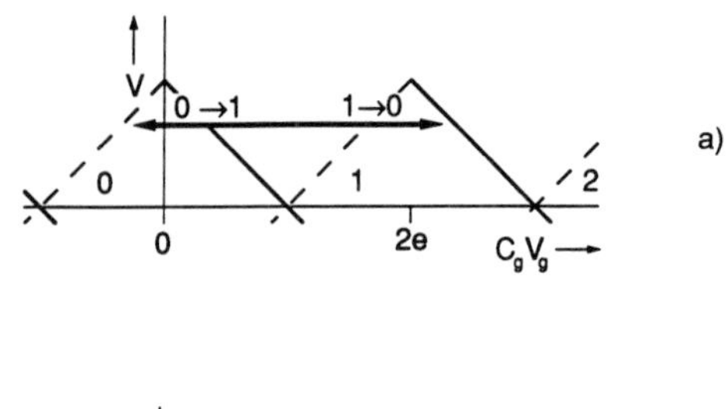

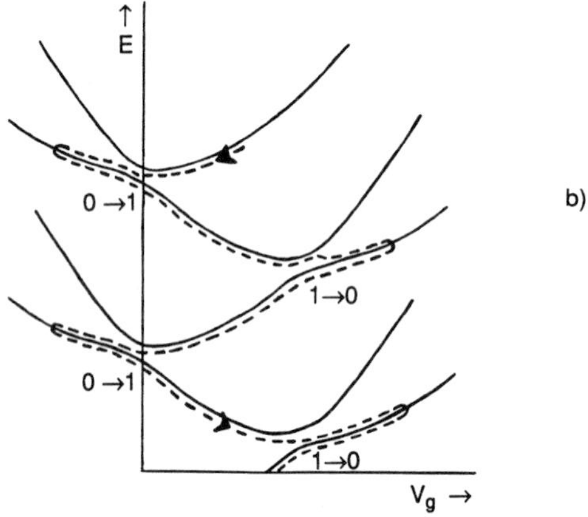

Figure 20: *(a) Resonance lines for* CPT *in a double junction. A Cooper pair can tunnel across the left junction on passing the solid lines, and across the right junction on passing the dashed lines. The double arrow is an example of a gate voltage modulation which will transport one Cooper pair per cycle. (b) Energy versus gate voltage, with a trajectory corresponding to the double arrow in (a).*

energy cannot be dissipated at a significant rate by quasiparticle tunnelling or the environment. This yields extra possibilities for the control of Cooper pair tunnelling.

The operation is based on the property that a Cooper pair will tunnel on sweeping through resonance only if this will result in lowering the electrostatic energy after resonance. More precisely, a downward transition across the energy gap opened by the Josephson coupling will occur when possible. This requires coupling to a dissipative environment.

In Figure 20a the resonance lines for CPT are plotted for a double junction, which yields of course a diagram very similar to the normal state. The indicated modulation should result in one Cooper pair being transferred per cycle in the direction of the applied voltage. Figure 20b gives the corresponding trajectory in energy versus gate voltage, clearly showing the necessity of controlled interband transitions. The d.c. gate voltage and the modulation amplitude are rather critical if stability against sporadic quasiparticle tunnelling is required. We will not consider this here.

Zorin (1991) has recently examined interband transitions in this system. He uses the conventional approximation of treating the dissipative environment for high frequencies as a transmission line with specific resistance ρ and capacitance c. The real part of the conductance at frequencies of order E_J/h is responsible for the interband transitions.

The total relaxation probability $P \approx 1 - \exp(-3\alpha B)$ where $\alpha = (4/R_K)(\rho/\omega_J c)^{1/2}$ and $B \approx E_J^2/hfE_C$. With B of order 10 this would require $\alpha \geq 0.1$ for a significant pumping action. This should be easily realised for typical ρ and c.

There are only preliminary results for this experiment (Geerligs 1990). These results indicate that a double junction can indeed work as a Cooper pair turnstile for quite low environmental impedances.

6 Semiconductor systems

6.1 Introduction

Charging effects in metal junctions and constrictions in a 2DEG are very similar. Effects like those in metal systems have also been observed in semiconductor structures. We will consider the differences which make semiconductors attractive for the study of charging effects, and mention some recent experimental and theoretical work.

The Fermi wavelength of the electrons in semiconductors is of the order of the limit of fabrication of artificial structures, 10–50 nm; in contrast, it is of the order of the lattice spacing in metals. Consequently, the coexistence of a discrete energy spectrum with the charging energy will be easier to study in semiconductors than in metal junctions. This coexistence expresses itself both in the $I(V)$ characteristics and in conductance oscillations as a function of gate voltage.

Even though the single particle level spacing Δ may be significant in STM-grain tunnelling experiments, it will be much smaller than the charging energy. Semiconductors offer the possibility to study the full range of $\Delta < E_C$ to $\Delta > E_C$. In semiconductors also the number of free electrons can be suppressed strongly, usually with the help of gate electrodes, to the point where only a few free electrons are present. The Coulomb island is usually named a quantum dot in this situation. Here deviations from a simple electrostatic energy following Equation (1) are expected. Equation (1) assumes capacitances independent of the number of electrons, whereas in semiconductors the potential profile (and thus the size) of the dot will strongly change during the charging of the dot with the first few free electrons.

6.2 $I(V)$ characteristics of quantum dots

Averin and Korotkov (1990) have calculated the changes in the $I(V)$ characteristics of Coulomb islands when the energy spectrum discreteness becomes significant. Their work was meant for metal systems with $\Delta \ll E_C$, but later work treated the general case (Averin, Korotkov and Likharev, 1991). Groshev (1990) has calculated the $I(V)$ curves of a quantum dot with one electron level, under the influence of charging energy.

In the usual calculations the tunnelling resistance is much larger than R_K, to avoid quantum fluctuations of the charge ($h\Gamma \ll E_C, \Delta$), but also so that the bare width $h\Gamma$ of the single electron levels is thermally broadened ($k_BT \gg h\Gamma$, which implies $k_BT/E_C \gg R_K/R$). In the latter limit, the levels can be given occupation numbers, and the tunnelling can be described by a master equation. Although tunnelling is still through a single resonant electron level, it cannot be distinguished from sequential tunnelling.

In the case of a discrete energy spectrum, the relevant free energy of a quantum dot becomes the sum of the electrostatic energy, Equation (1), and the sum of the single-particle energy levels $\sum_{p=1}^{N} E_p$. For a classical (*e.g.* metal) Coulomb island the electrostatic energy yielded the shift of the Fermi level and the electron energy spectrum due to voltages and charges. Likewise, in a quantum dot the electrostatic energy shifts the discrete energy spectrum of the island up or down with respect to the Fermi levels in the leads. Tunnelling rates to or from the island are usually calculated by the golden rule, *i.e.* neglecting the influence of the environment (as well as higher-order tunnelling).

The theoretical results for the $I(V)$ curve are not hard to understand with this in mind. If the charging energy is negligible and the Fermi level of the collector is below the bottom level of the quantum dot, there will be a one-to-one correspondence between $I(V)$ curve and energy spectrum of the quantum dot.

If the charging energy and average level separation are of comparable magnitude, the situation is somewhat more complicated. If the tunnel rate of the collector is large compared to that of the emitter (relatively thin collector barrier), there will be no charge accumulation in the quantum dot. Hence the $I(V)$ curve still shows only the discrete energy spectrum, be it offset by $e/2C_\Sigma$. In the opposite case of low collector tunnelling rate (relatively thick collector barrier), the quantum dot will have to be charged by a significant amount in order to obtain a significant (collector) current. Thus the discrete energy spectrum, or the periodicity of the resonances in the $I(V)$ curve, is expanded. The spin degeneracy of the discrete energy states is lifted, hence the periodicity is $\Delta + 2e/C_\Sigma$. Finally, for emitter and collector tunnelling rates about equal, a mixture of the two $I(V)$ curves is obtained. The smaller of the two quantities Δ or E_C gives a fine structure on the basic $I(V)$ curve for the larger one (Coulomb staircase or resonant tunnelling characteristic).

The calculations by Averin, Korotkov and Likharev assume thermal equilibrium in the quantum dot, *i.e.* that the electron relaxation rate τ_ε^{-1} is larger than the emitter and collector tunnelling rates. However, to avoid fine structure from being washed out, h/τ_ε should be small compared to Δ. A more detailed analysis (Averin and Korotkov, 1990) concluded that this is probably not fulfilled in very small metal particles.

6.3 Coulomb oscillations for a discrete energy spectrum

The differential conductance of the $I(V)$ curve of a quantum dot peaks where there is a resonance of emitter level with a single electron level in the dot. Similarly, by using a gate voltage, a single electron level in the dot can be brought between emitter and collector level even for very small applied bias voltage, *i.e.* in the linear response regime. The current results from alternating states with charge Ne and $(N-1)e$ (compare Figure 5). The resulting conductance peak, which is periodic in gate voltage, has received a rather large amount of attention in experiments and theory on semiconductor devices. The conductance oscillations are usually called Coulomb oscillations.

Experiments and theory are extensively presented in the recent review by Van Houten, Beenakker and Staring (1991) which also contains the theory of lineshape and temperature dependence of the oscillations.

The total free energy of a quantum dot [adding Equation (1) and the discrete energy levels] reduces in the linear response regime (vanishing V_l and V_r) to

$$F = \frac{Q_0'^2}{2C_\Sigma} + \sum_{p=1}^{N} E_p \,. \tag{15}$$

Using $Q_0' = Q_0 + C_g V_g$, the states of the quantum dot with charges $Q = -Ne$ and $Q = -(N-1)e$ will have equal free energy if

$$(N - \tfrac{1}{2})\frac{e^2}{C_\Sigma} + E_N = \frac{eC_g V_g}{C_\Sigma}. \tag{16}$$

This situation will result in a peak in conductance. The next conductance peak, where charges $-(N+1)e$ and $-Ne$ alternate, will be at a gate voltage V_g', related to V_g by

$$\frac{e}{C_\Sigma}\left[C_g(V_g' - V_g) - e\right] = E_{N+1} - E_N \,. \tag{17}$$

This confirms that the periodicity of Coulomb oscillations is e/C_g for systems with a nearly continuous energy spectrum. More interesting, the position of the conductance peaks versus gate voltage provides information about the energy spectrum if it is discrete. This was used by McEuen *et al.* (1991) in an experiment to measure the energy spectrum of a Coulomb island in the quantum Hall regime.

Acknowledgments

Many people have contributed to the work presented in these notes. I would like to especially thank V Anderegg, D Averin, U Geigenmüller, L Kouwenhoven, J Mooij, M Peters, G Schön and N van der Vaart in Delft, and M Devoret, D Esteve, H Pothier and C Urbina in Saclay. We have used facilities and research results of the Delft Institute for Microelectronics and Submicron Technology for fabrication of the devices. This work was supported by the Dutch Foundation for Fundamental Research on Matter. The notes were written during a NATO Science Fellowship, awarded by the Netherlands Organisation for Scientific Research, at the University of California, Berkeley.

References

Ambegaokar V, Eckern U, and Schön G, 1982. Quantum dynamics of tunnelling between superconductors, *Phys. Rev. Lett.* **48** 1745.

Averin D V, and Likharev V V, 1986. Coulomb blockade of single-electron tunnelling, and coherent oscillations in small tunnel junctions, *J. Low Temp. Phys.* **62** 345.

Averin D V, and Odintsov A A, 1989. Macroscopic quantum tunnelling of the electric charge, in small tunnel junctions, *Phys. Lett. A* **140** 251.

Averin D V, and Aleshkin, 1989. *Physica B* **165** & **166**.

Averin D V, and Schön G, 1990. Single electron effects in small tunnel junctions, to be published in the proceedings of the NATO Advanced Study Institute on Quantum Coherence in Mesoscopic Systems, (Les Arcs, April 1990).

Averin D V, and Nazarov Yu V, 1990. *Phys. Rev. Lett.* **65** 2446.

Averin D.V., Korotkov, A.N., and Nazarov, Yu.V., 1991. Transport of electron-hole pairs in arrays of small tunnel junctions, *Phys. Rev. Lett.* **66** 2818.

Averin D V, and K K Likharev, 1991. Mesoscopic Phenomena in Solids, chapter 6, B L Altshuler, P A Lee, and R A Webb (Elsevier, Amsterdam).

Bakhvalov N S, Kazacha G S, Likharev K K, and Serdyukova S I, 1989. Single-electron solitons in one-dimensional tunnel structures, *Zh. Eksp. Teor. Fiz.* **95** 1010 [*Sov. Phys. JETP* **68** 581].

Barner J B, and Ruggiero S T, 1987. Observation of the incremental charging of Ag particles by single electrons, *Phys. Rev. Lett* **59** 807.

Beenakker C W J, 1991. (preprint).

Ben-Jacob E, Mottola E, and Schön G, 1983. Quantum shot noise in tunnel junctions, *Phys. Rev. Lett.* **51** 2064.

Bol D W, Scheffer J J F, Giele W T, and Ouboter R de Bruyn, 1985. Thermal activation in the quantum regime and macroscopic quantum tunnelling in the thermal regime in a metabistable system consisting of a superconducting ring interrupted by a weak junction, *Physica B* **133** 196.

Brown R, and Simánek E, 1986. Transition to ohmic conduction in ultrasmall tunnel junctions, *Phys. Rev. B* **34** 2957.

Büttiker M, 1986. Quantum oscillations in ultrasmall normal loops and tunnel junctions, *Physica Scripta T* **14** 82.

Büttiker M, 1987. Zero-current persistent potential drop across small-capacitance Josephson junctions, *Phys. Rev. B* **36** 3548.

Cleland A N, Schmidt J M, and Clarke J, 1990. Charge fluctuations in Small-Capacitance junctions, *Phys. Rev. Lett.* **64** 1565.

Delsing P, Likharev K K, Kuzmin L S, and Claeson T, 1989a. Effect of high-frequency electrodynamic environment on the single-electron tunnelling in ultrasmall junctions, *Phys. Rev. Lett.* **63** 1180.

Delsing P, Likharev K K, Kuzmin L S, and Claeson T, 1989b. Time-correlated single- electron tunnelling in one-dimensional arrays of ultrasmall tunnel junctions, *Phys. Rev. Lett.* **63** 1861.

Delsing P, Haviland D B, Claeson T, Likharev K K, Korotkov A N, 1991, to be published in the proceedings of SQUID '91

Devoret M H, Esteve D, Grabert H, Ingold G-L, Pothier H, and Urbina C, 1990. Effect of the electromagnetic environment on the Coulomb blockade in ultrasmall tunnel junctions, *Phys. Rev. Lett.* **64** 1824.

Devoret M H, Martinis J M, and Grabert H (eds), 1991. Proceedings of the NATO ASI on single charge tunnelling, (Les Houches, France) (to be published as a special topics issue of *Z. Phys. B* and a regular volume in the series of Plenum Press).

Dolan G J, Dunsmuir J H, 1988, *Physica B* **152** 7.

Eckern U, Schön G, and Ambegaokar V, 1984. Quantum dynamics of a superconducting tunnel junction, *Phys. Rev. B* **30** 6419.

Esteve D, 1989, unpublished.

Fulton T A, and Dolan G J, 1987. Observation of single-electron charging effects in small tunnel junctions, *Phys. Rev. Lett.* **59** 109.

Fulton T A, Gammel P L, Bishop D J, Dunkleberger L N, and Dolan G J, 1989. Observation of combined Josephson and charging effects in small tunnel junction circuits, *Phys. Rev. Lett.* **63** 1307.

Geerligs L J, Anderegg V F, van der Jeugd C A, Romijn J, and Mooij J E, 1989. Influence of dissipation on the Coulomb blockade in small tunnel junctions, *Europhys. Lett.* **10** 79.

Geerligs L J, Anderegg V F, Holweg P A M, Mooij J E, Pothier H, Esteve D, Urbina C, and Devoret M H, 1990a. Frequency-locked turnstile device for single electrons, *Phys. Rev. Lett.* **64** 2691.
Geerligs L J, Averin D V, and Mooij J E, 1990b. Observation of macroscopic quantum tunnelling of the electric charge, submitted to *Phys. Rev. Lett.*
Geerligs L J, Anderegg V F, Romijn J, and Mooij J E, 1990c. *Phys. Rev. Lett.* **65** 377.
Geerligs L J, 1990. (thesis) Delft University of Technology.
Geerligs L J *et al.* , 1991. (to be published in *Z. Phys. B* special issue on charging effects.)
Geigenmüller U, and Nazarov Yu V, 1991. Coupling of tunnel junctions by quantum circuit modes (preprint).
Giaever I, and Zeller H R, 1968. Superconductivity of small tin particles measured by tunnelling, *Phys. Rev. Lett.* **20** 1504.
Gorter C J, 1951. A possible explanation of the increase of the electrical resistance of thin metal films at low temperatures and small field strengths, *Physica* **17** 777.
Grabert H, Ingold G-L, Grabert H, Devoret M H, Esteve D, Pothier H, and Urbina C, 1991,.*Z. Phys. B* **84** 143. Single electron tunnelling rates in multijunction circuits (preprint).
Josephson B D, 1962. Possible new effects in superconductive tunnelling, *Phys. Lett.* **1** 251.
Korotkov A N, Averin D V, Likharev K K, Vasenko S A, 1991, to be published in the proceedings of SQUID '91 (Springer, Berlin).
Kouwenhoven L P, Johnson, A T, Van der Vaart N C, Harmans C J P M, Foxon C T, 1991a, *Phys. Rev. Lett.*.**67** 1626
Kouwenhoven, L P, *et al.* , 1991b. Proceedings of the NATO ASI on Single Charge Tunnelling, Les House, France (to be published).
Kuzmin L S, Delsing P, Claeson T, and Likharev K K, 1989. Single-electron charging effects in one-dimensional arrays of ultrasmall tunnel junctions, *Phys. Rev. Lett.* **62** 2539.
Kuzmin L S, Haviland D B, 1991, proceedings of the Conference on Nanostructures and Mesoscopic Systems, May 1991, Santa Fe, to be published.
Lafarge, P, Pothier H, Williams E R, Esteve D, Urbina C, and Devoret M (to be published).
Lambe J, and Jaklevic R C, 1969. Charge-quantization studies using a tunnel capacitor, *Phys. Rev. Lett.* **22** 1371.
Likharev K K, 1987. Single-electron transistors: electrostatic analogs of the d.c. SQUIDS, *IEEE Trans. Magn.* **23** 1142.
Likharev K K, 1988. Correlated discrete transfer of single electrons in ultrasmall tunnel junctions, *IBM J. Res. Develop.* **32** 144.
Likharev K K, Bakhvalov N S, Kazacha G S, and Serdyukova S I, 1989. Single-electron tunnel junction array: an electrostatic analog of the Josephson transmission line, *IEEE Trans. Magn.* **25** 1436.
Maassen van den Brink A, Schön G, and Geerligs L J, 1991. (preprint submitted to *Phys. Rev. Lett.*.
Maassen van den Brink A, Odintsov A A, Bobbert P A, Schön G, 1991, to be published in *Z. Phys.* B, special topics issue on Single Charge Tunnelling
McEuen P, *et al.* 1991. *Phys. Rev. Lett.* **66** 1926.
Nazarov Yu V, 1989a. Coulomb blockade of tunnelling in isolated junctions, *Pis'ma Zh. Eskp. Teor. Fiz.* **49** 105 [*JETP Lett.* **49** 126].
Nazarov Yu V, 1989b. Anomalous current-voltage characteristics of tunnel junctions, *Zh. Eskp. Teor. Fiz.* **95** 975 [*Sov. Phys. JETP* **68** 561].
Nazarov Yu V, 1990. Influence of the electrodynamic environment on the electron tunnelling at finite traversal time (preprint).
Pothier H, *et al.* 1991. (to be published).

Schön G, and Zaikin A D, 1990. Quantum coherent effects, phase transitions, and the dissipative dynamics of ultra small tunnel junctions, *Physics Reports* **198** 237.

Urbina C, Pothier H, Lafarge P, Orfila P, Esteve D, Devoret M H, Geerligs L J, Anderegg V F, Holweg P A M, Mooij J E, 1991, *IEEE Trans. Magn.* **27** 2578.

Van Bentum P J M, van Kempen H, van de Leemput L E C, and Teunissen P A A, 1988a. Single-electron tunnelling observed with point-contact tunnel junctions, *Phys. Rev. Lett.* **60** 369.

Van Bentum, P J M, Smokers R T M, and van Kempen H, 1988b. Incremental charging of single small particles, *Phys. Rev. Lett.* **60** 2543.

Wellstood F, 1988. (thesis) University of California at Berkeley.

Wilkins R, Ben-Jacob E, and Jaklevic R C, 1989. Scanning-tunnelling-microscope observations of Coulomb blockade and oxide polarization in small metal droplets, *Phys. Rev. Lett.* **63** 801.

Williams, E, Martinis J, Devoret M, Esteve D, Urbina C, Pothier H, Larfarge P, Orfila P (to be published).

Zeller H R, and Giaever I, 1969. Tunnelling, zero-bias anomalies, and small superconductors, *Phys. Rev.* **181** 789.

Zorin A B, 1991. Controlled transfer of single Cooper pairs in the small-capacitance tunnel junction transistor, (to be published in *Proc. Third Int. Superconductive Electronics Conf. 1991*).

Zwerger W, 1991. preprint.

Spectroscopy of Semiconductor Nanostructures

C M Sotomayor Torres

University of Glasgow
Glasgow, U.K.

1 Introduction

Optical spectroscopy of semiconductors has been a lively subject for many decades. It has helped to shape the concepts of energy gaps, impurity states and resonant states among others. Of course spectroscopic techniques had been in use for some considerable time before semiconductors were developed, in the study of atomic spectra for example.

In spectroscopic terms, the size of the energy gap is the key difference between metals, insulators and semiconductors. In both intrinsic luminescence and near-gap absorption, the main features of the spectrum can be associated with the magnitude of the energy gap, since the photons reflect the energy spacing between the states taking part in the optical process. This will be discussed in more detail in Section 2.

The states involved in optical transitions must be allowed states, and hence an understanding of the density of allowed states (DOS) and its correlation with optical transitions is helpful. In the case of three-dimensional (3D) or bulk materials, the DOS follows an $E^{1/2}$ dependence on the energy of the carrier and the occupation of the energy levels is governed by Fermi-Dirac statistics. Lowering the dimensionality to 2D leads to a step-like DOS. In 1D the edges of the 2D DOS develop a spike with a tail decaying as $E^{-1/2}$ to high energies, and finally in zero dimensions (0D) the DOS is reduced to a set of δ-functions. See the chapter by Stern for a fuller discussion of the electronic states in low-dimensional semiconductors.

Not only must the states be allowed for a given optical process to occur, but they also need to be connected by selection rules. For example, the quantum numbers of the wavefunction of the electron and hole may have to be same, and for luminescence in 2D this would translate into the requirement $\Delta l = 0$ where l is the orbital angular momentum quantum number of the states.

The strength of optical transitions is usually described by the 'oscillator strength', which in 3D is dimensionless and of order unity. The concept of oscillator strength conveys the probability that the transition can occur, and is proportional to the number of **k**-states coupled to a given energy range. As discussed by Weisbuch and Vinter (1991), the oscillator strength per atom in 2D does not increase but the new form of the DOS arising from the 2D confinement results in a shared k_z for electrons and holes in the same subband. Having the same wave-vector k_z helps to concentrate the oscillator strength, when compared to 3D. Going to 0D, for the case where the dimensions of the dot are somewhat larger than the Bohr radius of the exciton, a giant oscillator strength develops, proportional to the ratio of the volume of the dot to the volume of the exciton. However, Weisbuch and Vinter caution us to consider the available data on weakly emitting dots and wires. These suggest that emission becomes weaker as the dimensions are reduced, contrary to expectations based on the enhancement of the oscillator strength predicted from a treatment of 'bottleneck' effects in energy and momentum relaxation. This is discussed in Section 5 on luminescence from free-standing nanostructures.

This chapter will concentrate on the spectroscopy of dry etched nanostructures, with special emphasis on the intensity of luminescence from wires and dots. It starts with a brief survey of optical techniques useful for the study of nanostructures (Section 2). Section 3 follows on the use of non-destructive techniques, such as Raman scattering and cathodoluminescence, for the study of damage induced by reactive ion etching. The regular geometry of fabricated nanostructures leads to new effects which are related to geometry rather than confinement, such as the observation of surface phonons (polaritons), discussed in Section 4. Studies of luminescence from free-standing GaAs-$Al_xGa_{1-x}As$ dots and wires are discussed in Sections 5 and 6, followed by Section 7 on the models currently available to determine the dependence of the efficiency of emission on the dimensions of nanostructures. Finally, Section 8 deals with recent studies on the effect of annealing and overgrowth on the emission characteristics of GaAs-$Al_xGa_{1-x}As$ dots and wires, since these are two necessary steps in likely fabrication sequences for potential devices. Confinement phenomena have been seen in dots and wires confined by other means, including direct growth and the use of stressors. A comprehensive survey of optical spectroscopy of wires and dots has been recently been published by Kash (1990).

2 Survey of some optical techniques

The optical characterisation of semiconductors has been reviewed several times. A recent publication edited by Stradling and Klipstein (1990) covers most techniques in a clear introductory manner. It includes luminescence, scanning electron microscopy and localised vibrational mode spectroscopy. A more advanced survey of techniques suitable for semiconductor quantum wells and superlattices is given by Fasol *et al.* (1989) which includes time-resolved spectroscopy, Raman scattering and the effects of a magnetic field.

2.1 Modulation techniques

The application of external perturbations such as electric fields and their impact on the absorption and emission spectra has been discussed in the chapter by Bastard. In general, a parameter which can be changed in a controllable manner and which affects at least one of the optical properties can be used to study the optical behaviour; with such a technique the background signal is subtracted from the processes under study. Pollak and Glembocki (1988) have given an excellent review of photoreflectance, electroreflectance, piezoreflectance, analysis of lineshape and interference effects in semiconductors and quantum wells. The information which can be obtained is usually directly related to the energy of the optically active electronic levels and the scattering mechanism involved. The application of modulation techniques to the study of nanostructures is in its infancy, but already some theoretical work is underway (Tang 1991). Although modulation techniques appear rather easy to use in principle, the analysis is non-trivial. In fact, the interpretation of photoreflectance from dot or wires arrays, for example, necessitates new models of the dielectric function of these nanostructured surfaces. This should not be a deterrent but encouragement to apply these modulation techniques to wires and dots.

2.2 Time-resolved spectroscopy

Time-resolved spectroscopy is used among other applications to measure the energy relaxation of quasiparticles such as excitons in quantum wells, to probe hot phonon distributions and the lifetime of electron-hole liquids and plasmas, and to measure trapping times of carriers by impurities or other crystalline defects. There are several techniques in use such as pump-and-probe, streak cameras, *etc.* A useful series of publications covering the applications of time-resolved spectroscopy to the study of 3D and 2D semiconductors has been edited by Alfano (1986, 1988, 1990). Techniques, experimental arrangements, tools for data analysis and results are covered.

2.3 Magneto-Raman and magneto-luminescence spectroscopy

In general, the application of large magnetic fields results in confinement of the electron and hole wavefunctions. Magnetic fields were use to obtain fingerprints of the 2DEG in the early days of low-dimensional semiconductor structures, confirming its existence by monitoring the positions of Landau levels as a function of the tilt angle between the magnetic field and the normal to the 2D layer. Similar oscillatory behaviour has been seen in magneto-photoluminescence (magneto-PL); in this case a Landau ladder appears associated with each occupied electronic subband. In undoped structures, magneto-PLE (photoluminescence excitation) has allowed the measurement of the binding energies of excitons in 2D and 1D semiconductors, as discussed in the chapter by Heitmann. Another parameter that can be obtained is the effective mass of the electrons, from the separation between Landau levels at a given field. A comprehensive coverage of the use of high magnetic fields can be found in the series on the proceedings of the conferences on applications of high magnetic fields in semiconductor physics. Magneto-Raman spectroscopy is also a very powerful technique; in doped structures it reveals

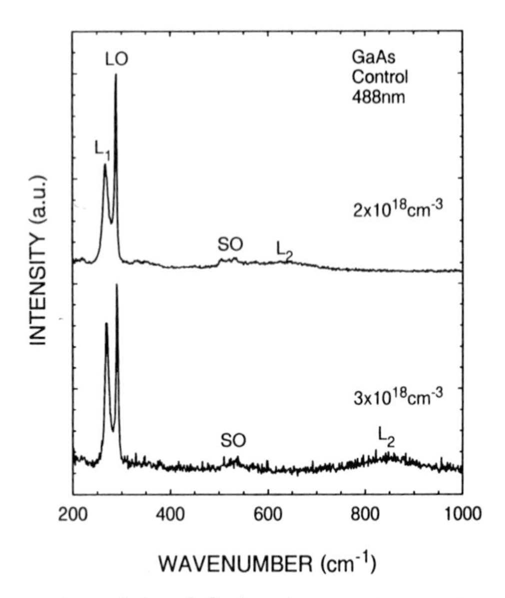

Figure 1: *Raman spectra of doped* GaAs *at room temperature using laser light of wavelength* 488 nm *for two carrier densities. The various peaks are discussed in the text and are identified as the low-frequency* LO *phonon-plasmon coupled mode L_1, the unscreened* LO *phonon, and L_2, the high-frequency phonon-plasmon coupled mode closer to the plasmon line (Wang et al. 1992).*

the spacing of energy levels, effective masses and allows probing of more exotic quasi-particles, such as the rotons recently demonstrated by Pinczuk *et al.* (1990).

3 Optical assessment of dry etching

The fabrication of nanostructures involves the removal of unwanted semiconductor material from specific areas defined by a pattern produced either holographically or with an electron beam. The nanostructures dealt with here are all fabricated using reactive ion etching (RIE) mostly with $SiCl_4$ and a few with CH_4-H_2. There is ample literature addressing the issue of RIE damage but some key questions remain to be answered. For a review the reader is referred to Wilkinson and Beaumont (1990). In this section the assessment of RIE damage with optical techniques comparable to those used to study optical transitions in dots and wires is described.

3.1 Raman scattering: LO phonon-plasmon coupled mode

Raman scattering has been used (Wang *et al.* 1992) to compare various etching methods, by monitoring the LO phonon-plasmon coupled mode in n^+-GaAs grown by molecular beam epitaxy (MBE). One form of damage known from conductivity measurements in n^+-GaAs wires is surface depletion enhanced by RIE. By monitoring the LO phonon-plasmon coupled mode for a given etching process, using light of different penetration

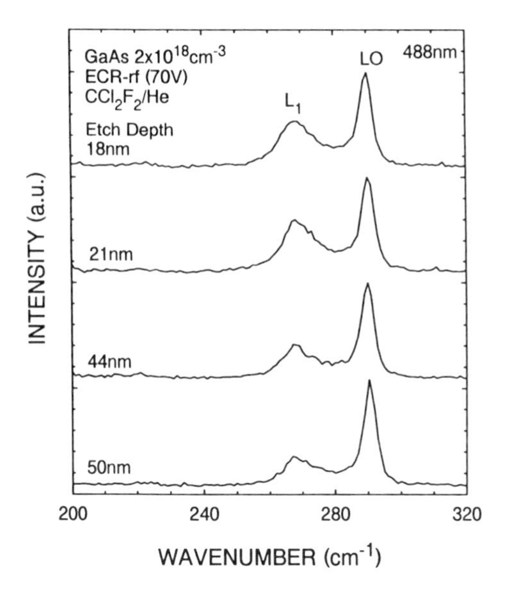

Figure 2: *Raman spectra at room temperature as a function of etch depth in* ECR *r.f. etching. The gas is* CCl_2F_2-He *(Wang et al. 1992).*

depths (different laser lines), a profile can be obtained of the carrier concentration as a function of depth in a non-destructive manner. Figure 1 shows the Raman spectra of unetched GaAs layers with different carrier concentrations and a given laser penetration depth, recorded at 300 K. The coupled LO phonon-plasmon mode L_1 can be seen at the low-frequency side of the unscreened LO phonon. L_2 is the higher-frequency LO phonon-plasmon branch, which is sufficiently close to the plasmon line to give the carrier density, and SO is a second-order Raman process. From the changes in the ratio of the integrated intensities of the LO and the L_1 lines, the change in the surface depletion region is extracted.

Samples were etched by Ne ion beam, wet etching, electron cyclotron resonance (ECR) in CCl_2F_2-He (see Figure 2), and RIE in CH_4-H_2. The depletion depth, obtained from the frequency of the plasmon line, is plotted as a function of etch depth for the various etching processes in Figure 3. The damage saturates after the removal of 20 nm of material for both CH_4-H_2 and ECR etching. These are the least damaging ionic processes, and are only surpassed by (isotropic) wet etching. At present, development work is in progress to optimise ECR etching to produce the high aspect ratios needed for nanostructures.

3.2 Phonon Raman scattering

In undoped GaAs, Raman scattering is also a useful technique in that severe RIE on a (100) surface manifests itself as a low-frequency broadening of the allowed LO phonon and the appearance of the TO phonon, forbidden in the normal back-scattering configuration. Watt *et al.* (1988a) studied phonon Raman scattering with various laser lines for flat surfaces etched with $SiCl_4$ at various powers, as shown in Figure 4. The exciting

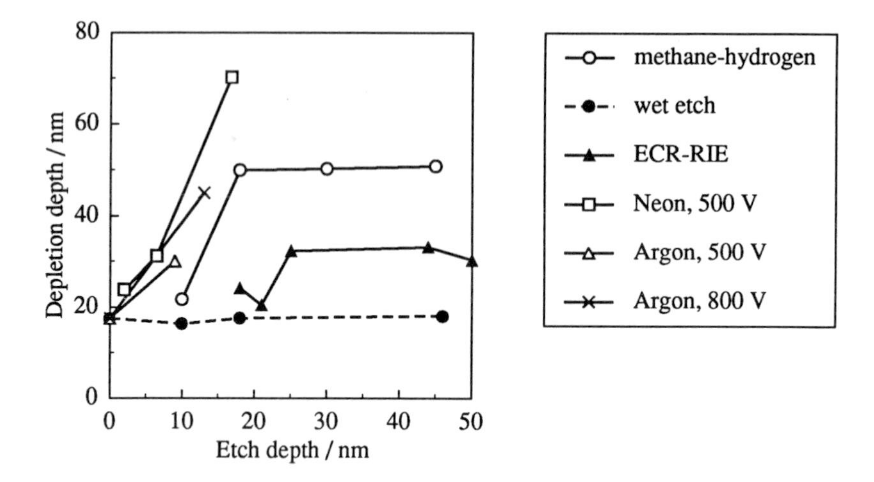

Figure 3: *Depletion depth calculated from the Raman scattering data as a function of etch depth, using various etching techniques. In both* ECR *and conventional* CH_4-H_2 *etching, saturation of damage is observed after a depth of about* 20 nm *has been etched (Wang et al. 1992).*

light was varied from 476.2 nm through 530.9 nm to 647.1 nm, changing the penetration depth from 70 to 130 to 290 nm. For the sample prepared using 100 W RIE power, the TO mode decreased in intensity from the 476.2 nm to the 530.9 nm spectrum and was not observable under 647.1 nm excitation. In the case of the 50 W sample, the TO phonon was not observed even with the 530.9 nm line. To confirm the selection rules, the sample was rotated about the axis parallel to the direction of the incident light, and the intensity of the LO mode was seen to follow a cosine dependence as expected. The intensity of the TO phonon remained unchanged with respect to rotation confirming that the damage is isotropic (see Figure 5). Concerning the low frequency LO phonon broadening, following the work of Tiong *et al.* (1984), the analysis suggests that the LO phonons exist in the crystal in undamaged regions of at least 25 nm in diameter.

3.3 Cathodoluminescence of etched nanostructures

It is common practice to examine the physical appearance of nanostructures using a scanning electron microscope (SEM). However, such a technique is not suitable for studying electrical or optical characteristics. Another non-destructive technique is PL, but it can hardly be used for homogeneous semiconductors since at low temperatures the excitons can migrate and the technique therefore only has a resolution in depth of around 1 μm in GaAs. In order to use luminescence for the assessment of damage, the layer has to have 'depth markers' such as quantum wells of different thickness so that their luminescence appears in different spectral regions. Samples from the wafers described in Section 5 have been used to assess damage. The multiple quantum well layer was etched to leave an array of dots or wires. Typically, the exciting laser beam was around 30–70 μm diameter which could be reduced to some 10 μm (with some

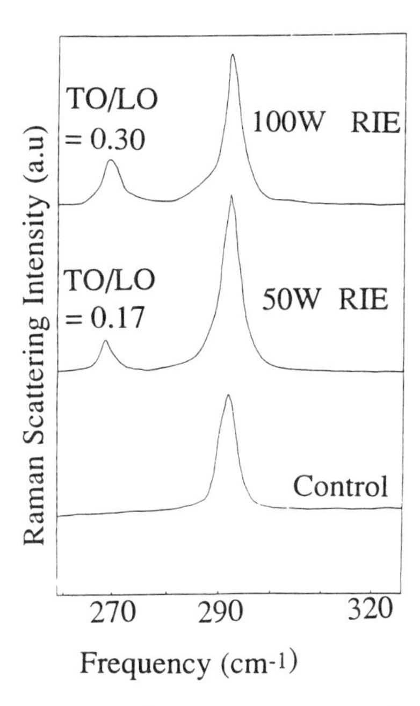

Figure 4: *Raman scattering spectra for two* RIE *powers and a control (unetched) sample made from undoped* GaAs *and measured using the* 482.5 nm *laser line. For both powers the symmetry-forbidden* TO *phonon is observed, as well as a low-frequency broadening of the* LO *phonon. The resolution is* 2.4 cm^{-1} *(Watt et al. 1988a).*

uncertainty in the measurement of the size of the beam). However, this means that at any time the luminescence signal was sampling between 15 000 and 20 000 60 nm dots. Thus standard PL only gives an average response.

The power of cathodoluminescence (CL) on the other hand is that it allows one to observe the physical structures from which the luminescence is obtained with a spatial resolution of approximately 1 μm or better. This comes about because the electrons which excite the electron-hole pair can be tightly focused. CL studies of poorly etched dots and of dots etched under optimised conditions from 1 μm to 60 nm diameter have been carried out (Williams *et al.* 1991, 1992). The scanning CL system used was based on a Cambridge Stereoscan 150 Mk 2 SEM with an electron source operated at 10 keV and giving a current of less than 45 nA in the beam with a diameter of approximately 200 nm. The CL spectra were recorded at 15 K.

The poorly etched sample had some areas (approximately 40 μm in extent) of unetched quantum well material in almost all the dot arrays. These areas were seen to luminesce very strongly in CL as a miniature mesa would. The sample etched under optimum conditions exhibited luminescence from dots down to 500 nm in diameter. Figures 6a and 6b show the CL at 15 K of the mesa and the 500 nm dots respectively. One of the most revealing aspects of these results is the ability to see monochromatic images of a given set of dots (in this case 15 dots), where it is clearly seen that the different wells luminesce with different strengths (see Figure 7), with the 6 nm well being the weakest. The intensities of the different quantum wells clearly reveal non-uniformities

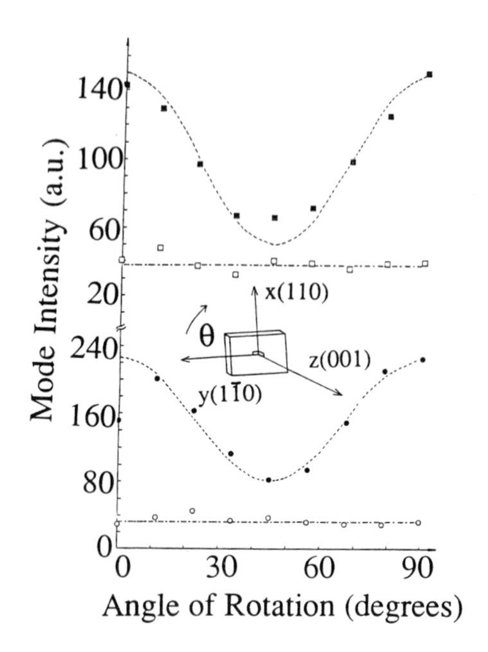

Figure 5: *Intensities of the* LO *and the* TO *phonon as a function of the angle of rotation as shown in the inset. The dashed lines are a sinusoidal fit to the* LO *points and the dot-dashed lines are a guide to the eye. Upper part:* 100 W RIE *sample with full squares denoting the* LO *and the open squares denoting the* TO *intensities. Lower part:* 50 W RIE *with full circles denoting the* LO *and open circles denoting the* TO *phonon intensities (Watt et al. 1988a).*

within each dot (cylinder) and from dot to dot. It is important to notice that the 15 dots were perfect in their appearance in SEM mode, and that the top right one appears optically inactive. The origin of the trend with depth is not clear. It may be associated with inhomogeneities in the material on the micrometre scale or with the distribution of non-radiative defects arising from RIE damage. It is clear that this method opens up the possibility of studying the depth profile of RIE damage in nanostructures in a manner so far unrivalled.

4 One-dimensional geometrical effects

4.1 Surface phonons

The fabrication of nanostructures enables optical scattering from faces other than the usual (100) surface to be studied. Effects of this type were clearly observed by Watt *et al.* (1990, 1988b) in arrays of undoped GaAs cylinders of 80 nm diameter and 310 nm height etched with $SiCl_4$ RIE, using phonon Raman scattering. They manifested themselves in the appearance of new phonons, interpreted as surface phonons (polaritons), between the LO and TO phonons. Their intensity increases as the angle of incidence of the

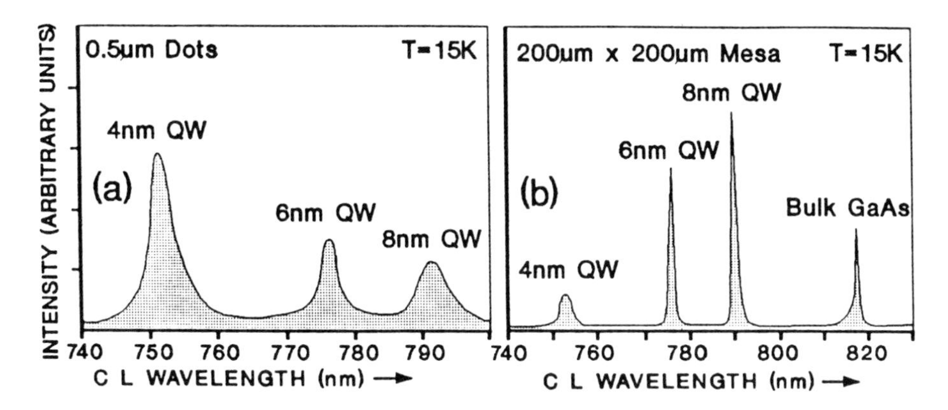

Figure 6: CL *at* 15 K *of (a) a control mesa* (200 × 200 μm²) *showing the three* GaAs-AlGaAs *quantum wells of thickness* 4 nm, 6 nm *and* 8 nm; *and (b) an array of* 500 nm *dots on the same sample (Williams et al. 1991).*

exciting laser with respect to the axis of the cylinder increases. In other words, they reflect directly the coupling to other wave-vectors. The Raman spectra are shown in Figure 8. The appearance of the TO phonon in etched flat surfaces and cylinders has already been documented by Kirillov *et al.* (1986), Watt *et al.* (1988b) and Semura *et al.* (1984) among others, and indicates the lack of smoothness of the unpatterned etched surfaces. In the spectra shown in Figure 8 the spectrum of the off-pattern area shows a negligible contribution from TO phonons. The spectra from the array of cylinders show clearly the TO phonon because this becomes allowed in first-order scattering when {110} surfaces are exposed. These surface phonons were modelled using an elastic continuum model of Ruppin and Englman (1970), which assumes that the material conserves its crystallinity over a small volume of well-defined geometry, with a well-defined dielectric constant inside the crystal and in the surrounding medium. The experimental frequencies of these new phonons are compared to the elastic continuum model in Figure 9. The implications of this observation are the following.

1. The cylinders of GaAs conserve their crystalline quality.

2. The cylinders are of sufficient uniformity to resemble a perfect semiconductor cylinder for phonons.

3. No evidence is found of an amorphous layer when the Raman signal is scanned over the frequency range of the phonon density of states. This is important in that recent and novel X-ray experiments on nanostructures (Tapfer *et al.* 1991) suggest the existence of a substantial volume of amorphous material. In dry etched wires of 150 nm width and approximately 300 nm in height, the crystalline core was reported to be only 60 nm wide (*i.e.* there is a thickness of some 45 nm on the outer part of the wire which is amorphous). However it is not clear that the dry etch process used was optimised to minimise damage.

4. These surface phonons probably participate in energy relaxation, as suggested by Stroscio *et al.* (1990) and Constantinou and Ridley (1990).

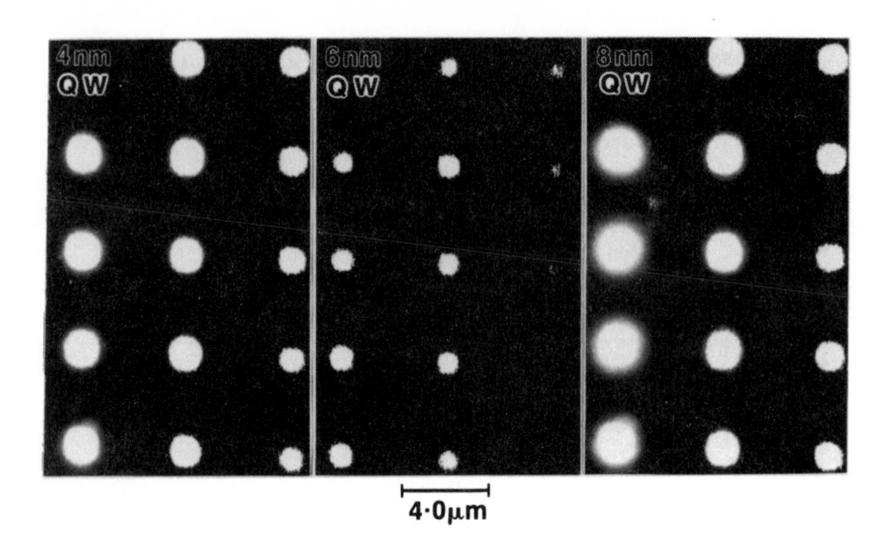

Figure 7: *Monochromatic images of free-standing quantum dots of* GaAs-AlGaAs *(sample B described in Section 5.1) showing the emission from the* 4 nm, 6 nm *and* 8 nm *wells from an array of 15 dots. The dot at the top right corner is not emitting light, although seen in* SEM *mode (Sotomayor Torres et al. 1991).*

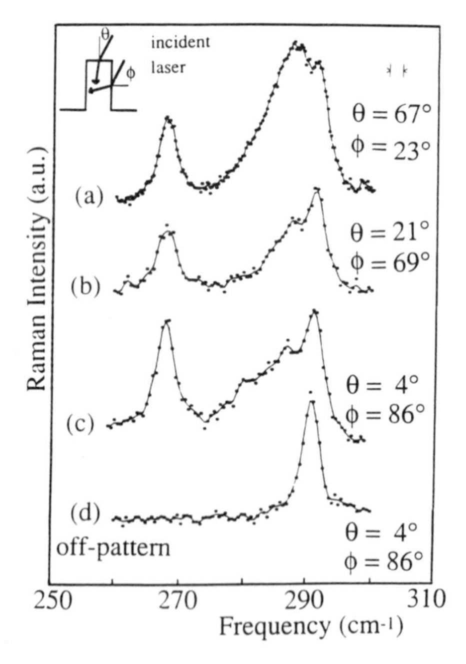

Figure 8: *Raman scattering from undoped* GaAs *cylinders of* 310 nm *height and* 80 nm *diameter for various angles of incidence. The bottom spectrum was obtained with the laser spot off the patterned area (Watt et al. 1990).*

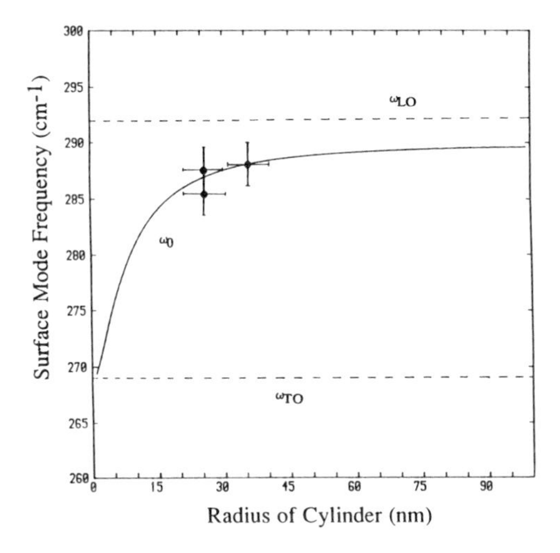

Figure 9: *Comparison of calculated frequencies of the surface mode with the frequencies observed as a function of the radius of the cylinder (Watt et al. 1990).*

4.2 Grating coupling effects

Patterned and etched samples, in which the exciting light wavelength becomes comparable to the dimensions of the structure, can act as grating couplers as discussed by Heitmann in this volume. The case of wires has been treated recently by Bockelmann (1991).

The dependence of the intensity of emission of an array of dots on the angle between the incoming light and the normal to the sample has been studied using 488 nm light impinging on an array of 300 nm diameter dots. It was found that the efficiency of luminescence depends strongly on this angle, yielding maximum intensity at around 45°. This form of grating-coupler effect is currently under discussion (Sotomayor Torres 1992).

5 Luminescence from free-standing dots

5.1 Luminescence from free-standing GaAs-AlGaAs dots

Many of the predicted properties resulting from a concentrated oscillator strength in semiconductors of one and zero dimensions (Hanamura 1988) have not yet been realised, partly due to technological barriers but most probably due to intrinsic phenomena limiting energy and momentum relaxation (Bockelmann and Bastard 1990, Izrael *et al.* 1990). Also, the prediction of Hanamura requires a coherent system for super-radiance, whereas monolayer fluctuations and the distribution of sizes are likely to invalidate this assumption in real systems. Therefore the present lack of experimental evidence is not surprising. In fact, what is usually observed is a decrease of the integrated intensity of luminescence as the dimensions of the structure decrease to sizes around 200 nm or, in some cases, even for larger structures. The rate of decrease is likely to reflect the ideality of the fabrication process employed, with the slowest drop in luminescence signalling the best route for fabrication.

A study of the intensity of emission as a function of the diameter of dots will now be described (Sotomayor Torres *et al.* 1991). Four samples were used, each of which contained a number of arrays of dots of different diameters. Three of these (samples A, B and C) were grown by solid source MBE using As_2 in a Varian Modular Gen II system, and consisted of three GaAs wells of 4 nm, 6 nm and 8 nm thickness separated by 20 nm of $Al_{0.30}Ga_{0.70}As$ with a superlattice buffer comprising 25 periods with 5 monolayers each of GaAs and AlAs. The fourth sample (D) was grown by MOCVD and had the same structure of three wells and a graded buffer layer of $Al_xGa_{1-x}As$ with 10%, 20% and 30% Al. Overgrowth of sample A was achieved by MOCVD with the deposition of 100 nm of $Al_{0.30}Ga_{0.70}As$ at 700°C. The emission intensity data of the overgrown dot are discussed in Section 7.

All samples were patterned by electron beam lithography using a modified JEOL 100CX II TEM/STEM operating at 100 keV. The masks were either negative resist (HRN) for samples A and D or SrF_2-AlF_3 for samples B and C. Dry etching in $SiCl_4$ completed the preparation of the samples. Dots of diameter 60 nm, 100 nm, 250 nm, 500 nm and 1000 nm were made in samples A, B, C and D covering areas up to $0.5 \times 0.5\,mm^2$ with a target density of 4%, *i.e.* a ration of dot diameter to interdot space of 1 : 4 for all sizes. The photoluminescence spectra at 5 K were obtained with He-Ne or argon lasers (maximum excitation power 30 mW) and either (a) a 1 m spectrometer and GaAs photomultiplier in current mode or (b) a 0.5 m spectrometer with Si detector and lock-in techniques.

Typical PL spectra from free-standing 60 nm dots and a control mesa in sample C are shown in Figure 10. There is negligible broadening of the line. Emission was obtained from all the quantum wells in all arrays of dots in sample C down to 60 nm diameter. In sample A emission from arrays of dots with diameter down to 250 nm was obtained, but not from all the quantum wells. In sample D emission was obtained from all arrays of dots down to 60 nm diameter. A summary of the dependence of emission intensity on the diameter of the dot for samples C and D is given in Table 1, where the data are corrected to take the filling factor into account. The blank spaces indicate that the emission for that particular array was below the limit of detection. It can be seen that the dots emit with a strength within two orders of magnitude of that of the mesa. Details of the data and its analysis will be published elsewhere (Lootens *et al.* 1992, Wang *et al.* 1992). For all samples used, a comparison of the intensity of emission from mesas of $0.2\times0.2\,mm^2$ and $0.5\times0.5\,mm^2$ with that from the starting material under the same experimental conditions showed that the emission from the mesa is an order of magnitude weaker than from the as-grown un-processed material.

This study suggests that the efficiencies of luminescence from the dots lie within two orders of magnitude of those of the adjacent mesas down to the smallest structures of around 60 nm diameter.

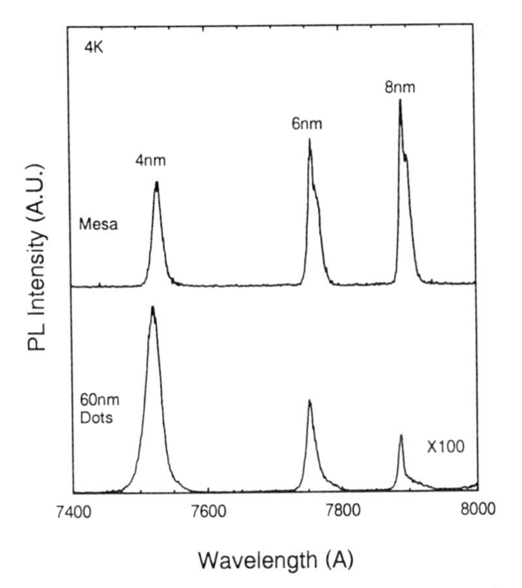

Figure 10: PL *spectra from sample C at* 5 K *under* 488 nm *laser light showing the three peaks associated with the* 4 nm, 6 nm *and* 8 nm *quantum wells. The top spectrum corresponds to a mesa and the bottom one to dots of* 60 nm *diameter (Sotomayor Torres et al. 1991).*

5.2 Luminescence from GaInAs-InP dots

GaInAs-InP is an interesting material system for nanostructures for at least two reasons: (a) the emission energies from GaInAs quantum wells fall in the range from about 1.0 to 1.5 μm; and (b) GaInAs and InP have no Fermi level pinning at the midgap level, and thus have a much slower surface recombination velocity than GaAs.

Dots have been fabricated in a single GaInAs-InP quantum well grown by MOCVD and lattice-matched to InP. The dots were fabricated using electron beam lithography and RIE in CH_4-H_2 and examined by PL at 5 K, excited with 2 mW of 633 nm light (MacLeod 1992). It was found that the mesas and dots luminesce detectably down to 100 nm in size. Emission from the 60 nm dots could not be observed with the excitation level used. Figure 11 shows the emission of a control mesa and of 100 nm dots. We also observed the high-energy tail reported by Patillon *et al.* (1990). We ascribe this tail tentatively to emission from higher bands, possibly including the light hole, in a mechanism akin to the 'bottlenecking' proposed by Reed *et al.* (1986), and suggest that a quasi-zero-dimensional effect sets in at 100 nm in GaInAs. Smaller dots in this material have been fabricated using wet etching by Notomi *et al.* (1991), yielding dimensions almost down to 10 nm. These appear to exhibit zero-dimensional quantisation.

Sample	Diameter of dot (nm)	4 nm well	6 nm well	8 nm well
C	1000	80	25	15
	500	22	4	1.7
	250	9	7	6
	100	2.7	1	6.5
D	500	113	90	
	250	24	22	28
	100	8	9	5
	60	5	5	7

Table 1: *Integrated* PL *at* 5 K *of dots as a percentage of the intensity from the control mesa (Sotomayor Torres et al. 1991).*

6 Luminescence from GaAs-AlGaAs wires

A few years ago the data in the literature concerning the dependence of the emission intensity on the size of nanostructures was markedly different for wires (Forchel *et al.* 1990) and dots (Arnot *et al.* 1989a), decreasing very rapidly with decreasing size by orders of magnitude for the former and very slowly for the latter.

Leitch *et al.* (1990, 1991) began a study of the strength of emission from wires of varying aspect ratio, which is now described. GaAs-$Al_xGa_{1-x}As$ multiple quantum wells with $x = 0.3$ were grown by solid source MBE using As_2 in a Varian Modular Gen II system. The material consisted of three GaAs wells of thickness 4 nm, 6 nm and 8 nm separated by 20 nm barriers, with a superlattice buffer of 25 periods (5 monolayers GaAs, 5 monolayers AlAs), and a 10 nm GaAs cap, grown on a semi-insulating substrate. The structures were defined by electron beam lithography, using a modified Philips PSEM 500 operating at 50 keV. Optical alignment marks were wet etched, and structures were exposed in 100 nm HRN resist to create patterned areas approximately 100 μm × 100 μm. Care was taken to avoid confusion associated with material trends across the wafer, and control mesas were placed close to the patterns. The packing densities of wires and dots were chosen to provide maximum coverage while still resolving the structures, resulting in a coverage of about 4%. Arrays of wires were made varying from 0.3 μm to 100 μm in length, and from 0.1 μm to 100 μm in width, thus permitting permutations of dimensions from 0.1 μm to 100 μm, enabling comparisons to be made simultaneously between dots, short wires, and long wires. The material was reactive ion etched in $SiCl_4$ in a Plasmatech RIE 80 system for one minute at a pressure of 10 mTorr with a r.f. power density of 0.65 W cm^{-2} and a d.c. self bias of 300 V. This resulted in an etch rate of 0.2 μm per minute.

Luminescence from the structures was characterised at 6 K by photoluminescence using an argon ion laser, a 1 m spectrometer and a GaAs photomultiplier in current mode. Low excitation powers were used to avoid exciting luminescence from areas away from the pattern under study. The laser beam was focussed onto the array using a 45° orientation, and an image of the array on the plane of the entrance slit allowed

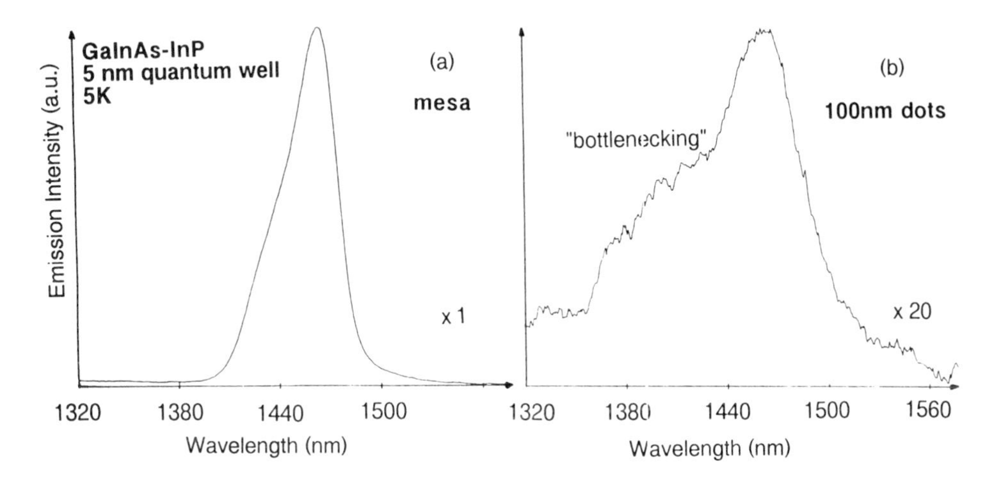

Figure 11: *Emission from (a) a mesa and (b)* 100 nm *dots in a* 5 nm *quantum well of* GaInAs-InP *(MacLeod et al. 1992).*

the signal to be maximised. The emission observed was only obtained from the patterns and mesas; no emission was observed from etched-back areas, and no stray emission from patterns other than those under observation was seen. This could be confirmed due to a slight gradient in the thickness of the wells across the wafer. The change in efficiency of the luminescence from the unpatterned starting material across the wafer was less than $\pm 50\%$ over 5 mm. Overall, the measurements of luminescence intensity were reproducible to within a factor of two; stray points resulted solely from imperfections in fabrication.

Emission from the structures was studied as a function of both the length and width of the wires (see Figures 12 and 13). It was found that emission from all the structures was within two orders of magnitude of that from the control mesas, which was itself of comparable intensity to that from the starting material. Considering the three quantum wells in all the arrays of wires examined, a distinct trend was noticeable in the emission from the 4 nm well (Figure 13), which fell off monotonically with reduced size of the structure. It was noticeable that the shorter structures gave more efficient luminescence than longer structures for a given width below 3 μm, as is discussed below. In the case of the 8 nm well the luminescence from the shortest wires showed an efficiency several times greater than that of the longer wires (0.3 μm to 3 μm) for a given wire width below 3 μm. Intermediate behaviour was seen in the 6 nm well. This behaviour was tentatively attributed to two factors. The damage to the surface is expected to be more significant for the narrow wells near the top of the structure due to their longer exposure to the etch gas. This reduces the emission, in line with current models of surface recombination and dead-layer effects. This behaviour is counteracted by an increase in the efficiency of the shorter wires for a given wire width, which is seen in the deeper wells.

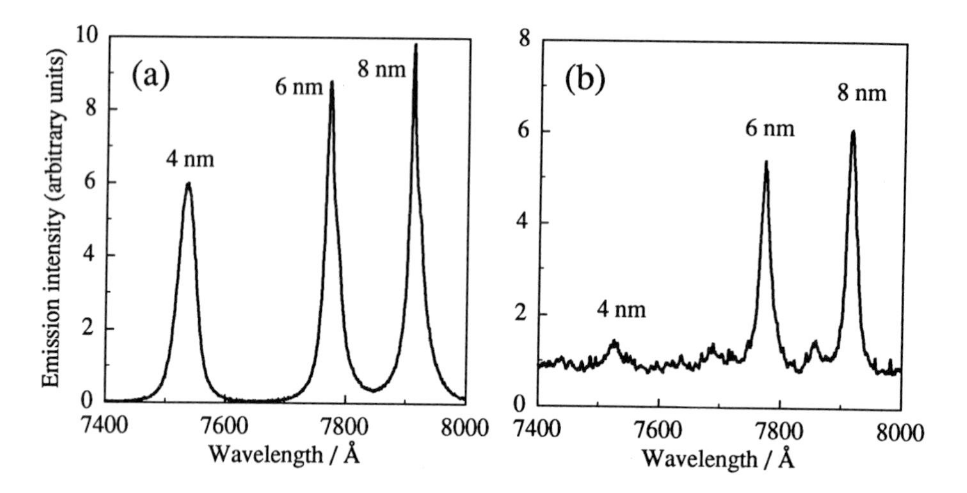

Figure 12: *Photoluminescence spectrum taken at* 6 K *showing the emission from the three quantum wells of (a) a* $100 \times 100\,\mu m^2$ *mesa, and (b)* $0.1 \times 0.3\,\mu m^2$ *wires (Leitch et al. 1991).*

7 Models governing the emission yield of nanostructures

7.1 'Extrinsic' — dead layer model

A model was described recently (Claussen *et al.* 1989) to illustrate the effects of recombination at the surface of etched dots. The diffusion equation was solved in cylindrical coordinates with recombination at the surface imposed as a boundary condition. Expanding the modified Bessel functions and neglecting powers of kR greater than two gives

$$\frac{N}{\pi c \tau R^2} = 1 - \frac{\left(1 + \frac{k^2R^2}{8}\right)}{\left[\left(1 + \frac{k^2R^2}{4}\right) - \frac{R}{2S\tau}\left(1 + \frac{k^2R^2}{8}\right)\right]},$$

where N is the total population of carriers, R is the radius of the dot, τ is the lifetime of the carriers, S is the surface recombination velocity, and k is the reciprocal of the mean free path l. Varying R from 0 to 500 nm, with $l = 500$ nm and $\tau = 10^{-9}$ s shows that the efficiency is insensitive to the value of S in the range from 10^6 to 10^{12} cm s^{-1}. This model does not account for the real cut-off in the emission seen by Claussen *et al.* (1989), and a 'dead layer' was invoked to account for the difference between the geometrical size of their structures and the cut-off predicted by the model.

A model invoking a similar concept of a 'dead layer' has been applied by Forchel *et al.* (1988) to the dependence of the intensity of emission on the width of the wire, with either the surface recombination velocity or the width of the optical dead layer as a fitting parameter.

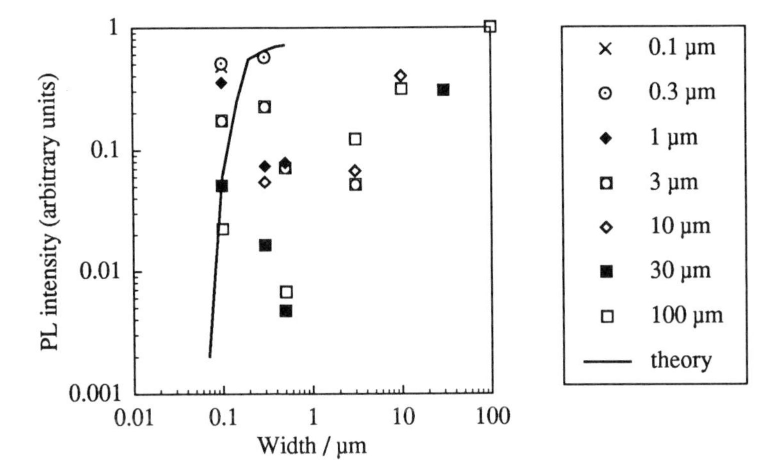

Figure 13: *Intensity of photoluminescence at* 6 K*, corrected for area coverage and normalised to the intensity of the mesa, as a function of the width of the wire for the* 8 nm *well. The solid line is the prediction of the theory of Weisbuch et al. (1991a) discussed in Section 7.2 (Sotomayor Torres et al. 1992).*

7.2 'Intrinsic' — slower relaxation model

Benisty *et al.* (1991) and Weisbuch *et al.* (1991a) put forward a new model which suggests that poor radiative efficiency in nanostructures can be attributed to an intrinsic effect, in which the relaxation of carriers from higher levels is inhibited due to quantisation and the orthogonality of states. This is supported by reference to magnetic confinement in quantum well lasers (Berendschot *et al.* 1989). This effect would reveal itself in a reduced efficiency of photoluminescence when exciting with light of energy well above the band gap. The model assumes that the hole thermalises to the bottom of the valence band in a time short compared to non-radiative lifetimes after an electron-hole pair has been created, whereas the electron undergoes a slowing-down process in energy and momentum relaxation because the inter-subband energies do not match that of an LO phonon. In addition, the scattering of electrons by acoustic phonons has been calculated to decrease substantially for wires narrower than 150 nm, compared to 2D confined carriers. The model takes into account irregularities in the size of the structures, modifications to the transition rates due to LA phonon scattering between all occupied levels, and the probability of the levels being occupied. It also assumes a typical non-radiative lifetime against recombination of 0.1 ns. The consequence is that conservation rules for energy and momentum are not satisfied simultaneously, since the non-radiative lifetime is shorter than the relaxation time for electrons. In Figure 13, the predicted intensity is compared with experimental data for the 8 nm well in wires of decreasing width. It can be seen that the model agrees well with the experimental points for wires whose width is around 100 nm. In addition, the model predicts that the emission intensity of a wire depends on its length L as $L^{-1/2}$ for a constant width. This trend can be clearly seen in Figure 14 for the 4 nm and 8 nm quantum wells in the 100 nm wide wires.

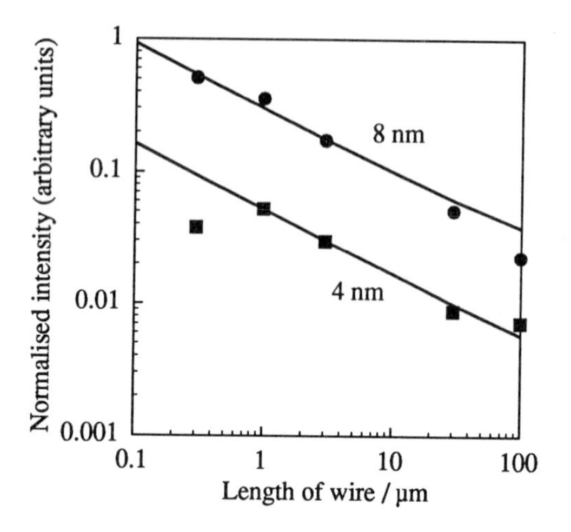

Figure 14: *Intensity of photoluminescence at 6 K, corrected for area coverage and normalised to the intensity of the mesa, as a function of the length L of wires of width* 0.1 μm. *The squares and the circles are the intensities from the* 4 nm *and the* 8 nm *quantum well, respectively. The solid lines are a fit to* $L^{-1/2}$ *(Leitch et al. 1991).*

7.3 Considerations of transport and trapping at defects

A model is currently being devised in which the efficiency of emission in the high-quality material of our structures is governed by the density of traps. The localisation of carrier due to interface roughness in the well is reduced in material with smooth interfaces. Thus the removal of 96% of the area of the well, together with a scaling of the remaining emission, could lead to an increase of the effective emission efficiency if non-radiative areas within a diffusion length of a trap were removed. This could even lead to an effective efficiency of emission above that of the mesa. The GaAs layers grown just prior to the series of MBE quantum well layers used in the work described in Section 5, exhibited record densities of free carrier and mobilities (Stanley *et al.* 1991). We expect our layers to have $4\times10^{13}\,\mathrm{cm}^{-3}$ free carriers, equivalent to a sheet density of ten ionised donors per μm^2. The implication is that there would be one ionised donor per wire in wires 1 μm long by 100 nm width. Thus there is a significant probability of obtaining wires containing no donors at all for these dimensions and smaller, and thus the possibility exists that this might give rise to more efficient emission for shorter wires and dots compared with the emission from longer wires of a given width.

8 Annealing and overgrowth of nanostructures

8.1 Annealed and overgrown GaAs-AlGaAs dots

After the pioneering work of Arnot *et al.* (1989b, 1989c) on the overgrowth of GaAs-$Al_xGa_{1-x}As$ dots it became clear that there was a need to separate the effect of annealing from that of overgrowth on the strength of emission from arrays of dots. The intensity of emission was seen to increase after overgrowth with respect to the free-standing dots in these early experiments, but a large blue shift suggested that serious intermixing of Ga and Al had occurred (Sotomayor Torres *et al.* 1990). A series of experiments on annealing and overgrowth was performed in order to understand the processes involved.

Sample A (described in Section 5.1) was examined after etching, after annealing and after overgrowth. The PL intensity was compared with that of the control mesas. The data are summarised in Table 2. The following four observations can be made.

1. The intensity of emission increases after annealing with respect to the as-etched levels for all sizes of dot and for all quantum wells in the dots (without significant linewidth broadening).

2. The intensity of emission increases after overgrowth, for all dot sizes and all quantum wells in the dots, to levels higher than that from the control mesas, indeed approaching the intensity of the as-grown material.

3. After overgrowth there is a red shift in the emission from all the quantum wells in the dots and mesas, which is largest for the narrowest well in the smallest emitting dots.

4. After overgrowth a significant broadening of the quantum wells occurs in dots and mesas, being larger for the narrowest well in the smallest emitting dots.

These observations are confirmed by CL spectra (Williams *et al.* 1992) recorded from the same sample. The increase in the intensity of emission after annealing is associated with a reduction in non-radiative traps located at the surface which are thermally released or relaxed. The red shift and the broadening are possibly associated with strong local strain surrounding the dots, although this does not explain the red shift in the mesas.

The hypothesis of strain was explored with photoluminescence excitation. (PLE). Spectra recorded at 15 K from the same sample after overgrowth of the dot arrays and mesas revealed substantial red shifts and a decrease of the E_1-HH_1 and E_1-LH_2 energy splittings. Red shifts in this overgrown sample were also observed independently in cathodoluminescence, confirming the presence of strain.

The broadening may be partly associated with some enhancement of interface diffusion which sets in the course of the second annealing period during overgrowth.

The importance of these experimental results is that no matter what is done to a dot sample, through annealing and/or overgrowth, at best the intensity of PL recovers to the as-grown levels in the range of dots that we normally have in our samples (60–1000 nm diameter).

It may be of interest to point out that emission from narrow dry etched wires (15–45 nm) of GaInAs-InP becomes observable only after overgrowth, as recently reported by Marzin *et al.* (1991).

Condition	Diameter of dot (nm)	4 nm well	6 nm well	8 nm well
etched	500	1	0.2	0.1
	250	1		
	100			
	60			
annealed	500	84	25	17
	250	115	18	13
	100			
	60	28	9	6
overgrown	500	336	166	288
	250	936	486	738
	100			
	60			

Table 2: *Integrated* PL *intensity at* 5 K *of various arrays of dots in sample A after reactive ion etching, after annealing at the growth temperature, and after overgrowth (see text) as a percentage of the emission from the control mesa. Blank spaces denote emitted light below the level of detection (Sotomayor Torres et al. 1991).*

8.2 Annealed GaAs-AlGaAs wires

Annealing tests were performed on wires fabricated in a similar quantum well structure to that used for dots, but grown by MOCVD. The 200 nm wires were fabricated using holography and reactive ion etching under conditions comparable to those described in Section 5, together with control mesas for comparison. The annealing was carried out in the MOCVD growth chamber under the same conditions as growth, but with the trimethyl-gallium source off, at 720°C for 30 minutes. The luminescence spectra were recorded before and after each of the annealing steps. An increased emission after the first annealing but a decrease in emission efficiency after the second annealing were observed. The results are summarised in Table 3.

The studies of annealing show that the etched surface is unstable under annealing, and further time-dependent annealing studies are required to establish the behaviour of the etched surfaces.

Condition	4 nm well	6 nm well	8 nm well
after etching	4.7	3.9	6.6
after anneal 1	23.0	18.6	10.6
after anneal 2	7.0	4.0	2.7

Table 3: *Intensity of photoluminescence from holographic wires as a percentage of the intensity from a control mesa (Lootens et al. 1991).*

9 Conclusions

The optical studies of semiconductor nanostructures have encompassed a variety of types of nanostructure, of which the deep etched ones have been discussed here. Clearly the need to reduce the regular lateral dimensions to sizes smaller than or equal to the diameter of excitons is still valid to ensure 1D and 0D quantization and to explore the potential benefits in optical activity. The size regime of the structures presented here is close to the confinement of the centre of mass of excitons, as discussed in the chapter by Heitmann. On the other hand, the decrease of the emission intensity with lateral size arising from slower energy and momentum relaxation suggests that innovative designs are required to ensure a continuous supply of excitons to the dots or wires from adjacent areas of the structure. In other words, novel laser and optoelectronic device designs will be essential. As shown here, overgrowth and annealing are non-trivial technological steps for deep etched nanostructures. The poor inter-band emission properties provide a strong incentive to search for intra-band transitions and devices based on these. We suspect this area of activity will see a surge in the next few years. The signs also point towards dot-to-dot and wire-to-wire communication between excitons as a topic that will play an essential role in the future research on theory and experiments on optical transitions in nanostructures. The application of modulation spectroscopy to the study of nanostructures is wide open to enterprising experimentalists and theorists and we expect imminent advance from that field. The much anticipated improvement in optoelectronic devices based on non-linear optical properties may have to wait until more regular structures, tightly packed, with truly atomic-like spectra, become available; semiconductor doped glasses, dots embedded in a solid matrix and porous silicon are some of the present possibilities.

Phonons in nanostructures have not yet attracted the attention they deserve. It is expected that the LO and TO phonons will become hybrids and that new phonon modes will arise, such as the interface polaritons predicted by the exciting theoretical models recently put forward (Ridley 1992, Constantinou and Ridley 1992) and currently undergoing experimental test in various laboratories.

Acknowledgments

This work was done in collaboration with W E Leitch, P D Wang, D Lootens, H Benisty, M Watt, S Thoms, G M Williams, H Wallace, M A Foad, C R Stanley, P Van Daele, P Demeester, C Weisbuch, G Armelles, A G Cullis, R W MacLeod, S P Beaumont, C D W Wilkinson, and F Briones. We are grateful to the technical staff of the Nanoelectronics Research Centre for expert assistance. This work is supported by the U.K. Science and Engineering Research Council and by the European Community's ESPRIT BRA 3133 (NANSDEV). In addition, the author wishes to acknowledge the support of the Nuffield Foundation.

References

Alfano R R, 1986, 1988, 1990. *Ultrafast Laser Probe Phenomena in Bulk and Microstructure Semiconductors*, Parts I, II and III, *SPIE*, **793, 942, 1282.**

Arnot H E G, Andrews S R, and Beaumont S P, 1989a. *Microcircuit Engineering*, **9**, 365.

Arnot H E G, Sotomayor Torres C M, Cusco R, Watt M, Glew R, and Beaumont S P, 1989b. *Optical Society of America, Technical Digest Series*, **10**, 83.

Arnot H E G, Watt M, Sotomayor Torres C M, Glew R, Cusco R, Bates J, and Beaumont S P, 1989c. *Superlatt. Microstruct.*, **5**, 459.

Benisty H, Sotomayor Torres C M, and Weisbuch C, 1991. To appear in *Phys. Rev. B.*

Berendschot T J M, Reinen H A J M, and Bluyssen H J A, 1989. *Appl. Phys. Lett.*, **54**, 1827.

Bockelmann U, 1991. *Europhys. Lett.*, in press.

Bockelmann U, and Bastard G, 1990. *Phys. Rev. B*, **42**, 8947.

Claussen E Jr, Craighead H G, Worlock J M, Harbison J P, Schiavone L M, Florez L, and van der Gaag B P, 1989. *Appl. Phys. Lett.*, **55**, 1427.

Constantinou N C, and Ridley B K, 1990. *Phys. Rev. B*, **41**, 10622, 10627.

Constantinou N C, and Ridley B K, 1992. To appear in *Phys. Rev. B.*

Fasol G, Fasolino A, and Lugli P (eds), 1989. *Spectroscopy of Semiconductor Microstructures*, Plenum Press, New York.

Forchel A, Maile B E, Leier R, Mayer G, and Germann R, 1990. In *Science and Engineering of One and Zero-Dimensional Semiconductors*, Eds. Beaumont S P and Sotomayor Torres C M, Plenum Press, New York, p. 277.

Forchel A, Leier H, Maile B E, and R German, 1988. *Festkörperprobleme*, **28**, 99.

Hanamura E, 1988. *Phys. Rev. B*, **37**, 12273.

Izrael A, Sermage B, Marzin J Y, Ougazzaden, Azoulay R, Etrillard J, Thierry-Mieg V, and Henry L, 1990. *Appl. Phys. Lett.*, **56**, 830.

Kash K, 1990. *J. Luminescence*, **46**, 69.

Kirillov F, Cooper C B, and Powell R A, 1986. *J. Vac. Sci. Technol.*, **B4**, 1316.

Leitch W E, Sotomayor Torres C M, Kean A H, and Beaumont S P, 1990 *Mat. Res. Soc. Extended Abstracts*, **EA-26**, 127.

Leitch W E, Sotomayor Torres C M, Lootens D, Thoms S, Van Daele P, Stanley C R, Demeester P, and Beaumont S P, 1991. To appear in *Surf. Sci.*.

Lootens D *et al*, 1992. To be published.

MacLeod R W *et al*, 1992. In preparation.

Marzin X J Y, Izrael A, Birotheau L, Roy N, Azoulay R, Robein D, Benchimol J-L, Henry L, Thierry-Mieg V, Ladan F R, and Taylor L, 1991. In *Proc. 5th Int. Conf. on Modulated Semiconductor Structures*, Nara, Japan, to appear in *Surf. Sci.*.

Notomi M, Naganuma M, Nishida T, Tamamura T, Iwamura H, Nojima S, and Okamoto M, 1991. *Appl. Phys. Lett.*, **58**, 720.

Patillon J N, Jay C, Iost M, Gamonal R, Andre J P, Soucail B, Delalande C, and Voos M, 1990. *Superlatt. Microstruct.*, **8**, 335.

Pinczuk A, Valladares J P, Heiman D, Pfeiffer L N, and West K W, 1990. *Surf. Sci.*, **229**, 384.

Pollak F H, and Glembocki O J, 1988. *SPIE*, **946**, 2.

Reed M A, Bate R T Bradshaw, K, Duncanm W M, Frensley W R, Lee J W, and Shih H D, 1986. *J. Vac. Sci. Technol. B*, **4**, 358.

Ridley B K, 1992. To appear in *Phys. Rev. B.*

Ruppin R, and Englman R, 1970. *Rep. Prog. Phys.*, **33**, 149.

Semura S, Saitoh H, and Asakawa K, 1984. *Appl. Phys. Lett.*, **55**, 3131.

Sotomayor Torres C M, 1992. In preparation.

Sotomayor Torres C M, Watt M, Arnot H E G, Glew R, Leitch W E, Kean A H, Cusco Cornet R, Kerr T M, Thoms S, Beaumont S P, Johnson N P, and Stanley C R, 1990. In *Science and Engineering of One and Zero-Dimensional Semiconductors*, Eds Beaumont S P and Sotomayor Torres C M, Plenum Press, New York, p. 297.

Sotomayor Torres C M, Leitch W E, Lootens D, Wang P D, Williams G M, Thoms S, Wallace H, Van Daele P, Cullis A G, Stanley C R, Demeester P, and Beaumont S P, 1991. To appear in *Nanostructures and Mesoscopic Systems*, Eds Kirk W P and Reed M A, Boston, Academic Press.

Sotomayor Torres C M, Wand P D, Leitch W E, Benisty H, and Weisbuch C, 1992. To appear in *Optics of Excitons in Confined Systems*, Ed Quattropani A, Institute of Physics Publishing, London.

Stanley C R, Holland M H, and Kean A H, 1991. *Appl. Phys. Lett.*, **58**, 478.

Stradling R A, and Klipstein P C, 1990. *Growth and Characterisation of Semiconductors*, Adam Hilger, Bristol.

Stroscio M A, 1990. *Phys. Rev. B*, **40**, 6428.

Tang Y S, 1991. *J. Appl. Phys.*, **69**, 8298.

Tapfer L, La Rocca G C, Lage H, Cingolani R, Grambow P, Fischer A, Heitmann D, and Ploog K, 1991. In *Proc. 5th Int. Conf. on Modulated Semiconductor Structures*, Nara, Japan, to appear in *Surf. Sci.*.

Tiong K K, Amirtharaj P M, Pollack F H, and Aspnes D E, 1984. *Appl. Phys. Lett.*, **44**, 122.

Wang P D *et al*, 1992. In preparation.

Wang P D, Foad M A, Sotomayor Torres C M, Thoms S, Watt M, Cheung R, Wilkinson C D W, and Beaumont S P, 1992. To be published in *J. Appl. Phys.*.

Watt M, Sotomayor Torres C M, Arnot H E G, and Beaumont S P, 1990. *Semicond. Sci. Technol.*, **5**, 285.

Watt M, Sotomayor Torres C M, Cheung R, Wilkinson C D W, Arnot H E G and Beaumont S P, 1988a. *J. Modern Optics*, **88**, 365.

Watt M, Sotomayor Torres C M, Cheung R, Wilkinson C D W, Arnot H E G and Beaumont S P, 1988b. *Superlatt. Microstruct.*, **4**, 243.

Weisbuch C, Sotomayor Torres C M, and Benisty H, 1991a. In *Proc. Int. Symposium on Nanostructures and Mesoscopic Systems, Santa Fe, New Mexico*, Ed. W P Kirk, Academic Press, in press.

Weisbuch C, and Vinter B, 1991b. *Quantum Semiconductor Structures: Fundamentals and Applications*, Boston, Academic Press.

Wilkinson C D W, and Beaumont S P, 1990. In *Science and Engineering of One and Zero-Dimensional Semiconductors*, Eds. Beaumont S P and Sotomayor Torres C M, New York, Plenum Press, p. 11.

Williams G M, Cullis A G, Sotomayor Torres C M, Thoms S, Beaumont S P, Stanley C R, Lootens D, and Van Daele P, 1991. *Microscopy of Semiconductor Materials*, Inst. Phys. Conf. Ser. **117**, Ed. Cullis A G (September 1991).

Williams G M *et al*, 1992. In preparation.

Collective and Single-Particle Excitations in Low-Dimensional Systems

D Heitmann

Max-Planck-Institut
Stuttgart, Germany

1 Introduction

Sophisticated epitaxial growth and lateral microstucturing techniques have made it possible to realise low-dimensional electronic systems with quantum-confined energy structure, *i.e.* quantum wells, quantum wires and quantum dots. Among the most fundamental and interesting physical properties of these systems are their dynamic electronic excitations. They are interesting by themselves and also give access to the microscopic properties of the systems. I will give an overview, starting from volume plasmons and surface plasmon polaritons in free-electron metals and show that the dynamic excitation spectrum of quantum wells, wires and dots develops with decreasing dimensions into an interesting mixture of collective low-dimensional plasmon and single-particle inter-subband (-level) excitations.

Currently there is great interest and progress in the preparation and investigation of low-dimensional electronic systems (LDESs). 'Low-dimensional' means here that electrons are confined on a very small scale; thus quantum mechanics becomes important and the original free motion of the electrons is quantised into discrete levels. Two-dimensional electronic systems (2DESs) have been widely studied, notably in semiconductor heterostructures and quantum wells, and in Si-MOS (metal-oxide-semiconductor) systems (Ando *et al.* 1982). The great success of physics in layered 2D semiconductor structures with 1D quantum-confined energy states has challenged many scientists to prepare and study systems with further reduced dimensionality, specifically quantum wires (Berggren *et al.* 1986, Smith *et al.* 1987, Hansen *et al.* 1987, Alsmeier *et al.* 1988, Brinkop *et al.* 1988, Demel *et al.* 1988a, Demel *et al.* 1988b, Kern *et al.* 1990, Egeler *et al.* 1990, Demel *et al.* 1991) and quantum dots (Reed *et al.* 1988, Hansen *et al.* 1989,

Sikorski and Merkt 1989, Liu *et al.* 1989, Demel *et al.* 1990, Lorke *et al.* 1990). In these systems, the original free motion of the electrons in the lateral directions is also quantised due to an ultrafine lateral confinement. A potential acting in the x-direction creates a 'quantum wire' (the direction of the layer growth is labeled z in the following.) With a confining potential in both the x and y-directions, quantum dots, *i.e.* artificial 'atoms' with a totally discrete energy spectrum, are formed. A reversed structure with respect to dots are 'antidots' where 'holes' are 'punched' into a 2D electron system (Ensslin and Petroff 1990, Weiss *et al.* 1990, Lorke *et al.* 1991, Kern *et al.* 1991b). For recent reviews on LDESs see, for example, the Proceedings of the 6th International Winter School on Localisation and Confinement of Electrons in Semiconductors (1990), Demel *et al.* (1989), Merkt *et al.* (1989).

There has been an enormous amount of fundamental research in this field of LDESs which brought many new, sometimes unexpected and unique results. These LDESs are also very important for many realised and proposed applications. 1DESs and 0DESs test the limits of conventional devices if these are reduced in size for higher integration, and also offer the possibility to realise totally new device concepts. Moreover, we can tailor the microstructured semiconductor systems during their vertical growth or lateral patterning, or at will in a controlled way with additional electric and magnetic fields, and prepare beautiful textbook systems to learn and understand physics. In this spirit I will use these lecture notes to teach dynamic excitations in electronic systems. I will start from collective plasmon excitations in 3D free-electron metals and end with atom-like transitions in quantum dots. I will show that the dynamic excitations exhibit an interesting mixture of collective and single-particle excitations with decreasing dimensions. In this overview I want to discuss general ideas, so I cannot explain the experiments in full detail and will refer instead to the original literature and more specific reviews.

2 Volume and surface plasmons

The fundamental dynamic excitations of a bulk 3D free-electron metal are volume plasmons (Pines and Nozières 1966). One can visualise these excitations, as sketched in Figure 1a, by considering planes of charges oscillating against the positively charged homogeneous background from the ionised atoms of the crystal. The restoring electric force on these planes is, at equal average volume charge density N_v, independent of the distance between the planes. This leads to the unique situation that the eigenfrequency in a 3DES, ω_{vp}^2, is independent of the wave-vector q, namely $\omega_{\mathrm{vp}}^2 = N_v e^2/\epsilon_0 m^*$, where m^* is the effective mass from the band structure. (In higher orders, an additional non-local term $\frac{3}{5}(qv_F)^2$ must be added to ω_{vp}^2 due to the finite compressibility of the system, the so-called Fermi pressure; v_F is the Fermi velocity.) Volume plasmons are purely longitudinal excitations which have no component of magnetic field and thus do not couple to light (except in very tricky experiments when the spatial dispersion becomes important; see for example Frostmann and Gerhardts 1986). Thus volume plasmons are mostly investigated with electron energy loss spectroscopy. Typical plasma frequencies are about 4 eV for Na and Ka, about 15 eV for Al and less than 1 eV in n^+-doped semiconductors.

Plasma oscillations also occur at the boundary between a semi-infinite metal and a dielectric material as shown in Figure 1b (Raether 1977, Broadman 1982, Raether 1980, Ritchie 1957). In this case the amplitude of the electron density oscillation decays exponentially from the surface, at large q proportional to $\exp(-|qz|)$ (I call these excitations '$2\frac{1}{2}$D' plasmons). In the following q is the plasmon wave-vector in the x-y plane, defined by the boundary. In the adjacent dielectric the oscillations are accompanied by electromagnetic fields which decay exponentially from the interface (Figure 1b). This weakens the restoring force and leads to a reduced frequency of the surface plasmon resonance, $\omega_{\rm sp} = \omega_{\rm vp}/\sqrt{1+\epsilon_2}$ at large q ($q \gg \omega/c$ without the non-local term) for a free-electron gas that is covered by a dielectric material of dielectric constant ϵ_2. The electrodynamic forces (retardation) become increasingly important with decreasing q, leading to a decrease of the surface plasmon frequency to $\omega_{\rm sp} \approx qc$ for small q. In MOS and heterostructures it is possible to confine electrons in a very narrow channel of width $d_{\rm in}$ = 1–10 nm in the z-direction. Thus $d_{\rm in}$ is very small compared to $1/q$ for typical wave-vectors in usual experiments. In such a system plasma oscillations can be visualised as oscillations of charged wires in the x-y plane. In this case the restoring forces are inversely proportional to the distance of the wires which leads, without retardation effects, to a characteristic dependence of the 2D plasmon frequency on the square root of the wave-vector, $\omega_{\rm 2Dp} \propto \sqrt{q}$. We can derive this dispersion if we consider a thin metallic film of thickness d. The surface plasmons at the two boundaries couple if d becomes comparable to $1/q$ with decreasing thickness. This leads to a splitting of the surface plasmon dispersion into two branches, $\omega_{\rm sp}^+$ and $\omega_{\rm sp}^-$, as shown in Figure 1d. In the limit $qd \ll 1$, the $\omega_{\rm sp}^-$ mode developes into the 2D plasmon mode with a $\sqrt{q}$ dispersion (Ritchie 1957). Formally, the $\omega_{\rm sp}^+$ mode develops into the inter-subband transition of the 2DES.

The dispersion of plasmons is very often calculated within the random phase approximation, *i.e.* a global dielectric function $\epsilon(\omega, q)$ is calculated whose zeros determine the dispersion (Pines and Nozières 1966, Stern 1967). The plasmons appear then as quasiparticles—the quanta of the collective excitation—with energy $\hbar\omega_p$ and momentum $\hbar q$. However, there is also another electrodynamic approach which considers the same excitations as electrodynamic surface resonances. These calculations start from macroscopic models for the dielectric function and conductivities. The response, *i.e.* the reflected or transmitted electric and magnetic field amplitudes for plane waves, are determined using Maxwell's equations and boundary conditions. Since many experiments on surface plasmons and 2D plasmons are performed by optical spectroscopy, this approach is advantagous in explaining directly the observed quantities, such as retardation, oscillator and various coupling strengths. This electrodynamic approach is reviewed in detail in Heitmann (1987).

Surface plasmons and 2D plasmons are similar in the sense that they are both electrodynamic resonances which have electric and magnetic field components and thus interact with electromagnetic fields, in contrast to volume plasmons (thus more accurately they are called 'plasmon polaritons'). However, since the wave-vector q of these excitations is larger than ω/c, their fields decay exponentially from the boundary and do not couple directly with freely propagating waves. A freely propagating wave, incident with an angle θ with respect to the surface normal, has a component $k_x = (\omega/c)\sin\theta$ of the wave-vector parallel to the surface. Two different arrangements can be used to establish wave-vector or phase matching. In the prism arrangements (Otto geometry and

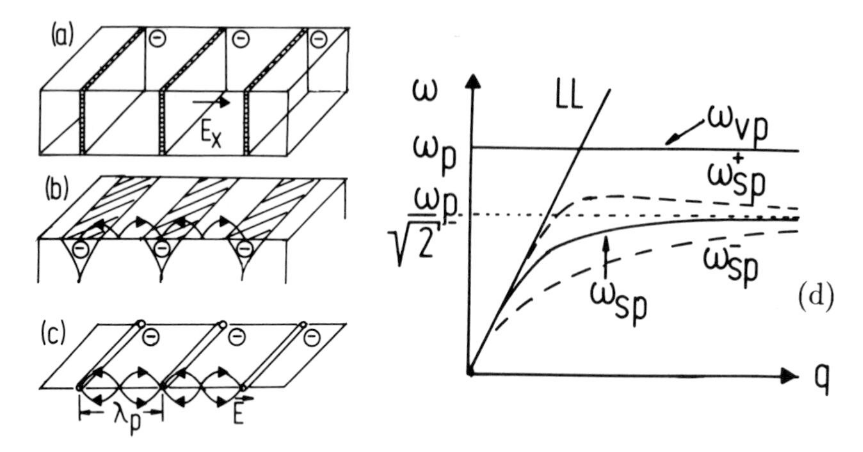

Figure 1: *Plasma oscillations in* 3D, $2\frac{1}{2}$D *and* 2D. *(a) Volume plasmons, plasma oscillations in the bulk of a metal, can be considered as planes of charges oscillating against the positively charged homogeneous background. (b) Surface plasmons, plasma oscillations at a boundary between a metal (the lower half-space) and a dielectric material. In this case the amplitude of the electron density oscillation decays exponentially from the surface. In the adjacent dielectric the oscillations are accompanied by electromagnetic fields (indicated by the field lines and arrows) which decay exponentially from the interface. (c) Plasma oscillations in a* 2DES *can be visualised as oscillations of charged wires in the x-y plane. They are also accompanied by exponentially decaying electromagnetic fields. (d) The dispersion of volume plasmons (ω_{vp}), surface plasmons at a semi-infinte metal (ω_{sp}) and in a thin metal film of thickness d with $q\omega_{\mathrm{sp}}/cd \approx 0.6$ (ω_{sp}^+ and ω_{sp}^-).* LL *is the light line, $\omega = cq$.*

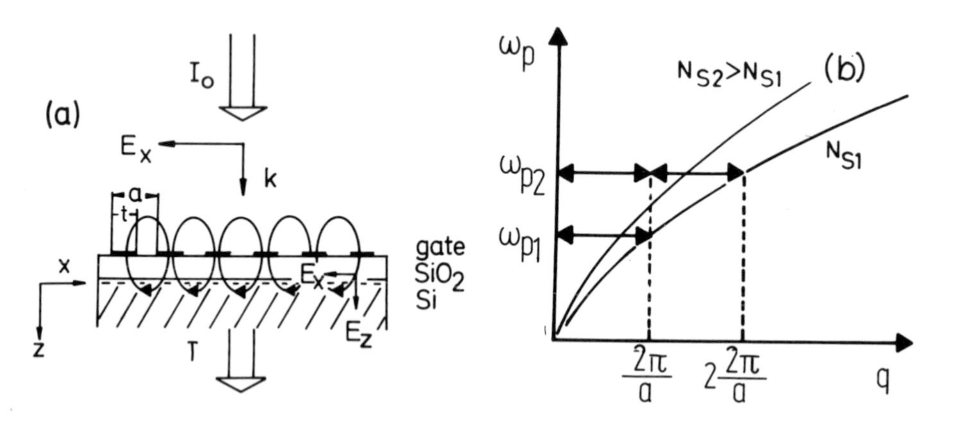

Figure 2: *(a) The action of a periodic structure in a* 2DES *as a grating coupler. Periodic metal stripes of width t short-circuit the E_x component of the incident electric field and induce spatially modulated parallel, $e_x(\omega, q)$, and perpendicular, $e_z(\omega, q)$, components of electric field in the near field. (b) Coupling of* FIR *radiation to 2D plasmons of wave-vectors $q_1 = 2\pi/a$ and $q_2 = 2(2\pi/a)$.*

Kretschmann-Raether geometry) the wave-vector is enhanced by the refractive index n of the prism, $k_x = n(\omega/c)\sin\theta$ (Raether 1977, Broadman 1982, Raether 1980). The radiation is coupled via an optical gap to the plasmons, and resonant coupling occurs if $k_{xp} \equiv n(\omega/c)\sin\theta_p = q$.

A second method, which is emphasised here, is the grating coupler. Any periodic variation of the dielectric properties in the x-y plane, such as a periodic surface corrugation or periodic metal stripes on a 2DES as shown in Figure 2a, will spatially modulate an incident plane wave with electric field $\mathbf{e} \propto \exp(ik_x^0 x)$ of wave-vector component k_x^0. The response is then given by a Fourier series of induced electric field components (and similarly the magnetic field components)

$$\mathbf{e} = \sum_{n=-\infty}^{+\infty} (e_x^n, 0, e_z^n)\exp\left[\mathrm{i}\left(k_x^n x - \omega t\right)\right] \tag{1}$$

where $k_x^n = k_x^0 + n(2\pi/a)$, $k_x^0 = (\omega/c)\sin\theta^0$ and a is the periodicity. If $|k_x^n|$ is smaller than ω/c this leads to the emission of diffracted waves at angle θ^n with $k_x^n = (\omega/c)\sin\theta^n$. If $|k_x^n|$ is larger than ω/c the induced Fourier components decay exponentially and exist only in the near field of the grating coupler. Here they can couple to plasmons of wave-vector $q = k_x^n$, leading to a resonant decrease of the reflected or transmitted radiation. Vice versa, a grating coupler can also be used to induce the radiative decay of plasmons which are excited by different means, *e.g.* by fast electrons (Teng and Stern 1967, Heitmann 1977, Heitmann *et al.* 1987), tunnelling electrons (Heitmann *et al.* 1987, Kirtley *et al.* 1981) or optically (Heitmann and Raether 1976). Another interesting aspect of the surface plasmon resonance is the strong field enhancement that is associated with this excitation and which gives rise to interesting applications, *e.g.* in non-linear optics for the giant Raman effect (Broadman 1982) or in light-emitting devices (Heitmann *et al.* 1987, Kirtley *et al.* 1981).

Experiments on the radiative decay of surface plasmons induced by a grating coupler (Heitmann 1977) are shown in Figure 3a. Fast electrons are incident onto a sinusoidally corrugated silver surface with periodicity $\Lambda = a = 885\,\mathrm{nm}$. It is known from measurements of energy loss that fast electrons excite surface plasmons very efficiently (Raether 1980). Several emission peaks are observed for different directions θ of the emitted radiation. The positions of the peaks shift systematically with decreasing wavelength λ of detection. From the experimental resonance positions, shown in Figure 3b, we find that the emission results from the radiative decay of surface plasmons by the $n = \pm1, \pm2$ and ±3 reciprocal lattice vectors of the grating, $g_n = n(2\pi/a)$, both from the left and right-hand branch of the dispersion. Due to the non-normal incidence of the electrons ($\alpha = 70°$ with respect to the surface normal), plasmons propagating in the direction of the normal component of the electron velocity v_x are more strongly excited leading to a correspondingly stronger emission, *i.e.* plasmons with positive $q = k_x$ from the right hand dispersion branch. The efficiency of the grating coupler for different 'diffraction' orders n depends on the Fourier components of the grating coupler and decreases in general with increasing $|n|$.

Grating couplers can be realised for a 2DES in MOS or heterostructures by fabricating periodic metal stripes (see Figure 2) on top of the structure (Allen *et al.* 1977). For periodicities $a \approx 500\,\mathrm{nm}$ the corresponding wave-vector is about $10^5\,\mathrm{cm}^{-1}$. 2D plasmons of this wave-vector have very small energies of about $5\,\mathrm{meV}$ for a typical 2D charge

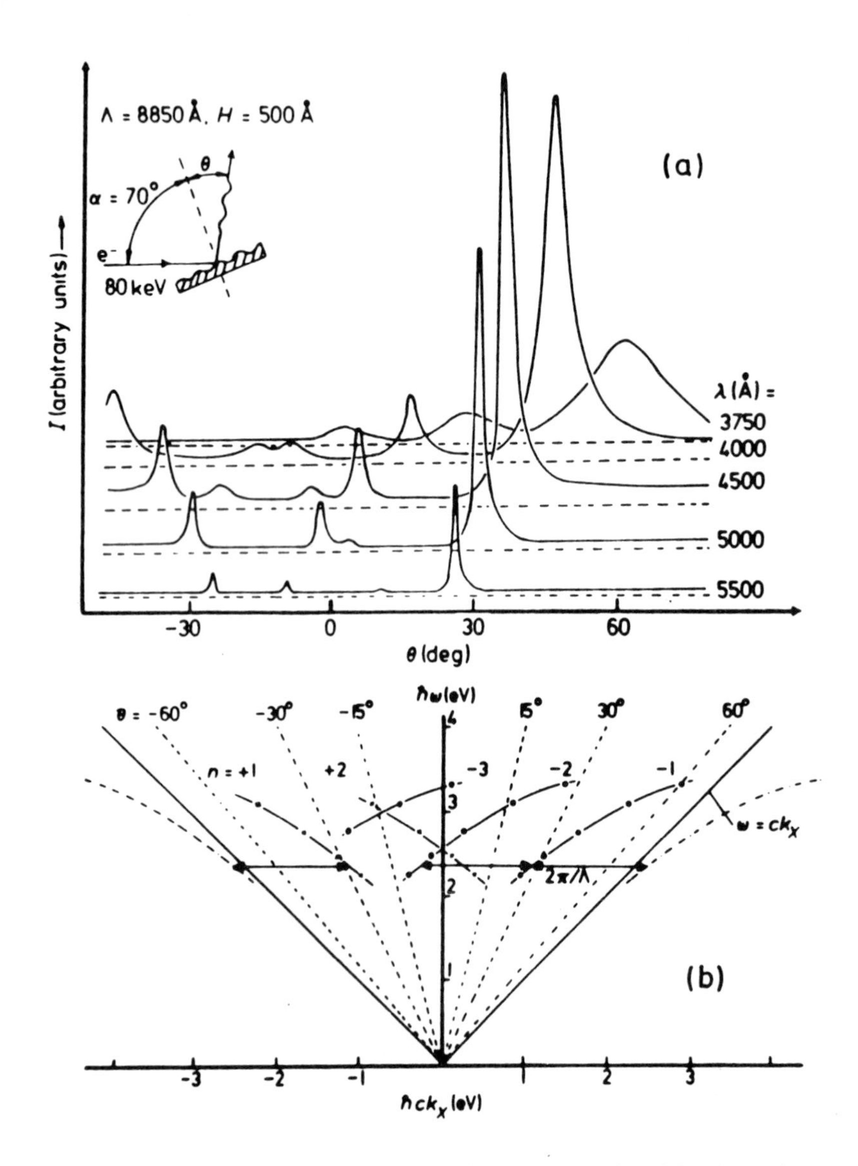

Figure 3: *Angular dependence of the emitted radiation induced by the radiative decay of surface plasmons on a sinusoidally modulated* Ag *surface with grating constant* $\Lambda =$ 8850 Å *and a modulation height of about* 500 Å *for various wavelengths. The surface plasmons are excited by* 80 keV *electrons which are incident with an angle of incidence* $\alpha = 70°$ *onto the sample. The zero intensity of each curve (shown as horizontal broken lines) has been shifted vertically for clarity. (b)* ω-k_x *diagram of the emission peak positions. The broken lines indicated by angles* θ *are related to the angles of observation. The left and right-hand branches of the surface plasmon dispersion (chain curves) are derived by adding reciprocal grating vectors* $2\pi/\Lambda$ *to the experimental peak positions (Heitmann 1977).*

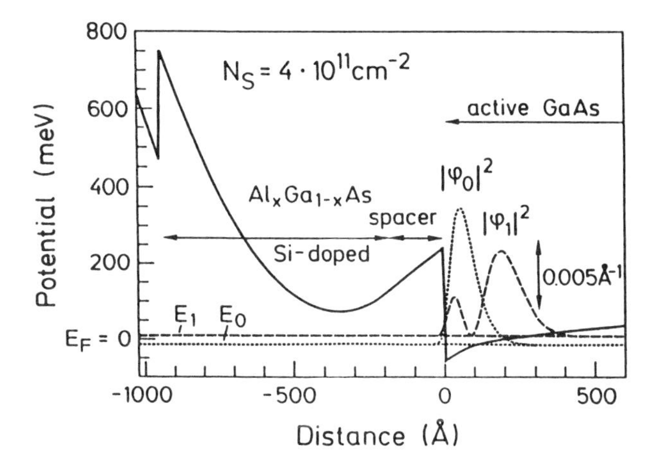

Figure 4: *Self-consistently calculated potential, band structure, wave-functions and energy levels in a* GaAs-Al_xGa_{1-x}As *heterostructure (Ensslin 1989).*

density $N_s = 7 \times 10^{11}\,\mathrm{cm}^{-2}$. Thus far-infrared (FIR) spectroscopy at low temperatures must be applied to investigate 2D plasmons. The wave-vector q is much larger than ω/c (about 100–400 times), in contrast to the surface plasmons discussed above. Thus retardation effects can be totally ignored and the light line ω/c is practically at $q = 0$. This means that q cannot be changed by the angle of incidence; rather one has to use the higher Fourier components of the grating coupler to vary q in discrete steps.

3 Two-dimensional electronic systems

In modulation-doped semiconductor heterostructures or quantum wells, *e.g.* in GaAs-Al_xGa_{1-x}As or in Si-MOS, it is possible to realise 2DESs of high mobility where electrons are one-dimensionally confined in the z-direction. The extent of the wave-functions in the z-direction is as small as 1 nm in Si-MOS at a high gate voltage and correspondingly high electron density $N_s = 10^{13}\,\mathrm{cm}^{-2}$, and 3 to 30 nm in GaAs-Al_xGa_{1-x}As depending on the width of the well and density. The one-particle energy spectrum consists of a set of quantum-confined 2D subbands ($i = 1, 2, 3, \ldots$):

$$E^i(k_x, k_y) = \frac{\hbar^2 k_x^2}{2m^*} + \frac{\hbar^2 k_y^2}{2m^*} + E_z^i\,, \qquad (2)$$

where the electrons have a free dispersion only in the x-y plane (k_x and k_y are the wave-vectors of the electron). Electron wavefunctions, energy levels, and densities have to be determined by a self-consistent calculation of the Poisson and the Schrödinger equation (Ando *et al.* 1982). An example for a typical GaAs-Al_xGa_{1-x}As heterostructure is given in Figure 4 (Ensslin 1989).

The elementary dynamic excitations of a homogeneous 2DES are 2D plasmons (collective intra-band excitations) and, in a magnetic field **B**, cyclotron resonances (CR) (transitions between Landau levels). In addition to the collective plasmon excitation that we have discussed for free-electron systems in a 3DES, we have now a different type of excitation, *i.e.* inter-subband resonances, resonant transitions between the quantised energy levels of the 2DES. These transitions are predominantly single-particle excitations but, as we will see below, are also subject to collective effects. Under typical experimental conditions these resonances occur in the FIR frequency range (in particular regarding the resonance condition $\omega\tau > 1$). Thus FIR spectroscopy (transmission, reflection and emission) is a very powerful tool to study these systems [for reviews on spectroscopy see, for example, Batke and Heitmann (1984), Heitmann *et al.* (1990)]. In transmission spectroscopy FIR radiation is transmitted through the sample and the transmission T is measured. The spectra are normalised to a reference spectrum to eliminate the frequency-dependent sensitivity of the spectrometer. It is very convenient if one can deplete the carrier density with a gate voltage. Then one can perform differential spectroscopy and evaluate $\Delta T = [T(N_s) - T(0)]/T(0)$. This expression is, for small signals, proportional to the real part of the dynamic conductivity and can thus be directly related to the microscopic properties of the electronic system.

4 Two-dimensional plasmons

Two-dimensional plasmons were first observed for electrons on the surface of liquid He (Grimes and Adams 1976) and for the Si (100) MOS system (Allen *et al.* 1977, Theis *et al.* 1977, Theis *et al.* 1978, Theis 1980). The first calculations of the 2D plasmon dispersion were performed by Ritchie (1957), Stern (1967) and Chaplik (1972). [For extended reviews on 2D plasmons see Theis (1980) and Heitmann (1986) (experiment) and Chaplik (1985) (theory).] As discussed above, 2D plasmons have the dispersion

$$\omega_p^2 = \frac{N_s e^2}{2\bar{\epsilon}\epsilon_0 m^*} q \tag{3}$$

where N_s is the 2D charge density, m^* the effective mass, $\bar{\epsilon} \equiv \bar{\epsilon}(\omega, q)$ is the effective dielectric function (which also depends on the surroundings) and q is the wave-vector in the x-y plane.

Two-dimensional plasmons are non-radiative modes in the sense that $q > \omega/c$ (taking into account retardation, also at small q and ω). Thus no direct coupling to transmitted radiation is possible. Grating couplers are an effective tool to excite these nonradiative modes, as discussed above. A grating coupler may consist of periodic stripes (periodicity a) of materials of high and low conductivity (Figure 2a). (In principle, any periodic variation of the dielectric properties in the vicinity of the 2DES, *e.g.* a modulation of the 2DES itself, acts as a grating coupler). Then the incident field, which has a homogeneous e_x-component only, is short-circuited in the region of high conductivity. Thus, in the near field of the grating, the electromagnetic fields are spatially modulated and consist of a series of Fourier components (Equation 1). Fourier components with wave-vector $k_x^n = n(2\pi/a)$ will couple to 2D plasmons if $q = k_x^n$ and ω satisfies the dispersion relation (Figure 2b). Another important feature is that the grating induces e_z-components of the electric field in the near field. It is possible to excite inter-subband resonances with these components (Heitmann *et al.* 1982, Heitmann and Mackens 1986). A full general calculation of the dielectric response in the

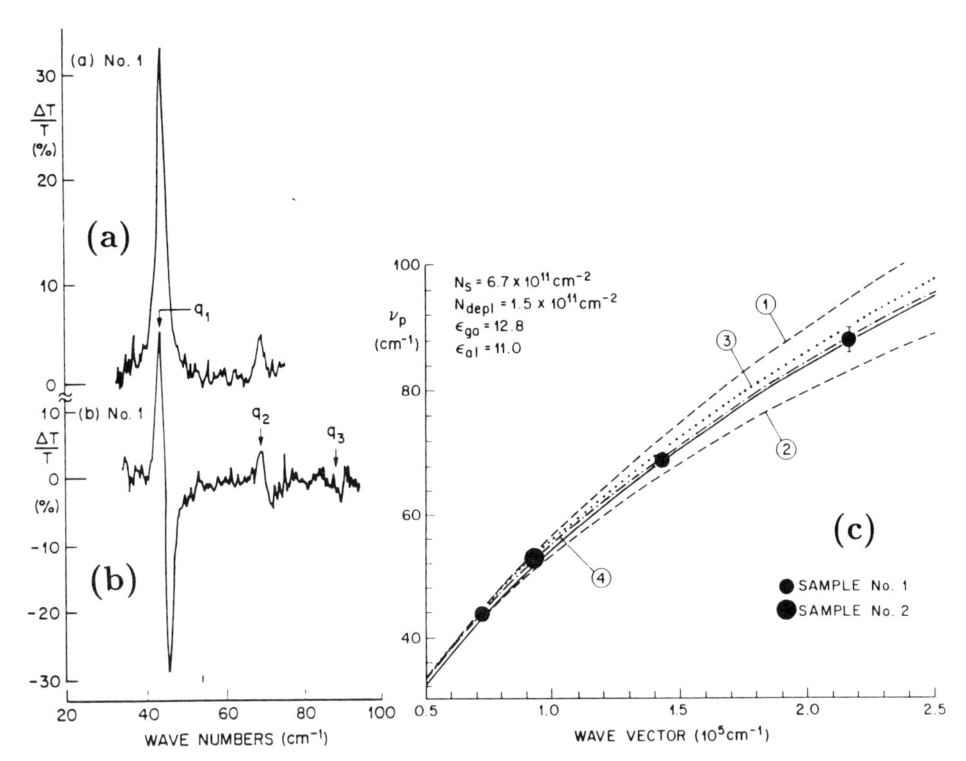

Figure 5: *Experimental plasmon resonances for a space-charge layer in* GaAs *with* $N_s = 6.7 \times 10^{11}\,\text{cm}^{-2}$ *and* $q_n = n \times (0.72 \times 10^5\,\text{cm}^{-1})$. *The arrows mark the positions of plasmon resonances. (a) The ratio of the transmissions,* $T(\nu)$, *at* $B = 0$ *and* $B \geq 8\,\text{T}$ *determine* $\Delta T/T$. *Trace (b), qualitatively the derivative of (a), is obtained by changing* N_s *slightly with the persistent photo-effect. (c) Theoretical and experimental plasmon dispersions. The solid line is the classical local plasmon dispersion. The curves marked 1–4 are defined as follows: curve 1, plasmon dispersion including non-local correction; curve 2, plasmon dispersion including finite-thickness effect; curve 3, plasmon dispersion including both non-local and finite-thickness corrections; and curve 4, plasmon dispersion including all correction terms (from Batke et al. 1986).*

presence of a grating is a complicated numerical problem. Approximate treatments for the conditions here are given in Theis (1980) and Tsui *et al.* (1978). The effectiveness of the grating coupler depends on its design. The nth Fourier components of the fields decay with $\exp(-|k_x^n|z)$ for $|k_x| \gg \omega/c$. Thus the distance d_z of the 2DES from the grating should be small compared with the periodicity [roughly $d_z < 0.1(a/n)$].

High-mobility modulation-doped GaAs-$\text{Al}_x\text{Ga}_{1-x}$As heterostructures are an ideal system in which to study 2D plasmons. In Figure 5a we show 2D plasmon excitations induced by a grating coupler (Batke *et al.* 1986). The periodicity of the grating coupler is $a = 880\,\text{nm}$. The transmission spectrum, T, of the FIR radiation is normalised to a spectrum taken at $B = 0\,\text{T}$. A magnetic field shifts the plasmon resonance, such that the effective conductivity at $B = 8\,\text{T}$ is low in the frequency range of Figure 5a. Two resonances are observed which are excited by the first and second Fourier components of the grating and correspond to wave-vectors $q_1 = 2\pi/a$ and $q_2 = 2(2\pi/a)$. Three plasmon resonances are resolved in the differential spectrum in Figure 5b (here $\Delta T/T$ is evaluated for two slightly different densities). This allows a detailed study of the dependence of the dispersion on q, as depicted in Figure 5c. It is shown in Batke *et al.* (1986) that several corrections to the simple Equation (3) become effective at high wave-vectors q. They arise (a) from non-local properties of the 2DES, *i.e.* the finite compressibility of the Fermi gas (Fermi pressure); (b) from virtual coupling to inter-subband transitions; and (c) from the finite thickness of the 2DES. The finite thickness becomes important if q approaches $1/d_{\text{in}}$, the inverse spatial extent of the electron wavefunctions in the z-direction. It is shown that these effects nearly cancel each other, in agreement with the experimental findings.

Two-dimensional plasmons have also been studied in the Si-MOS system. An interesting aspect of the plasmon excitation is that one can measure the dispersion for different directions in the surface. For Si (110) the plasmon dispersion is found to be anisotropic for different directions $\mathbf{q}$ with respect to the [001] direction (Batke and Heitmann 1983). This arises from the anisotropic bandstructure $E(\mathbf{k})$ of a Si (110) surface. Also 2D plasmons in hole space-charge layers of Si (110) have been investigated (Batke *et al.* 1983). Here the plasmon dispersion reflects the non-parabolicity of the surface band structure for holes and the anisotropy of the Si (110) surface.

5 Inter-subband resonances in a 2DES

Besides 2D plasmons, inter-subband resonances (ISRs) are characteristic dynamic excitations of heterostructures and quantum wells. ISRs represent oscillations of the carriers perpendicular to the interface. Thus in highly symmetrical systems, *e.g.* electrons in Si (100) MOS systems or in GaAs heterostructures, an e_z-component of the incident electric field is necessary to excite these transitions. Strip-line and prism coupler arrangements have been used to study ISR on Si (100) (Kneschaurek *et al.* 1976, McCombe *et al.* 1979). Another very powerful method is the grating coupler technique (Heitmann *et al.* 1982, Heitmann and Mackens 1986). As we have discussed above, a grating coupler excites e_z-components of the electric field in the near field (Figure 2a). Experimental spectra measured on Si (100) samples are shown in Figure 6a. The period of the grating is $a = 1800\,\text{nm}$. Two resonances, $\tilde{E}_{01}$ and $\tilde{E}_{02}$, are observed for $N_s = 3.3\times10^{12}\,\text{cm}^{-2}$, which correspond to resonant transitions from the lowest subband to the first and second excited subband. With increasing N_s the resonances shift to higher frequencies, corresponding to a larger separation between the subbands in the steeper potential well at larger surface electric fields. An additional resonance $\tilde{E}'_{01}$ is observed for $N_s > 8\times10^{12}\,\text{cm}^{-2}$, which can be attributed to resonant transitions in the primed subband system (Ando *et al.* 1982). The primed subband system arises from the projection of four volume energy ellipsoids of Si onto the Si (100) surface and is separated in $\mathbf{k}$-space by $0.86(2\pi/A)$ in [001] and equivalent directions, where A is the lattice constant of the crystal. It is known (Ando *et al.* 1982) that these subbands are occupied for $N_s > 7.5\times10^{12}\,\text{cm}^{-2}$.

The resonant energy measured in an ISR spectrum is exactly the spacing between

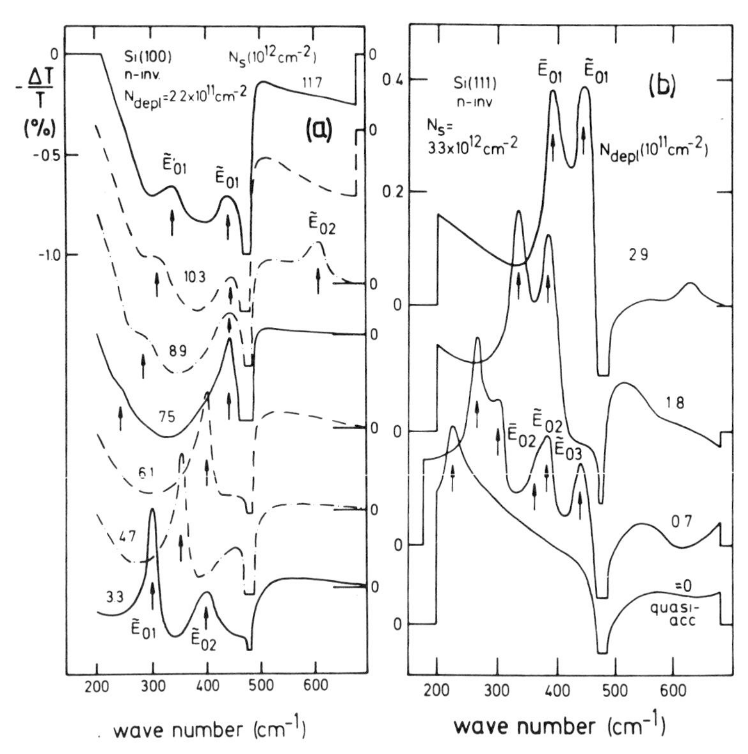

Figure 6: *Inter-subband resonances induced by a grating coupler for* Si (100) *at different charge densities* $N_s = 3.3\text{–}11.7 \times 10^{12}\,\text{cm}^{-2}$. *Resonances* $\tilde{E}_{01}$ *and* $\tilde{E}_{02}$ *in the lower subband system and* $\tilde{E}'_{01}$ *in the second subband system are observed. A resonant coupling to polaritons is measured in the regime of the optical phonon frequency of* SiO_2 *(about* $480\,\text{cm}^{-1}$*), which is not fully shown here for clarity. (b) Excitation of inter-subband resonances on a* Si (111) *sample with a grating coupler for different* N_{depl}. *Directly parallel excited* $\bar{E}_{oi}$ *resonances and depolarisation-shifted resonances induced by the grating coupler* $\tilde{E}_{oi}$ *are present (from Heitmann and Mackens 1986).*

subbands, $E_{01} = E_1 - E_0$, only in a strictly 2D system, which implies that the spacing between subbands must be very large. The real systems are quasi-2D and have a finite extent of the wavefunction. This means, in a hand-waving argument, that there is still a little bit of free-electron-like, plasmon-type behaviour left in the z-direction. Thus, if we try to excite a single electron from the lowest into the next subband, all the other electrons screen this excitation which shifts the actual resonance energy with respect to the single-particle subband spacing. A phenomenological model for this so-called depolarisation shift has been given by Chen *et al.* (1976). This collective influence in the quasi-2DES is usually calculated within the random phase approximation (RPA). In this model one distinguishes two effects which shift the observed resonance with respect to the subband spacing, $\tilde{E}_{01} = E_{01}\sqrt{1 + \alpha - \beta}$. The first, the so-called exciton shift (Ando 1977), results from the energy renormalisation when an electron is transferred to the first excited subband E_1 leaving a 'hole' in the E_0 subband. This exciton effect, characterised in a two-band model by β, shifts the resonance to smaller energies. A second effect, which is stronger than the exciton shift at medium or higher densities, is characterised by α and is called the depolarisation shift and increases the resonance energy. It arises from the resonant screening of the microscopic one-particle dipole excitation by the collective effect of all other electrons in the potential well (Chen *et al.* 1976, Allen *et al.* 1976). To include this effect in a calculation it is necessary to take into account the full time-dependence of the Hartree potentials (Allen *et al.* 1976) beyond a purely static calculation of subband energies. In typical GaAs-$Al_xGa_{1-x}As$ or Si-MOS sytems the depolarisation shift is, as we will see in the following, well observable but nevertheless a small effect. Then it is possible to calculate α and β in a two-subband approximation, where α and β are Coulomb matrix elements involving only the wave-functions of the lowest and first excited subbands. We will see below that collective effects are much more important in the currently available quantum wires and dots. Here it is vital to include all occupied subbands and virtual transitions to all unoccupied subbands in a RPA approach, which requires a very detailed knowledge of the self-consistent confining potential and the wavefunctions.

For Si (111) and (110), the band structure on the surface results from the projection of ellipsoids in the bulk band structure which are tilted with respect to the surface (Ando *et al.* 1982). Due to the anisotropic energy contours, the x and z-components of the dynamic surface current are coupled and thus ISR can be excited with a parallel field component e_x (Heitmann and Mackens 1986, Ando *et al.* 1977). Whereas parallel excited ISR (labeled $\bar{E}_{01}$) are not affected by the depolarisation shift, $\bar{E}_{01} = E_{01}\sqrt{1-\beta}$, the perpendicular excited resonance ($\tilde{E}_{01}$) is affected by both effects, $\tilde{E}_{01} = E_{01}\sqrt{1+\alpha-\beta}$. Thus both resonances, parallel resonances excited directly and perpendicular resonances excited by the grating coupler (and thus shifted by depolarisation) are observed (Figure 6b) when a grating coupler is used on Si (111). We find in the spectra in Figure 6b that the depolarisation shift slightly increases with the depletion field which is characterised by N_{depl}. It becomes much smaller for transitions to higher subbands ($\tilde{E}_{02}$ and $\tilde{E}_{03}$). Very detailed investigations of ISR induced by a grating coupler for different surface orientation on Si are reported in Heitmann and Mackens (1986). Grating couplers have also been used to investigate a resonant coupling of inter-subband resonances and 2D plasmons, leading to an anti-crossing of the dispersions, in a very tricky experiment with a Si (100) MOS structure under external stress (Oelting *et al.* 1986). They have also been applied to investigate the GaAs system (Helms *et al.* 1989). Another very powerful tool to study ISR is the method of resonant subband–Landau level coupling (Schlesinger *et al.* 1983, Rikken *et al.* 1986, Wieck *et al.* 1987, Ensslin *et al.* 1989, Heitmann and Ensslin 1991).

6 Systems with modulated charge density

A first step in the direction of lower-dimensional electronic systems is a system whose density is modulated (Mackens *et al.* 1984, Heitmann *et al.* 1985). Figure 7 shows measurements on Si (100) MOS systems with periodically modulated oxide thickness. The thickness of the oxide is d_1 in the region t_1 and d_2 for the remainder $t_2 = a - t_1$ of the period a (except for the small region of the slopes). A continuous layer of NiCr, 3 nm thick, is evaporated at varying angles onto the structured oxide. If a gate voltage

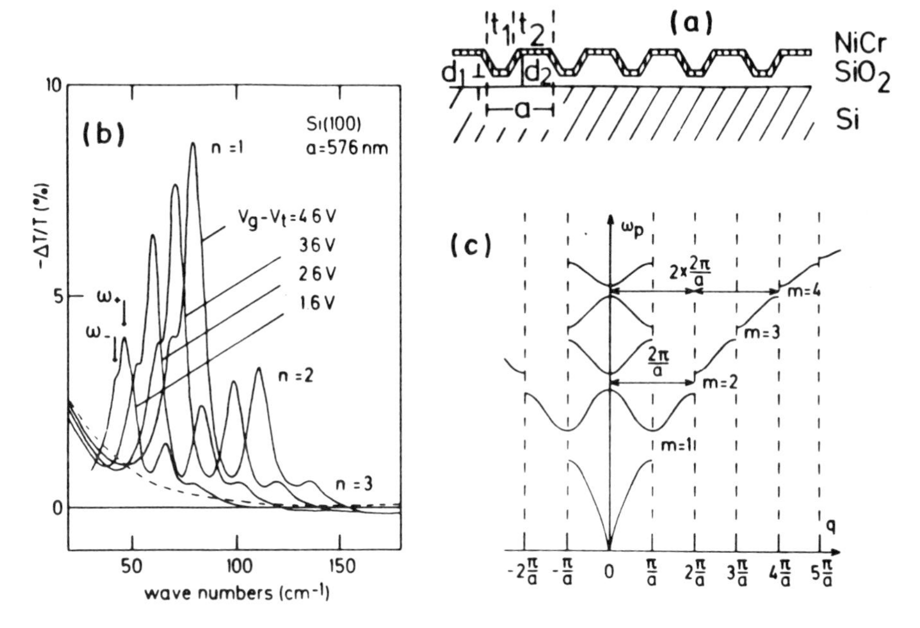

Figure 7: *Schematic geometry of a MOS sample with modulated oxide thickness (a), excitation of 2D plasmons with split resonances (b) due to the superlattice effect of the charge density modulation on the 2D plasmon dispersion (c) (from Mackens et al. 1984, Heitmann et al. 1985).*

V_g is applied between this gate and the substrate, an electron gas with a low charge density N_{s2} in the region t_2 and with high densities N_{s1} in the region t_1 is induced. Typically d_1 and d_2 are 20 and 50 nm.

FIR transmission spectra are shown in Figure 7b. Superimposed on the Drude background, well pronounced plasmon resonances are observed which shift with increasing gate voltage and corresponding charge densities to higher wave-numbers, roughly as expected from the plasmon dispersion (Equation 3). Three resonances labeled $n = 1, 2, 3$ are excited by n reciprocal lattice vectors of the grating, $q_n = n(2\pi/a)$. A characteristic of these systems with modulated charge density is that the plasmon excitations are split into two resonances, ω_- and ω_+. To discuss the origin of this splitting, we show in Figure 7c the plasmon dispersion in a system with modulated charge density. The superlattice effect of the periodically modulated charge density creates Brillouin zones with boundaries $q = m\pi/a$, $m = \pm 1, \pm 2, \ldots$. If we fold the plasmon dispersion back into the first Brillouin zone, $-\pi/a < q < \pi/a$, we expect a splitting of the plasmon dispersion at the zone boundaries and at the centre of the zone, $q = 0$. The plasmon dispersion forms bands with minigaps at $q = 0$ and $q = \pi/a$. Since we use the same grating that produces the Brillouin zones also for the coupling process $[q_n = n(2\pi/a)]$, we can only observe the gaps at $m = 2, 4, \ldots$. The resonances observed in Figure 7b are thus the lower and upper branches of this dispersion at $q = 0$.

The plasmon dispersion in a system with modulated charge density has been calculated by Krasheninnikov and Chaplik (1981). It is found in this approach using perturbation theory that the splitting of the mth gap is proportional to N_{sm}, the Fourier coefficient of the charge density:

$$N_s(x) = \sum_{m=-\infty}^{+\infty} N_{sm} \exp\left(\mathrm{i}\frac{2\pi m}{a}x\right).$$

Thus a system with a large second Fourier component N_{s2} in the charge density is needed to observe a large splitting of the plasmon resonance for $q = 2\pi/a$. This explains a strong influence of the sample geometry on the amount of the experimentally observed splitting and also the fact that the splitting observed for higher gaps ($m = 4, 6, \ldots$) is generally smaller (Mackens *et al.* 1984, Heitmann *et al.* 1985). Another characteristic feature of plasmons in a periodically modulated system is that they have the character of standing waves, where the two branches have different symmetry. Because of the symmetry, the upper branch ω_+ for the configuration in Figure 7c has a radiative character, whereas the ω_- branch is less radiative. Thus FIR radiation can excite the ω_+ branch with higher efficiency. This is observed in Figure 7b.

7 One-dimensional electronic systems

A one-dimensional electronic system is used here for a system whose energy spectrum consists of a set of quantum-confined 1D subbands ($i, j = 0, 1, 2, 3, \ldots$):

$$E^{ij}(k_y) = \frac{\hbar^2 k_y^2}{2m^*} + E_x^i + E_z^j, \tag{4}$$

The electrons have a free dispersion only in the y-direction, E_z^j represents the quantised energy levels in the original 2DES, and E_x^i represent the quantum-confined energy states due to the additional lateral confinement acting in the x-direction. We have assumed that the x and the y-directions are decoupled. To achieve a lateral quantisation of 2 meV in the GaAs system with $m^* = 0.065\, m_0$, for instance, one has to confine the electrons into a width w of some 100 nm, as we will see below. Some examples of realised 1DESs in the GaAs-$\mathrm{Al}_x\mathrm{Ga}_{1-x}$As system are sketched in Figure 8. The preparation of such lateral arrays of quantum wires has been reviewed in Heitmann (1990); etching techniques are described in Grambow *et al.* (1989) and Grambow *et al.* (1990).

Self-consistent band structure calculations show that the external confinement potential for electrons has a nearly parabolic shape (Laux *et al.* 1988). We can see this from the simplified arrangement sketched in Figure 8d. Due to the strong original 2D confinement the electrons can only move in the x-y plane. A positively charged layer of 2D density N_s and width w from donors in the AlGaAs layer or, in a gated structure, from a gate voltage, holds the electron in an equilibrium position. Moving the electron in the x-direction produces a force approximately linear in the displacement which gives rise to a parabolic confinement potential. We can calculate the force constant and thus the eigenfrequency by integrating the x-component of the electric field from the charged layer at the position of the electron and find approximately $\Omega_0^2 = N_s e^2 \pi / 2m^* \bar{\epsilon}\epsilon_0 w$ where we have assumed that the distance d in Figure 8d is small and the medium surrounding is described by an effective dielectric constant $\bar{\epsilon}$. This model is correct for a small number of electrons; self-consistent effects become important for a larger number of

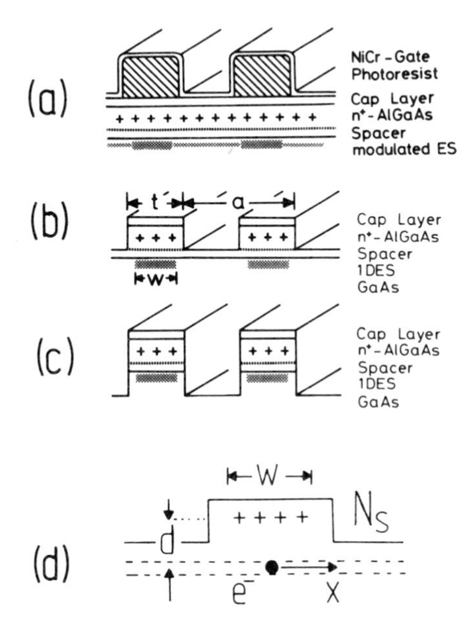

Figure 8: *Some examples of realised periodic arrays of quantum wire structures in the* GaAs-Al$_x$Ga$_{1-x}$As *system. (a) is a so-called split-gate configuration where carriers are depleted leaving isolated quantum wires by a gate voltage and a varying distance between gate and channel. The gate-distance modulation is achieved with a modulated photoresist layer. (b) and (c) sketch examples for mesa etched quantum wire structures: (b) shows schematically a 'shallow' mesa etched structure, (c) a 'deep' mesa etched structure (from Demel 1989). (d) sketches a simple model for the calculation of the lateral confinement potential of electrons.*

electrons. The most sophisticated calculations for a 1DES so far have been performed for confinement by the lateral field effect (Figure 8b), sometimes called the split-gate configuration, by Laux *et al.* (1988). It is found that the potential in the channel is indeed nearly parabolic for a small number of electrons. However, self-consistent effects flatten the potential and the separation between subbands decreases drastically with increasing 1D charge density N_l.

In most studies so far a 1DES is characterised by magnetic depopulation of the 1D subbands, which occurs if the 1DES is exposed to a perpendicular magnetic field **B** (Berggren *et al.* 1986). The underlying physics can be explained without loss of generality if one assumes a parabolic confinement potential $V(x) = \frac{1}{2}m^*\Omega_0^2x^2$. In this case the Schrödinger equation can be solved analytically (Berggren *et al.* 1986). The magnetic field induces an additional potential $V_B(x) = \frac{1}{2}m^*\omega_c^2(x - x_0)^2$, where $\omega_c = eB/m^*$ is the cyclotron frequency, $x_0 = k_y l_c^2$ is the central coordinate and $l_c = (\hbar/eB)^{1/2}$ is the magnetic length. For this model the energy levels are given by

$$E^i(k_y, B) = \hbar\Omega\left(i + \tfrac{1}{2}\right) + \frac{\hbar^2 k_y^2}{2m_y^*(B)} \tag{5}$$

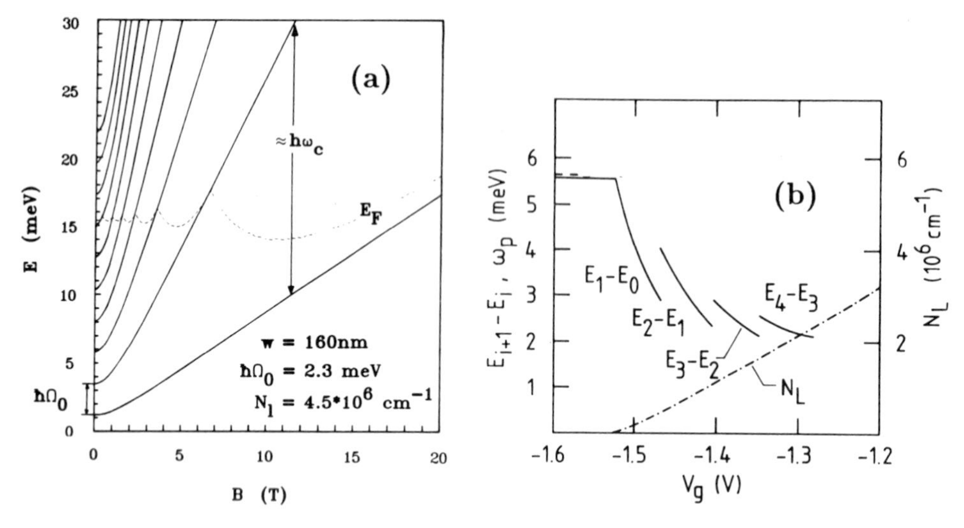

Figure 9: *(a) Single-particle energy spectrum for an electron confined in a harmonic parabolic potential and in a magnetic field B. The dotted line shows the oscillations of the Fermi energy for the indicated linear charge density N_l (screening effects are neglected). The hybrid 1D subband-Landau levels become successively depopulated with increasing B. (b) Self-consistently calculated spacing between 1D subbands in a quantum wire confined by the field effect (Figure 8a). The separation between subbands decreases with increasing linear density N_L due to self-consistent screening (Laux et al. 1988, Shikin et al. 1990).*

with $\Omega^2 = \omega_c^2 + \Omega_0^2$, and $m_y^*(B) = m^*\Omega^2/\Omega_0^2$. This energy spectrum is shown in Figure 9. Since the 1D density of states, $D_{1D}(E, B)$, and the separation between subbands increase with increasing B, the 1D subbands become successively depopulated giving rise to oscillations of the Fermi energy. In a transport measurement this leads to the Shubnikov–de Haas (SDH) type of oscillations. In a 2DES the number of occupied Landau levels increases with decreasing B, leading ideally to an infinite number of SDH oscillations periodic in $1/B$. In a 1DES, however, only a finite number of 1D subbands are occupied at $B = 0$; thus only a finite number of SDH oscillations occur, which are no longer linear in $1/B$.

Within this model we can determine the confinement energy Ω_0 and the total 1D carrier density N_l from the observed depopulation of the 1D subbands by a magnetic field. For a shallow etched sample (Figure 8b) we find (Demel *et al.* 1988a) typical values of $\hbar\Omega_0 = 2.3\,\text{meV}$ and $N_l = 4.5\times10^6\,\text{cm}^{-1}$. For these values, there are 6 occupied 1D subbands at $B = 0$. Defining the width w of the electron channel by the amplitude at the Fermi energy, $E_F(B = 0) = V(w/2) = \frac{1}{2}m^*\Omega_0^2(w/2)^2$, we find $w \approx 160\,\text{nm}$ which is smaller than the geometrical width of $t = 250\,\text{nm}$. In deep mesa etched structures as shown in Figure 8c we determine, for example, $\hbar\Omega_0 = 1\,\text{meV}$, $N_l = 12\times10^6\,\text{cm}^{-1}$ and 12 occupied 1D subbands. Comparing the geometrical channel width of $t = 550\,\text{nm}$ with the determined electronic width $w = 390\,\text{nm}$ we deduced that there is a lateral depletion length $w_{\text{dl}} = 80\,\text{nm}$ on either side of the wire (Demel *et al.* 1988a).

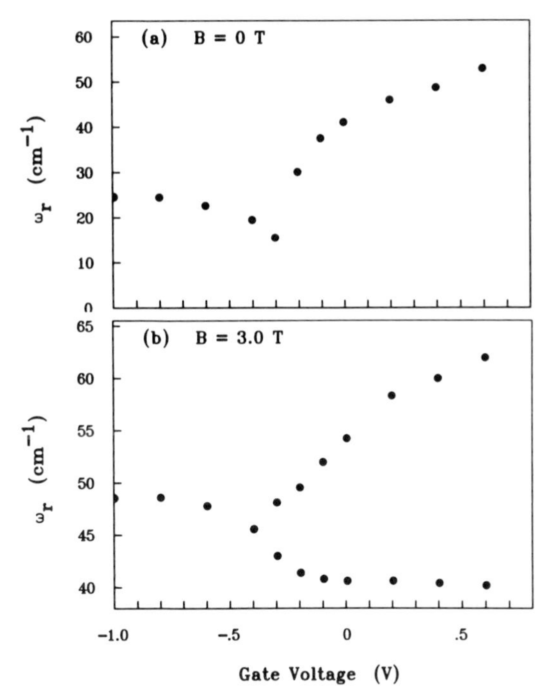

Figure 10: *Experimental* FIR *resonance positions measured on an* GaAs-Al$_x$Ga$_{1-x}$As *heterostructure with periodically modulated gate distance (Figure 8a) at $B = 0$ and $B = 3\,\mathrm{T}$. For $V_g \geq -0.4\,\mathrm{V}$ the system is density modulated; isolated 1D quantum wires wires are formed below this (from Demel et al. 1988b).*

For the 1DES we expect a strongly anisotropic dynamic response. With an exciting electric field in x-direction, that of the lateral confinement, we expect in principle to observe 1D inter-subband resonances, *i.e.* resonances reflecting the separation between 1D subbands. However we will find that these resonances are strongly governed by collective effects since the confinement energies are small and due to the special shape of the confining potential. We will discuss this behaviour in the rest of this section. With an exciting field in y-direction, along the length of the wire, we can excite 1D plasmons propagating freely along the wire. However, grating couplers (Demel *et al.* 1991) or Raman spectroscopy (Egeler *et al.* 1990) are required to produce the wave-vector of the 1D plasmons. We will discuss 1D plasmons in the next paragraph. With an exciting field in z-direction we expect to observe the inter-subband resonance of the original 2D confinement which is much stronger than the lateral confinement. This has indeed been observed in Si-MOS systems of modulated density (Mackens *et al.* 1986). However, a strong interaction of 1D and 2D energy states was observed in recent studies of the resonant subband-Landau level coupling in tilted magnetic fields for quantum wire structures (Kern *et al.* 1991a). We will not elaborate on this point here, but refer to the original paper.

In Figure 10 we show experimental FIR resonance positions (Demel *et al.* 1989) for a GaAs-Al$_x$Ga$_{1-x}$As heterostructure with periodically modulated gate distance as sketched in Figure 8a. The incident electric field is polarised in the x-direction in the original 2D plane and perpendicular to the periodic structure. For $V_g = 0$ and $B = 0$ one observes the 2D plasmon resonance induced by the grating coupler at $\omega_p = 40\,\text{cm}^{-1}$, for $B = 3\,\text{T}$ the cyclotron resonance at $\omega_c = 40\,\text{cm}^{-1}$ and the magnetoplasmon resonance ω_{mp} at $53\,\text{cm}^{-1}$. The latter is shifted with respect to ω_p according to the well-known magnetoplasmon dispersion $\omega_{\text{mp}}^2 = \omega_p^2 + \omega_c^2$ (Chaplik 1972, Theis *et al.* 1978, Heitmann 1986). With decreasing V_g we induce a density-modulated system with a decreased average density. Correspondingly the plasmon frequency shifts to lower frequencies. At $V_g = -0.4\,\text{V}$ there is a sudden change in the slope and ω_{mp} and ω_c merge for $B = 3\,\text{T}$. From d.c. transport one finds that this singular behaviour occurs in the range of gate voltages where isolated wires are formed. Such a behaviour was first observed by Hansen *et al.* (1987). One finds that the electron system forms a quantum-confined 1DES in the regime $V_g < -0.4\,\text{V}$; however the FIR response reflects only indirectly the spacing between 1D subbands.

The most surprising result is that the resonant frequencies ω_r in the FIR spectra are significantly higher in energy than one would expect from the separation between the 1D subbands, $\hbar\Omega_{01}$, which was determined from the measurements of d.c. magnetotransport. For example, at $V_g = -0.7\,\text{V}$ in Figure 10 we determine a 1D subband spacing of $\hbar\Omega_{01} = 1.5\,\text{meV}$ from magnetic depopulation, and measure a resonant frequency of $\hbar\omega_r = 3\,\text{meV}$ in the FIR experiment. For the deep mesa etched quantum wire system shown in Figure 8c it was found that $\hbar\Omega_0 = 1\,\text{meV}$ and $\hbar\omega_r = 4\,\text{meV}$ (Demel *et al.* 1988b). This result, puzzling at a first glance, can be explained by the strong influence of collective effects, which occur in a unique way in systems with parabolic confinement. The energy spectrum in Equation (5) was calculated for one electron. If we fill the external potential Ω_0 with more electrons this external potential is screened. Self-consistent Hartree calculations (Laux *et al.* 1988) show that electron-electron interaction, *i.e.* the self-consistent screening, strongly decreases the separation of the 1D subbands. This is demonstrated in Figure 9b. However, this difference in the Hartree energy, $\hbar\Omega_{01} = E_1 - E_0$, cannot be observed in an optical dipole excitation.

The Hamiltonian for N electrons, including electron-electron interaction, is

$$H_N = \sum_{j=1}^{N} H(\mathbf{r}_j, \mathbf{p}_j) + \sum_{i<j} v(\mathbf{r}_i - \mathbf{r}_j), \tag{6}$$

where

$$H(\mathbf{r}, \mathbf{p}) = \frac{1}{2m^*}\left[\mathbf{p} + \frac{e}{c}\mathbf{A}(\mathbf{r})\right]^2 + \tfrac{1}{2}m^*\Omega_0^2 x^2 \tag{7}$$

is the single-particle Hamiltonian. It has been shown for quantum wells (Ruden and Döhler 1983, Brey *et al.* 1989) and quantum dots (Maksym and Chakraborty 1990) with parabolic confinement, that H_N separates into two parts: a centre-of-mass (CM) motion, $(\mathbf{R}, \mathbf{P})$, and relative internal motion, $(\mathbf{r}_{j,\text{rel}}, \mathbf{p}_{j,\text{rel}})$:

$$H_N = H(\mathbf{R}, \mathbf{P}) + H_{\text{rel}}(\mathbf{r}_{j,\text{rel}}, \mathbf{p}_{j,\text{rel}}). \tag{8}$$

The CM motion obeys precisely the single-particle Hamiltonian and is the only allowed optical dipole excitation. Thus the optical response of quantum wires with parabolic confinement represents the collective CM excitation at the frequency of the unscreened external potential, *i.e.* for $B = 0$, the response is at $\hbar\Omega_0$ and *not* at the screened one-particle energy separation $\hbar\Omega_{01} = E_1 - E_0$ in Figure 9b. Thus dipole excitations are insensitive to electron-electron interaction for parabolic confinement; they are independent of the electron density in the wire and reflect the energy spectrum of the 'empty' wire. Shikin *et al.* (1990) have drawn the same conclusion from a classical description of an electron system in a parabolic confinement. The quantum mechanical CM motion corresponds exactly to the classical dipole plasma mode of the quantum wire. The result that the dipole excitation in a parabolic confinement is not affected by electron-electron interaction is a generalisation of Kohn's famous theorem (Kohn 1961), which states that the cyclotron frequency in a translationally invariant system is not influenced by electron-electron interaction.

In the RPA model, which we introduced above for a 2DES, we would interpret the observation that the resonant frequency is independent of the electron density in a parabolic confinement potential as arising from an exact cancellation of the decreasing one-particle energy (Figure 9b) by the increasing depolarisation shift. RPA calculations for quantum wires have been performed without (Que and Kirczenow 1988) and with (Wulf *et al.* 1990) self-consistent potential treatments. Actually this polarisation shift is very large. If we take, for example, the experimental results from Demel *et al.* (1988) and define the depolarisation shift ω_d as in the case of a 2DES, we find $\omega_d^2 = \omega_r^2 - \Omega_{01}^2 = (4\,\text{meV})^2 - (1\,\text{meV})^2 = (3.9\,\text{meV})^2$ for the deep mesa etched quantum wires (Figure 8c). Thus here the resonant frequency is strongly dominated by collective effects. Under such conditions, and also motivated by the fact that the dipole excitation in a parabolic confinement is a rigid CM motion of all electrons, it is also helpful to consider the excitations as 'local' plasmons in the following sense. We consider the confined charge oscillation in the quantum wire as a 'plasmon in a box'. The continuous 2D plasmon dispersion (Equation 3) of a homogeneous system with a free wave-vector q is now quantised in fixed values of $q = \pi/w_e$ and correspondingly

$$\omega_{pl}^2 = \frac{N_s e^2}{2\bar{\epsilon}\epsilon_0 m^*}\,\frac{\pi}{w_e}. \tag{9}$$

Here $\bar{\epsilon}$ is the effective dielectric constant. The effective width w_e is given by $w_e = w(1+\varphi)$, where φ takes account of the phase when the plasmon is 'reflected' at the walls of the box. This is of course a rather simplified model which, however, can explain many of the observed experimental observations qualitatively and also gives a quantitative estimate of the observed resonant frequency. For example, this model explains nicely the occurance of several resonances in multilayered wire structures which can be interpreted as local optical and acoustical plasmon resonances, *i.e.* resonances where the electrons in the different layers oscillate in or out of phase, respectively (Demel *et al.* 1988b). A more accurate model for a classical localised plasma oscillation in a parabolic confinement has be presented by Shikin *et al.* (1990).

8 One-dimensional plasmons

The excitations that we have discussed so far in the quantum wire were a mixture of single-particle excitations and confined plasmon oscillations in the direction perpendicular to the wire. It is however also possible to excite 1D plasmons propagating freely

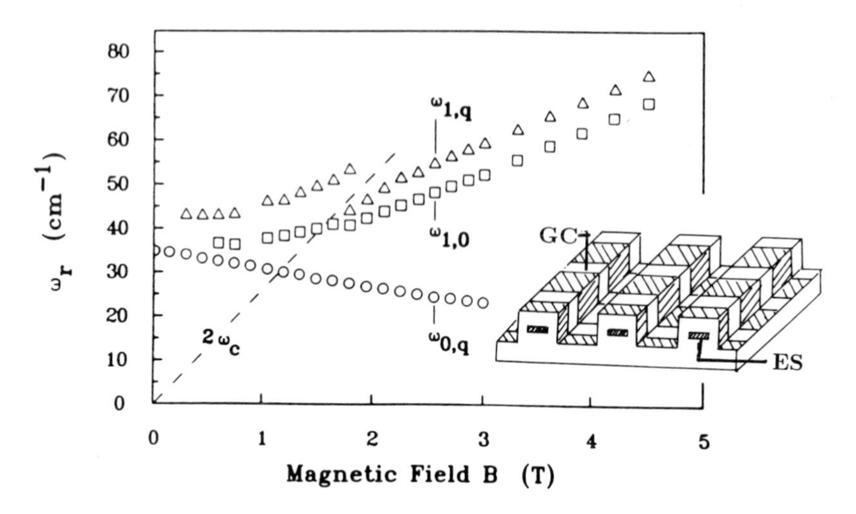

Figure 11: *Experimental dispersion in magnetic field B of 1D plasmons and combined plasmon modes in a deep mesa etched quantum wire sample with a perpendicularly oriented grating coupler as sketched in the inset. GC denotes the metal stripes of the grating coupler, and ES the 1D electronic system. The dash-dotted line is the calculated harmonic* $2\omega_c$ *of the cyclotron resonance and illustrates the anticrossing of* $\omega_{1,q}$ *with* $2\omega_c$. *From Demel et al. (1991).*

along the direction of the wires (Egeler *et al.* 1990). We have recently prepared single layered quantum wire structures where it was possible to couple 1D plasmons to FIR radiation which gave access to the interesting magnetic properties of these excitations (Demel *et al.* 1991). Periodic arrays of quantum wires have been fabricated from GaAs-$Al_xGa_{1-x}As$ heterostructures, consisting of a 10 nm thick cap layer, a 50 nm thick n-doped AlGaAs layer and a 25 nm thick spacer layer. With magnetic depopulation we determined that the quantum confinement in these narrow wires induces a 1D subband spacing of 1.7 meV (Berggren *et al.* 1986, Demel *et al.* 1988a). The linear charge density in the dark was $N_l = 4\times10^6\,\mathrm{cm}^{-1}$, the width of the electronic system was $w = 200\,\mathrm{nm}$ and six 1D subbands were occupied at $B = 0$. The density in the wires could be increased with short light pulses from a red light-emitting diode, leading simultaneously to an increase of the width w and to a decrease of the subband spacing. The critical point of the sample fabrication was the preparation of an additional array of periodic metal stripes, which were oriented perpendicular to the wires and acts as a grating coupler, *i.e.* couples FIR radiation to 1D plasmons propagating along the wires (inset to Figure 11). The deep mesa etched quantum wires discussed here have a geometrical width $t = 540\,\mathrm{nm}$ and a period $a = 1000\,\mathrm{nm}$, with a perpendicularly oriented metal grating coupler with period $b = 1000\,\mathrm{nm}$. We made the open space s between the metal stripes as small as possible ($s/b = 0.2$) to obtain a sufficiently large coupling strength to the 1D plasmon (Allen *et al.* 1977, Theis 1980, Tsui *et al.* 1978).

The experimental spectra show a number of modes. The dispersions for a sample with a quasi-2D density $N_s = 4.4\times10^{11}\,\mathrm{cm}^{-2}$ are shown in Figure 11. We explain these excitations in the following way. The mode labeled $\omega_{1,0}$ corresponds to a confined plasma oscillation perpendicular to the wire. The frequency of this mode is given approximately by the local plasmon frequency (Equation 9). $\omega_{1,0}$ is the same mode which has been observed in structures without metal grating couplers [see above and Demel *et al.* (1988), Hansen *et al.* (1987)] and shows the same linear dependence in a plot of ω^2 against B^2. Actually this is the only mode that we observed in the same wire sample before the preparation of the metal grating coupler.

The most interesting resonance in the spectrum is $\omega_{0,q}$, which can only be observed with the grating coupler. This resonance is a 1D plasmon which propagates with a wave-vector $q = 2\pi/b$ along the wire. The most prominent feature of this 1D plasmon is the negative dispersion in B, a behaviour which so far has only been observed for edge plasmons whose frequencies were determined by the circumference of the samples (Glattli *et al.* 1985, Mast *et al.* 1985, Allen *et al.* 1983). We find experimentally that we can vary the slope $\Delta\omega$ of the dispersion by changing the density profile in the wire. This is expected from calculations (Cataudella and Iadonisi 1987) and indicates that the plasmons observed on our structures are not just pure isolated edge magnetoplasmons, but are strongly governed by the finite width of the wire and density profile and thus represent real 1D plasmons. Another unique feature of 1D plasmons is a logarithmic term in the dispersion which is predicted for small qw (Das Sarma and Lai 1985, Li and Das Sarma 1989, Wu *et al.* 1985). So far our values of qw are too large to extract such effects from the experiments.

At $B = 0$ the plasma modes of a 2DES bound by two edges can be described approximately by an extension of the local plasmon model discussed above (Cataudella and Iadonisi 1987, Eliasson *et al.* 1986),

$$\omega_{n,q}^2 = \frac{N_s e^2}{2m^*\bar{\epsilon}\epsilon_0}\left[\left(\frac{n\pi}{w}\right)^2 + q^2\right]^{1/2} \qquad (n = 0, 1, 2, \ldots). \tag{10}$$

We have used n and q to label our modes. This dispersion relation is obtained from the 2D plasmon dispersion by replacing the 2D wave-vector by a combination of the wave vector q for the excitation parallel to the wires and a 'wave-vector' $n\pi/w$ for a plasma oscillation perpendicular to the wires, which can be thought of as a standing wave. In this model we can interprete the $\omega_{1,q}$ mode in Figure 11 as a confined plasma oscillation perpendicular to the wires, combined with a freely propagating plasmon along the wires. This mode shows a positive dispersion in B, in agreement with theory (Cataudella and Iadonisi 1987, Eliasson *et al.* 1986).

9 Quantum dots

In Demel *et al.* (1990) quantum dot samples were prepared starting from modulation-doped heterostructures. An array of photoresist dots (with a period of $a = 1000\,\text{nm}$ in both x and y-directions) was prepared by a holographic double exposure. Using an anisotropic plasma etching process, rectangular grooves 200 nm deep were etched all the way through the 10 nm thick GaAs cap layer, the 53 nm thick Si-doped AlGaAs layer, and the 23 nm thick undoped AlGaAs spacer layer into the active GaAs, leaving rectangular dots with rounded corners and geometrical dimensions of about $600 \times 600\,\text{nm}^2$.

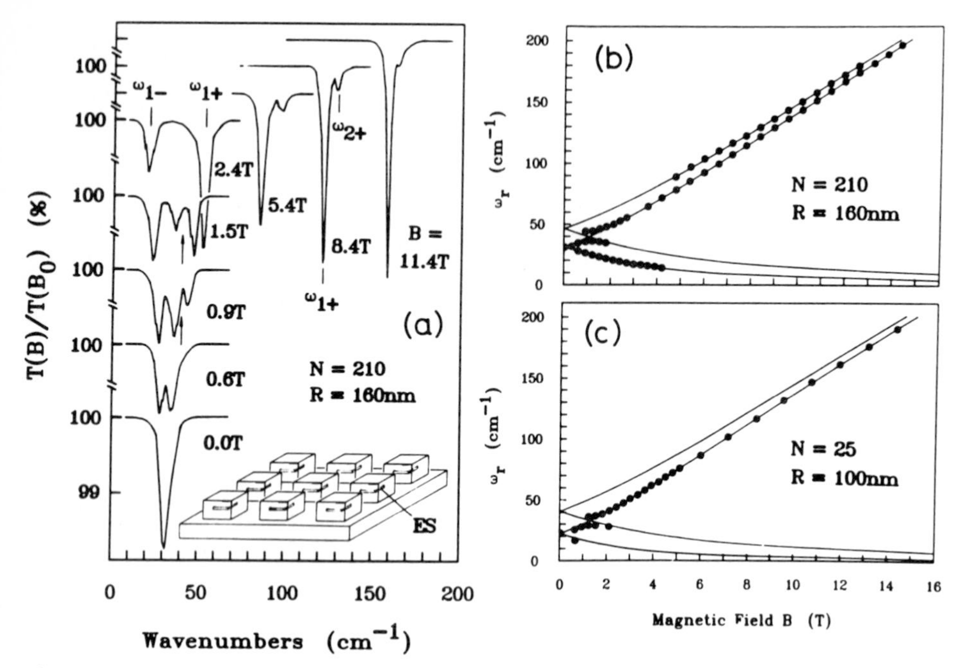

Figure 12: *(a) Normalised transmission of unpolarised* FIR *radiation for a* GaAs-$\mathrm{Al}_x\mathrm{Ga}_{1-x}$As *quantum dot structure with radius* $R = 160\,\mathrm{nm}$ *and* $N = 210$ *electrons per dot. The inset shows the structure of the dot schematically. In (b) and (c) we have plotted the experimental dispersion in B of resonant absorption in quantum dot structures with* $R = 160\,\mathrm{nm}$ *and* $N = 210$ *and with* $R = 100\,\mathrm{nm}$ *and* $N = 25$*, respectively. The full lines are fits with the theoretical dispersions (Equation 14). From Demel et al. (1990).*

These quantum dot samples are sketched in the inset to Figure 12a. Electrons are separated from the edge of the etched dots by lateral depletion regimes. We have estimated the actual radius of the confined electron system from the observed resonance frequency via Equation (13) which will be explained below. The number of electrons per dot, N, can be tuned with the persistent photo-effect.

The transmission $T(B)$ of unpolarised FIR radiation through the sample was measured at fixed magnetic fields, B, oriented normal to the surface of the sample. The spectra were normalised to a spectrum $T(B_0)$ with a flat response. The temperature was 2.2 K. The active area of the sample was $3 \times 3\,\mathrm{mm}^2$ containing 10^7 dots. Experimental spectra for a sample with dots of radius $R = 140\,\mathrm{nm}$ and $N = 210$ electrons per dot are shown in Figure 12a. For $B = 0$ one resonance is observed at $\omega_{01} = 32\,\mathrm{cm}^{-1}$. With increasing B the resonance splits into two. One, ω_{1-}, decreases in frequency while the other, ω_{1+}, increases. For $B > 4\,\mathrm{T}$ a second resonance, ω_{2+}, can be resolved whose position also increases with B. Experimental resonance positions for two situations where each dot contains $N = 210$ and only $N = 25$ electrons are shown in Figure 12b and c. An interesting observation is the resonant anticrossing at $\omega \approx 1.4\,\omega_{01}$.

The FIR resonances observed here are, except for the resonant splitting and the higher modes, very similar to earlier observations on larger finite-sized 2DESs in GaAs-$Al_xGa_{1-x}As$ heterostructures (dots with $R \approx 1.5\,\mu m$) by Allen *et al.* (1983). Similar dispersions and higher modes, but no anticrossing, were observed for electron systems on liquid He with large radii, $R \approx 1\,cm$ (Glattli *et al.* 1985, Mast *et al.* 1985). These experiments have been interpreted as depolarisation or, equivalently, as edge magnetoplasmon modes (Shikin *et al.* 1990, Fetter 1985, Fetter 1986, Sandormirskii *et al.* 1989, Shikin *et al.* 1991). Also, Sikorski and Merkt (1989) observed resonances with the same dispersions, except for the anticrossing, as our ground modes $\omega_{1\pm}$ for quantum dots in InSb systems containing only a very small number of electrons per dot ($N \approx 5$). An even more complex mode structure was found for coupled quantum dots (Lorke *et al.* 1990). For the quantum-confined systems one might expect, at a first glance, that an adequate description of the FIR response can be given in terms of transitions between the discrete energy levels of the 'atoms'. However, we have already seen from the experiments on 1DESs with quantum-confined energy states that the optical response of such low-dimensional systems is strongly influenced by collective effects with increasing number of electrons. To elucidate this behaviour for quantum dot structures, and to explain the experimentally observed excitation spectra, we will approach this problem from two limits: (i) transitions in an 'atom' including collective effects, and (ii) plasmon-like of excitations in a 2DES of finite size.

We have seen already that the external confinement potential for electrons has a nearly parabolic shape for the quantum wire structures; see also the self-consistent calculations for quantum dots by Kumar *et al.* (1990). The one-electron energy eigenvalues in a magnetic field B for a potential $\frac{1}{2}m^*\Omega_0^2 r^2$ have been calculated by Fock (1928):

$$E_{n,l} = (2n + |l| + 1)\sqrt{(\hbar\Omega_0)^2 + \left(\frac{\hbar\omega_c}{2}\right)^2} + l\left(\frac{\hbar\omega_c}{2}\right), \tag{11}$$

where $n = 0, 1, 2, \ldots$ and $l = 0, \pm1, \pm2, \ldots$ are the radial and azimuthal quantum numbers respectively, and ω_c is the cyclotron frequency.

From calculations of the matrix elements one finds that allowed dipole transitions have energies

$$\Delta E^{\pm} = \sqrt{\hbar^2\Omega_0^2 + \left(\frac{\hbar\omega_c}{2}\right)^2} \pm \left(\frac{\hbar\omega_c}{2}\right). \tag{12}$$

These are shown in Figure 13b. This result has been derived for one electron. However, we can use the same arguments as in case of the quantum wires, *i.e.* the generalisation of Kohn's theorem, that the one-particle excitation spectrum gives also directly the allowed dipole transitions for N electrons in dots with parabolic confinement (Maksym and Chakraborty 1990). Actually, this dispersion (Equation 12) is, except for the anticrossing, exactly the behaviour that we observe in our experiments for the strongest mode in Figure 12. Sikorski *et al.* (1989) have observed very nicely that the resonant frequency is indeed independent of the number of electrons in the dot.

For a small number of electrons per dot it is possible to calculate accurately, without any approximation, wavefunctions and the many-electron energy states in a parabolic or square well confinement potential (Maksym and Chakraborty 1990, Bryant 1987, Merkt *et al.* 1991). This gives us an interesting insight into many-body effects in

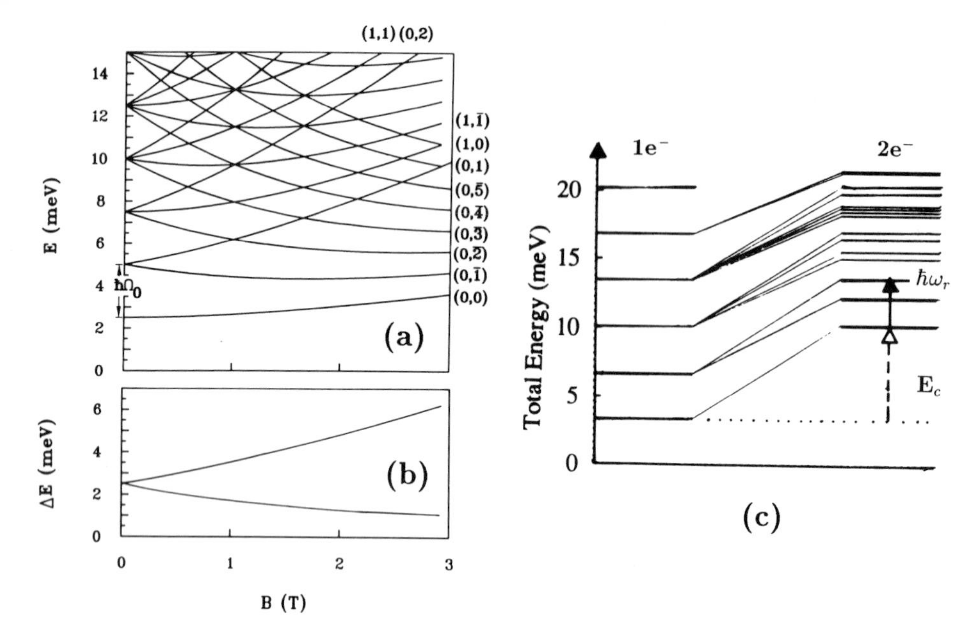

Figure 13: *(a) One-particle energy spectrum of quantum dots with a parabolic confinement potential and (b) allowed dipole transition energies. (c) shows, at $B = 0$, the one-particle energy spectrum and the two-particle excitation spectrum ('quantum-dot He') in a parabolic confinement potential. The indicated allowed dipole transition ($\hbar\omega_r$) for quantum-dot He has exactly the same energy as for one electron. (With special thanks to D Pfannkuche and R R Gerhardts.)*

spectroscopy. In Figure 13 we compare the one-particle energy spectrum with the two-particle energy spectrum in the same parabolic confinement. The many-body energy spectrum is governed by two energies: the quantum confinement energy, $E_q = \hbar^2/m^* l_0^2$, and the Coulomb energy, $E_c = e^2/4\pi\epsilon\epsilon_0 l_0$, which represents the electron-electron interaction. Here l_0 is the harmonic oscillator length. The Coulomb energy becomes increasingly important with respect to the confinement energy as the confinement length l_0 increases. The scale is set by the effective Bohr radius a^* which is 10 nm in GaAs. In Figure 13 with $E_q = 3.4$ meV and correspondingly $l_0 = 20$ nm, the Coulomb energy is $E_c = 7$ meV. This reflects the energy which is required to 'squeeze' the second electron into the quantum dot and shows up in Figure 13 in the increase of the lowest energy level with respect to the one-particle spectrum. Some of the degeneracies of the one-particle spectrum are now lifted due to the electron-electron interaction, giving rise to a complex excitation spectrum. However the only allowed dipole transition ($\hbar\omega_r$ in Figure 13) has exactly the same energy as in the one-particle case, as expected from the generalised Kohn's theorem. In recent experiments it was possible to realise quantum dot arrays confined by the field effect with well-defined numbers of electrons in each individual dot, $N = 1, 2, 3$ and 4. It was shown that the high Coulomb energy in a small quantum dot stabilises a well-defined number of electrons (Meurer *et al.* 1992).

Since the FIR response of the quantum dots reflects dominantly the rigid CM motion, we can also approach the dynamic excitations of dots from a classical plasmon type of excitation in a finite geometry, which gives us insight into the microscopic nature of the excitations. A simple way is to assume a disk-like 2DES with 2D density N_s and radius R and start from linear edge plasmons, *i.e.* excitations which exist at an edge of a semi-infinite 2DES (see, for example, Shikin *et al.* 1990, Fetter 1985, Fetter 1986, Sandormirskii *et al.* 1989, Shikin *et al.* 1991). These excitations have the dispersion $\omega_{ep}^2 = 0.81\,\omega_p^2(q)$ where ω_p is the 2D plasmon frequency (Equation 3). For a disk the circumference quantises the q-vectors into values $q = i/R$ ($i = 1, 2, \ldots$). For $B = 0$ we thus have

$$\omega_{0i}^2 = 0.81 \frac{N_s e^2 i}{2m^* \epsilon_0 \bar{\epsilon} R}. \tag{13}$$

In a magnetic field one calculates a set of double branches

$$\omega_{i\pm} = \sqrt{\omega_{0i}^2 + \left(\frac{\omega_c}{2}\right)^2} \pm \left(\frac{\omega_c}{2}\right). \tag{14}$$

The prefactor 0.81 in Equation 13 and the spacings of higher modes depend slightly on the modeling of the electron density profile (Shikin *et al.* 1990, Fetter 1985, Fetter 1986, Sandormirskii *et al.* 1989, Shikin *et al.* 1991). Note that the dispersion in B of the lowest mode agrees exactly with the quantum mechanical result. These classical treatments also give the higher modes which are so far not available from quantum mechanical treatments, *e.g.* in Maksym and Chakraborty (1990). In the classical model and with increasing B the ω_{i+}-resonances correspond to cyclotron-type excitations of the electrons. The ω_{i-}-modes represent edge magnetoplasmon modes where the individual electrons within the collective excitation perform skipping orbits along the edge of the dot. Within this classical model one can also explain the FIR response of antidots which has recently been studied by Kern *et al.* (1991).

Within the limited space of this contribution we cannot discuss in detail the interesting anti-crossing behaviour that we observe in the different systems, mode anti-crossing for the dots (Figure 12) and anti-crossing with $2\omega_c$ for the quantum wires (Figure 11). If we adopt the plasmon model, this anti-crossing resembles very similar observations in 2D magnetoplasmon excitations (Batke *et al.* 1985), where it was shown to arise from non-local interactions. The 'non-local interactions' for the microstructures here depend strongly on the exact shape of the potential which shows that a microscopic quantum theoretical description of the system, which is automatically a non-local theory, is necessary to explain the anti-crossing quantitatively.

10 Summary

We have reviewed dynamic excitations in electronic systems with different dimensionality. Free-electron 3D metals show dominantly a collective plasmon-like behaviour. In low-dimensional quantum-confined systems we have both collective plasmon excitations and single-particle excitations. The latter are, however, always influenced by collective contributions due to the finite strength of the confinement. Currently available modulation-doped quantum wires and dots have a nearly parabolic confinement potential. This leads to a unique situation that the dipole excitations represent the rigid CM oscillation of all electrons. Thus, even with only two electrons in quantum-dot He, the excitation is, in a certain sense, a localised plasmon in the dot.

Acknowledgments

In these lecture notes I have reported investigations which were only possible due to excellent cooperation with many colleagues. I would like very much to thank all of my colleagues who have worked with me on these different subjects, as listed in the References. I also acknowledge financial support from the BMFT.

References

Allen S J, Tsui D C, and Vinter B, 1976. *Solid State Commun.*, **20**, 425.
Allen S J, Tsui D C, and Logan R A, 1977. *Phys. Rev. Lett.*, **38**, 980.
Allen S J, Störmer H L, and Hwang J C, 1983. *Phys. Rev. B*, **28**, 4875.
Alsmeier J, Sikorski Ch, and Merkt U, 1988. *Phys. Rev. B*, **37**, 4314.
Ando T, Fowler A B, and Stern F, 1982. *Rev. Mod. Phys.*, **54**, 437.
Ando T, 1977. *Z. Phys. B*, **26**, 263; **58**, 263 (1976).
Ando T, Eda T, and Nayakama M, 1977. *Solid State Commun.*, **23**, 751.
Batke E, and Heitmann D, 1983. *Solid State Commun.*, **47**, 819.
Batke E, and Heitmann D, 1984. *Infrared Phys.*, **24**, 189.
Batke E, Heitmann D, Wieck A D, and Kotthaus J P, 1983. *Solid State Commun.*, **46**, 269.
Batke E, Heitmann D, and Tu C W, 1986. *Phys. Rev. B*, **34**, 6951.
Batke E, Heitmann D, Kotthaus J P, and Ploog K, 1985. *Phys. Rev. Lett.*, **54**, 2367.
Berggren K-F, Thornton T J, Newson D J, and Pepper M, 1986. *Phys. Rev. Lett.*, **57**, 1769.
Brey L, Johnson N, and Halperin P, 1989. *Phys. Rev. B*, **40**, 10647.
Brinkop F, Hansen W, Kotthaus J P, and Ploog K, 1988. *Phys. Rev. B*, **37**, 6547.
Broadman A D (ed), 1982. *Electromagnetic Surface Modes*, Chichester, Wiley.
Bryant G W, 1987. *Phys. Rev. Lett.*, **59**, 1140.
Cataudella V, and Iadonisi G, 1987. *Phys. Rev. B*, **35**, 7443.
Chaplik A V, 1972. *Sov. Phys. JETP*, **35**, 395.
Chaplik A V, 1985. *Surf. Sci. Rep.*, **5**, 289.
Chen W P, Chen Y J, and Burstein E, 1976. *Surf. Sci.*, **58**, 26.
Das Sarma S, and Lai Y W, 1985. *Phys. Rev. B*, **32**, 1401; *Surf. Sci.*, **170**, 43.
Demel T, Heitmann D, Grambow P, and Ploog K, 1990. *Phys. Rev. Lett.*, **64**, 788.
Demel T, Heitmann D, and Grambow P, 1989. In *Spectroscopy of Semiconductor Microstructures*, Eds. Fasol G, Fasolino A, and Lugli P, Nato ASI Series B: Physics **206**, New York, Plenum Press, p. 75.
Demel T, Heitmann D, Grambow P, and Ploog K, 1988a. *Appl. Phys. Lett.*, **53**, 2176.
Demel T, Heitmann D, Grambow P, and Ploog K, 1988b. *Phys. Rev. B*, **38**, 12732.
Demel T, Heitmann D, Grambow P, and Ploog K 1991. *Phys. Rev. Lett.*, **66**, 2657.
Egeler T, Abstreiter G, Weimann G, Demel T, Heitmann D, and Grambow P, 1990. *Phys. Rev. Lett.*, **65**, 1804.
Eliasson G, Wu J-W, Hawrylak P, and Quinn J J, 1986. *Solid State Commun.*, **60**, 41.
Ensslin K, and Petroff P M, 1990. *Phys. Rev. B*, **41**, 12307.
Ensslin K, Heitmann D, and Ploog K, 1989. *Phys. Rev. B*, **39**, 10879.
Ensslin K, 1989. PhD thesis, Stuttgart.

Fetter A L, 1985. *Phys. Rev. B*, **32**, 7676.
Fetter A L, 1986. *Phys. Rev. B*, **33**, 5221.
Fock V, 1928. *Z. Phys.*, **47**, 446.
Forstmann F, and Gerhardts R R, 1986. *Metal optics near the plasma frequency*, in *Springer Tracts on Modern Physics*, **109**, Berlin, Springer.
Glattli D C, Andrei E Y, Deville G, Poitrenaud J, and Williams F I B, 1985. *Phys. Rev. Lett.*, **54**, 1710.
Grambow P, Vasiliadou E, Demel T, Kern K, Heitmann D, and Ploog K, 1990. *Microelectronic Engineering*, **11**, 47.
Grambow P, Demel T, Heitmann D, Kohl M, Schüle R, and Ploog K, 1989. *Microelectronic Engineering*, **9**, 357.
Grimes C C , and Adams G, 1976. *Phys. Rev. Lett.*, **36**, 145.
Hansen W, Smith T P, Lee K Y, Brum J A, Knoedler C M, Hong J M, and Kern D P, 1989. *Phys. Rev. Lett.*, **62**, 2168.
Hansen W, Horst M, Kotthaus J P, Merkt U, Sikorski Ch, and Ploog K, 1987. *Phys. Rev. Lett.*, **58**, 2586.
Heitmann D, 1977. *J. Phys. C: Solid State Phys.*, **10**, 397.
Heitmann D, 1986. *Surf. Sci*, **170**, 332.
Heitmann D, 1987. *Physics and applications of quantum wells and superlattices*, in Nato ASI Series B: Physics, **170**, Eds. Mendez E E, and von Klitzing K, New York, Plenum Press, p. 317
Heitmann D, 1990. *Electronic Properties of Multilayers, and Low-Dimensional Semiconductor Structures*, in Nato ASI Series B: Physics, **231**, Eds. Chamberlain J M, Eaves L, and Portal J C, New York, Plenum Press, p. 151.
Heitmann D, Demel T, Grambow P, and Ploog K, 1990. *Advances in Solid State Physics*, **29**, Ed. Rössler U, Braunschweig, Vieweg, p. 285.
Heitmann D, and Ensslin K, 1991. *Quantum coherence in mesoscopic systems*, in Nato ASI Series B: Physics **254**, Ed. Kramers B, New York, Plenum Press, p. 3.
Heitmann D, Kotthaus J P, and Mohr E G, 1982. *Solid State Commun.*, **44**, 715.
Heitmann D, Kotthaus J P, Mackens U, Beinvogl W, 1985. *Superlatt. Microstruct.*, **1**, 35.
Heitmann D, Kroo N, Schulz Ch, and Szentirmay Zs, 1987. *Phys. Rev. B*, **35**, 2660.
Heitmann D, and Mackens U, 1986. *Phys. Rev. B*, **B33**, 8269.
Heitmann D, and Raether H, 1976. *Surf. Sci.*, **59**, 17.
Helms M, Colas E, England P, DeRosa F, and Allen S J, 1989. *Appl. Phys. Lett.*, **53**, 1714.
Höpfel R A, and Gornik E, 1984. *Surf. Sci.*, **142**, 412.
Kern K, Heitmann D, Gerhardts R R, Grambow P, Zhang Y H, and Ploog K, 1991a. *Phys. Rev. B*, **44**, 1139.
Kern K, Heitmann D, Grambow P, Zhang Y H, Ploog K, 1991b. *Phys. Rev. Lett.*, **66**, 1618.
Kern K, Demel T, Heitmann D, Grambow P, Ploog K, Razeghi M, 1990. *Surf. Sci.*, **229**, 256.
Kirtley J, Theis T N, and Tsang J C, 1981. *Phys. Rev. B*, **24**, 5650.
Kneschaurek P, Kamgar A, and Koch J F, 1976. *Phys. Rev. B*, **14**, 1610.
Kohn W, 1961. *Phys. Rev.*, **123**, 1242.
Krasheninnikov M V, and Chaplik A V, 1981. *Sov. Phys. Semicond.*, **15**, 19.
Kumar A, Laux S E, and Stern F, 1990. *Phys. Rev. B*, **42**, 5166.
Laux S E, Frank D J, and Stern F, 1988. *Surf. Sci.*, **196**, 101.
Li Q, and Das Sarma S, 1989. *Phys. Rev. B*, **40**, 5860.
Li Q, and Das Sarma S, 1990. *Phys. Rev. B*, **41**, 10268.
Liu C T, Nakamura K, Tsui D C, Ismail K, Antoniadis D A, and Smith H I, 1989. *Appl. Phys. Lett.*, **55**, 168.

Lorke A, Kotthaus J P, and Ploog K, 1990. *Phys. Rev. Lett.*, **64**, 2559.
Lorke A, Kotthaus J P, and Ploog K, 1991. *Superlatt. Microstruct.*, **9**, 103.
Mackens U, Heitmann D, and Kotthaus J P, 1986. *Surf. Sci.*, **170**, 346.
Mackens U, Heitmann D, Prager L, Kotthaus J P, and Beinvogl W, 1984. *Phys Rev. Lett.*, **53**, 1485.
Maksym P, and Chakraborty T, 1990. *Phys. Rev. Lett.*, **65**, 108.
Mast D B, Dahm A J, and Fetter A L, 1985. *Phys. Rev. Lett.*, **54**, 1706.
McCombe B D, Holm R T, and Schafer D E, 1979. *Solid State Commun.*, **32**, 603.
Merkt U, Sikorski CH., and Alsmeier J, 1989. In *Proc Spectroscopy of Semiconductor Microstructures*, Eds. Fasol G, Fasolino A, and Lugli P, Nato ASI Series B: Physics **206**, New York, Plenum Press, p. 89.
Merkt U, Huser J, and Wagner M, 1991. *Phys. Rev. B*, **43**, 7320.
Meurer B, Heitmann D, and Ploog K 1992. *Phys. Rev. Lett.*, in press.
Oelting S, Heitmann D, and Kotthaus J P, 1986. *Phys. Rev. Lett.*, **56**, 1846.
Pines D, and Nozières Ph, 1966. *The Theory of Quantum Liquids*, New York, Benjamin.
Proc. 6th Int. Winter School on Localisation and Confinement of Electrons in Semiconductors, Mauterndorf, Austria, 1990. Eds Bauer G, Heinrich H, and Kuchar F, Berlin, Springer.
Que W, and Kirczenow G, 1988. *Phys. Rev. B*, **37**, 7153.
Que W, and Kirczenow G, 1989. *Phys. Rev. B*, **39**, 5998.
Raether H, 1977. In *Physics of Thin Films*, **9**, New York, Academic Press, p. 145.
Raether H, 1980. In *Springer Tracts in Modern Physics*, **88**, Berlin, Springer.
Reed M A , Randall J N, Aggarwal R J, Matyi R J, Moore T M, and Wetsel A E, 1988. *Phys. Rev. Lett.*, **60**, 535.
Reed M A, and Kirk W P (Eds), 1989. *Nanostructure Physics and Fabrication*, Boston, Academic Press.
Rikken G L J A, Sigg H, Langerak G J M, Myron H W, and Perenboom J A A J, 1986. *Phys. Rev. B*, **34**, 5590.
Ritchie R H, 1957. *Phys. Rev.*, **106**, 874.
Ruden P, and Döhler G H, 1983. *Phys. Rev. B*, **27**, 3547.
Sandomirskii V B, Volkov V A, Aizin G R, Mikhailov S A, 1989. *Electrochim. Acta*, **34**, 3.
Schlesinger Z, Hwang J C M, and Allen S J, 1983. *Phys. Rev. Lett.*, **50**, 2098.
Shikin V, Nazin S, Heitmann D, and Demel T, 1991. *Phys. Rev. B*, **343**, 11903.
Shikin V, Demel T, and Heitmann D, 1990. *Surf. Sci.*, **229**, 276.
Sikorski Ch. , and Merkt U, 1989. *Phys. Rev. Lett.*, **62**, 2164.
Smith T P, Arnot H, Hong J M, Knoedler C M, Laux S E, and Schmid H, 1987. *Phys. Rev. Lett.*, **59**, 2802.
Stern F 1967. *Phys. Rev. Lett.*, **18**, 546.
Teng Y Y, and Stern E A, 1967. *Phys. Rev. Lett.*, **19**, 511.
Theis T N, 1980. *Surf. Sci.*, **98**, 515.
Theis T N, Kotthaus J P, and Stiles P J, 1977. *Solid State Commun.*, **24**, 273.
Theis T N, Kotthaus J P, and Stiles P J, 1978. *Solid State Commun.*, **26**, 603.
Tsui D C, Allen S J, Logan R A, Kamgar A, Coppersmith S N, 1978. *Surf. Sci.*, **73**, 419.
Weiss D, von Klitzing K, and Ploog K, 1990. *Surface Sci.*, **229**, 88.
Wieck A D, Maan J C, Merkt U, Kotthaus J P, Ploog K, and Weimann G, 1987. *Phys. Rev. B*, **35**, 4145.
Wu J-W, Hawrylak P, and Quinn J J, 1985. *Phys. Rev. Lett.*, **55**, 879.
Wulf U, Zeeb E, Gries P, Gerhardts R R, Hanke W, 1990. *Phys. Rev. B*, **41**, 3113; **42**, 7637.

Experiments on Lateral Superlattices

W Hansen

Universität München
München, Germany

1 Introduction

Man-made modifications of the electronic properties of semiconductors by imposing an artificial superlattice on the crystal offers intriguing possibilities for physical investigations as well as technology (Esaki and Tsu 1970). Well established are one-dimensional (1D) superlattices in multilayered heterostructures as well as so-called *nipi*-structures (Esaki 1987). There a vertical sequence of alternating layers of semiconductor material with different bandgap, workfunction and/or doping is grown on a substrate crystal by means of epitaxial methods. The bandstucture of formerly three-dimensional electrons in the semiconductor conduction band is modified in the vertical dimension, *i.e.* normal to the surface of the multilayer structure. More recently, a large effort has been devoted to the generation of potential gradients that perturb the potential of a formerly homogeneous two-dimensional electron system (2DEG) residing at an interface close and parallel to the crystal surface (Ando *et al.* 1982). Rapid progress in refined lithographic methods has made feasible the fabrication of masks that vary in a dimension parallel to the crystal surface, *i.e.* in a lateral dimension, on a scale of about 100 nm. The geometry of such masks can be transformed into a spatially modulated potential landscape so that an electron system is formed whose density is modulated if the Fermi energy is still above the potential maxima or, furthermore, electron systems of lower dimension are generated if the lateral potential forms channels or pockets with walls that reach well above the Fermi energy.

The widespread interest in the study of such laterally patterned systems stems from the advantageous properties of the two-dimensional systems (2DEGs) from which they are derived. The electrons in a 2DEG are strongly confined in the direction normal to the interface and may be well separated from donors. Therefore a degenerate electron plasma with high density can be formed without severely degrading the mobility by

Coulomb scattering from the potentials of the donors. Thus large electronic mean free paths at low temperatures are among the most prominent advantages of 2DEGs. Since we expect an artificial superlattice potential to modify the electronic states of the system significantly only if the mean free path in the unperturbed system is much larger than the period of the potential, long mean free paths are vital for the observation of superlattice effects. More generallly we expect that the electron system changes its physical properties each time its extension becomes comparable to a length scale tied to a distinct physical process. Typical length scales are the phase coherence length, the elastic mean free path or the de Broglie wavelength. The fact that we can control the electron density in 2DEGs and study their electronic properties as a function of the Fermi energy is therefore highly desirable. Another important property is the reduced dimensionality. A superlattice potential has an increasing impact on the density of states the lower the dimensionality of the underlying system. This becomes immediately plausible if we remember that only in a 1D system is a true gap in the density of states created by a 1D superlattice potential.

From an experimentalist's point of view there is another aspect that makes superlattices invaluable for the investigation of nanometre systems, if experimental methods other than mere resistance measurements are to be applied. The lateral periodic repetition of nanostructures in a superlattice offers the possibility to extend the investigation of nanostructures to methods which inherently probe large areas of samples and/or for which the signal strength has to be boosted up above noise level by measuring in parallel a large number of nanometre systems that are as equal as possible. In an optical transmission or reflection experiment the focal area of the probe is much larger than a single mesoscopic device. In far-infrared spectroscopy, for instance, the wavelength of the probe is of order $100\,\mu$m and, accordingly, the areas probed in the focus of a transmission experiment are in the mm^2 range. As another example the capacitance of a single quasi zero-dimensional (0D) electron dot is, according to the small space it occupies, extremely small. In order to get reasonable signal strengths in measurements of quantum dot capacitances, a large number of dots is therefore measured in parallel, so that with sufficient similarity of the dots the contributions of the individual dots to the signal add coherently. In this article, I would like to focus on two different properties of nanometre systems to which experimental access is possible by taking advantage of the periodic repetition of nanometre systems: the dynamic conductivity and the capacitance. The measurements of these properties have given us a great deal of new insights into the physics of nanometre systems. Since the currently complex field of static transport measurements in 2DEGs with a weak superlattice potential is covered in a special contribution to this workshop, I will address static transport only very briefly at those points where cross links between different experiments should be pointed out. The discussion of interband optical spectroscopy on such systems, which has gained vast interest in the last three years, is completely omitted. This topic is also covered in different contributions to this volume.

2 Fabrication of superlattices

Usually the fabrication of narrow 1D electron wires or electron dots in semiconductor systems starts from a 2DEG, because of its advantageous properties. Within the

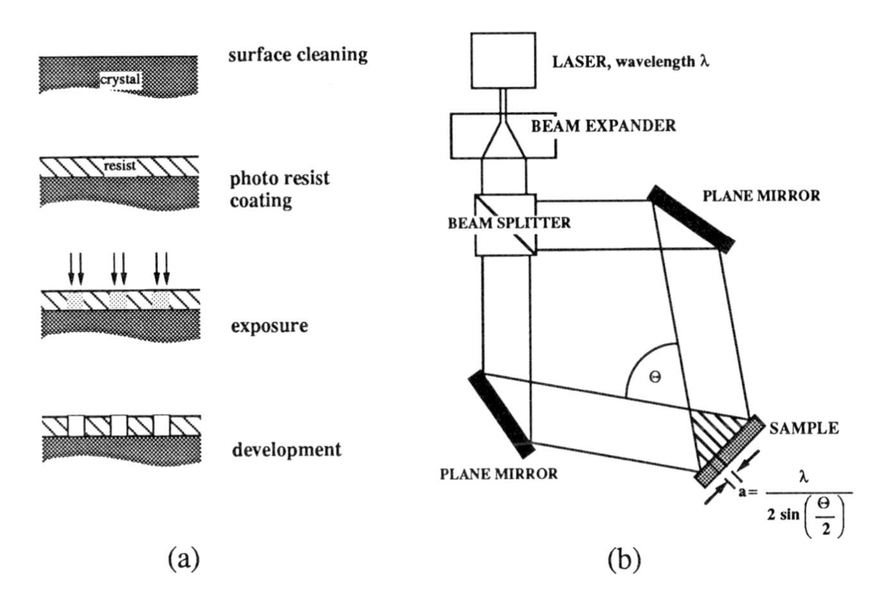

Figure 1: *(a) Process steps for definition of a lithography mask; (b) typical optical setup for holographic lithography.*

last decade a large number of methods has been invented to impose a lateral potential modulation on 2DEGs. They may be divided naturally into maskless techniques and in patterning methods that utilise lithography. Maskless techniques take advantage of special properties of the material under investigation. For instance, the persistent photo-effect in GaAs-$Al_xGa_{1-x}As$ heterojunctions can be utilised to modulate the electron density spatially with a light beam of spatially modulated intensity (Tsubaki *et al.* 1984, Weiss *et al.* 1988). Other maskless techniques of great future potential are regrowth on tilted surfaces (Tanaka and Sakaki 1989, Petroff *et al.* 1984) and focused ion beam exposure (Nieder *et al.* 1990).

The experiments discussed in the following sections are, however, performed on samples of nanostructures that have been fabricated with more conventional lithographic techniques. There the crystal surface is first covered with a resist mask which is subsequently exposed to light, electron or ion beams. The exposing beam writes the lateral pattern into the resist, so that a resist pattern is left on the crystal surface after a development process, and can be used as a mask for subsequent fabrication processes. The most important steps of lithographic mask production are depicted in Figure 1a. As an example, an exposure setup is sketched in Figure 1b that is of specific importance for the fabrication of periodic superlattices. Here the resist is sensitive to optical radiation and it is exposed in an interferometer. An expanded laser beam of wavelength λ is split into two beams of equal intensity which are then brought together to interfere with each other. The interference pattern is a field of light like a grating to which the resist on the sample surface is exposed. After development a grating of stripes in resist is left on the sample with period $a = \lambda/[2\sin(\frac{1}{2}\theta)]$, where θ is the angle between the two interfering laser beams. A matrix of resist dots can be generated if the sample is rotated

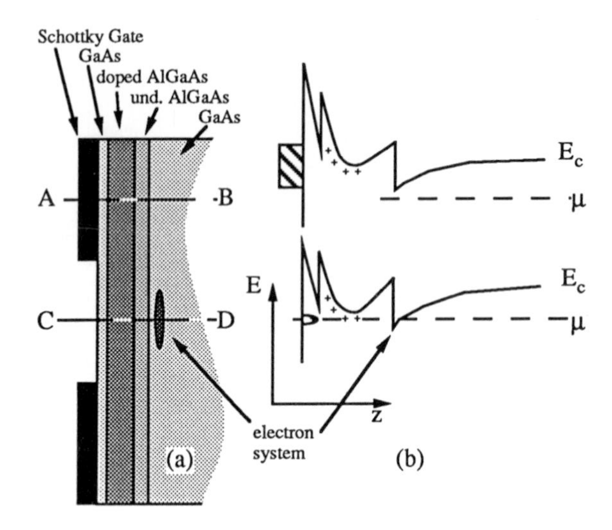

Figure 2: *(a)* GaAs-Al_xGa_{1-x}As *heterojunction device with a split Schottky gate, and (b) sketch of the heterojunction conduction band beneath the gate at negative bias and at the open heterojunction surface between the metal patterns.*

after the first exposure and exposed a second time to the interference pattern. This technique of holographic lithography is especially suited for the generation of periodic microstructures on large areas with precise homogeneity of the lattice constant.

Using the resist as a mask, its pattern is transformed into a potential in the next step in fabrication. Here again, a large number of possible methods is available. In the most important ones, which are used in the experiments discussed in the following three sections, either external fields have been applied via properly patterned field-effect (gate) electrodes or the surface has been modified by etching. As an example, the resist mask has been used to evaporate a patterned Schottky gate on an GaAs-Al_xGa_{1-x}As heterostructure in Figure 2a. The effect of the patterned Schottky gate on the band bending at the open heterostructure surface in the gap between the metal patterns strongly depends on a large density of surface states which is believed to exist at the surface of the GaAs cap around mid-gap position. The corresponding band bending of the conduction band in the heterostructure is sketched in the lower part of Figure 2b. Here it is assumed that the donor density in the doped GaAs-Al_xGa_{1-x}As layer is sufficiently high that there are electrons in the potential minimum at the heterojunction. Due to the different work functions, the band bending beneath the metal patterns can be different, even without application of a voltage between the Schottky gate and the electron system. The top band structure of Figure 2b depicts the situation with a negative gate voltage applied so that the conducting band at the heterojunction interface is well above the electrochemical potential in the GaAs. Consequently the heterojunction is no longer occupied in the areas beneath the biased gate patterns, and the electron system is confined in the lateral direction to the space beneath the gaps in the patterns. In this manner quasi-one and zero-dimensional electron systems can be generated with novel and intriguing electronic properties.

Figure 3 shows a series of different possible gate configurations apart from that in

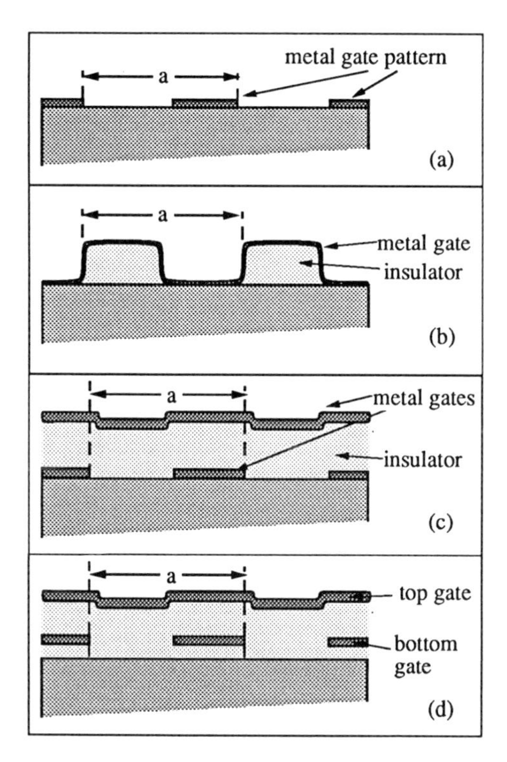

Figure 3: *Different gate configurations that are used to impose lateral potential gradients on* 2DEGs.

Figure 2. Each of them has successfully been applied to generate a superlattice of electron density in a 2DEG or electron systems of lower dimensionality. The configuration depicted in Figure 3b is a modification of the 'split gate' in Figure 3a. The distance of a metal layer from the 2DEG is modulated by an insulator that has been microstructured. In the areas beneath the patterned insulator the electron density is less affected by the gate voltage than in those where the gate directly contacts the semiconductor surface. The important advantage of this method is that the gate consists of a layer that still covers the whole sample surface. Thus even patterns that are not singly connected can be realised. If, for instance, small holes — so-called 'antidots' — are to be punched into the electron system, the split gate pattern in Figures 2 and 3a would have to consist of individual tiny discs that are not easily connected to a voltage source.

In Figure 3c a combination of a Schottky gate with a field-effect gate is shown. This method has been applied to generate quantum wires and quantum discs in InSb-MOS devices. In those devices the floating Schottky gate is made of NiCr and it is thought that it pins the Fermi energy in the band gap above the edge of the valence band. The top gate is positively biased so that electrons in the inversion layer are induced at the semiconductor-insulator interface in the gaps between the Schottky gate patterns. The most versatile gate configuration is the 'stacked gate' design shown in Figure 3d. This configuration can be used in *p*-type MOS devices either to confine electrons in the gaps between or, in a reversed mode, directly beneath the bottom gate patterns. In the first mode the bottom gate is biased negative and the top gate is positive, *vice versa* in the reversed mode. The 'stacked gate' configuration is the only one that allows us to control both the lateral extent and the charge density of confined electron systems by proper adjustment of the top and bottom gate voltages.

Alternative methods to generate 1D as well as 0D electron systems employ physical and/or chemical alteration of the surface of the heterostructure by etching methods. Here the patterned resist serves as a mask that protects those areas of the sample which are not to be etched. The distance between the electron system at the heterojunction and the surface is decreased by physical removal of material from the surface of the heterostructure and thus, due to surface states, the impact of the conduction band bending increases. Clearly, in so-called deep mesa etched samples, where the heterostructure surface is removed down to a depth beyond the heterojunction, the electrons are confined within the unetched patterns. Carriers trapped in surface states of the exposed sidewalls of the etched patterns cause large potential gradients, so that the effective width to which the electron system is found to be confined is significantly smaller than the width of the etched pattern. Depletion widths of several 100 nm were found in deep mesa etched GaAs-$Al_xGa_{1-x}As$ heterojunction wires. Thus the lateral depletion lengths are much larger than the widths anticipated for homogeneous depletion layers at semiconductor junctions of similar doping, where the potential varies only in one dimension, *i.e.* normal to the junction. Because of the large influence of the surface potential it turns out that it is sufficient to etch the surface to a depth which is only a fraction of the distance to the electron system in order to achieve the confinement of carriers. It is believed that the damage introduced unintentionally by etching into the masked areas is reduced in such a so-called shallow etching process so that the mobility in the confined electron system is degraded less. The resulting cross sections of deep and shallow mesa etched GaAs-$Al_xGa_{1-x}As$ heterostructures are sketched in Figure 4(a) and (b). The difference between the lithographic width of the etched patterns and the effective width of the confined electron system is significantly reduced in shallow etched samples. In Figure 4c a combination of a shallow-etched surface with a field effect electrode is depicted. The confinement potential originates from the Schottky barrier at the etched surface. The electron density and the lateral extent of the confined electron system are altered with the potential of the gate electrode.

The technologies for the realisation of artificial superlattice potentials with small periods are rapidly developing because of the interest of both researchers as well as device technologists in structures on a nanometre scale. Only those techniques that are important for the experiments discussed in the following paragraphs have been described here. They represent, however, only a fraction of existing possibilities, which for instance include regrowth on patterned structures (Kapon *et al.* 1989), ion bombardment (Scherer *et al.* 1987, Nieder *et al.* 1990), strain induced by patterned surface layers (Kash *et al.* 1988), and growth on tilted surfaces or heterostructure cross-sections (Gershoni *et al.* 1991, Kash *et al.* 1991).

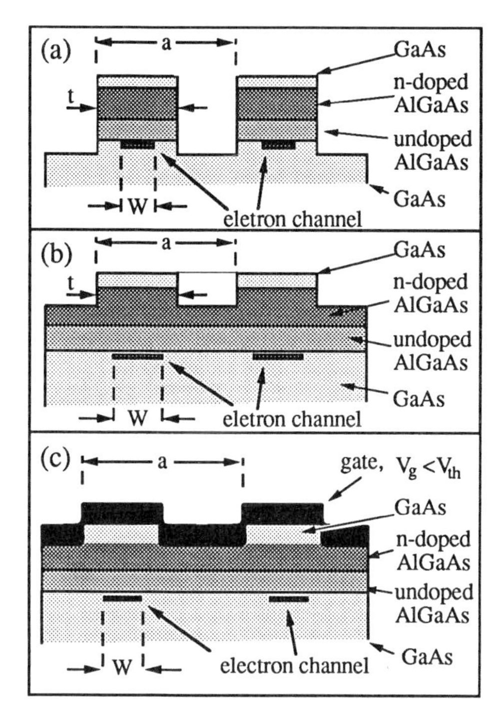

Figure 4: *Cross-sections of etched* GaAs-Al_xGa_{1-x}As *heterojunctions: (a) deep mesa etched patterns, (b) shallow-etched patterns and (c) combination of shallow-etched patterns with a front gate electrode.*

3 Electron density superlattices

A 1D superlattice of period a at the crystal surface defines a so-called minizone in reciprocal space with boundaries at $\pm\pi/a$. The energy dispersion of single particles as well as elementary excitations can be thought of being folded back into the first minizone, and interaction with the potential of the superlattice causes energy gaps at points where branches of the energy dispersions intersect. Thus so-called minibands and minigaps are formed.

The first lateral superlattices were experimentally realised in MOS-structures prepared on so-called vicinal planes (Cole *et al.* 1977, Matheson and Higgins 1982, Evelbauer *et al.* 1986). In such samples the semiconductor surface at the semiconductor-oxide interface is slightly misoriented by an angle Θ with respect to the surface of high symmetry. Small angles Θ of a few degrees cause a surface superlattice potential with periods of typically 100 Å and a new minizone is formed in momentum space with reciprocal lattice vector $G = Q \sin\Theta$, where Q is the extension of the first Brillouin zone in the direction normal to the interface. Minigaps have been observed in measurements of resistance (Cole *et al.* 1977, Matheson and Higgins 1982, Evelbauer *et al.* 1986) as well as far-infrared transmission (Kamgar *et al.* 1980, Sesselmann and Kotthaus 1979) as a function of carrier density and magnetic field.

Lithographically defined surface superlattices which have been investigated experimentally to date have periods typically an order of magnitude larger than those found on vicinal planes. Thus the first minigap is expected at densities which are roughly two orders of magnitude smaller, and experiments on such artificial superlattices are much more difficult (Warren *et al.* 1985, Bernstein and Ferry 1986, Bernstein and Ferry 1987, Ismail *et al.* 1989a). However, the important advantage of these devices is that in principle both the carrier density as well as the strength of the modulation potential can be controlled. Furthermore, it is possible to combine lateral confinement into a 1D wire with a 1D superlattice potential along the wire. By lowering the dimensionality of the electron system, the impact of minizones in the energy dispersion on the density of states and thus on transport properties is enhanced. For instance, evidence for the formation of a miniband in a finite 1D superlattice has been found in a corrugated electron channel defined by a split gate (Kouwenhoven *et al.* 1990, Ulloa *et al.* 1990, Brum 1991). However, a magnetic field of $B = 2\,\mathrm{T}$ had to be applied normal to the channel in order to make the observations. Similar observations were made by Haug *et al.* (1990) on a channel with fewer periods even in absence of magnetic fields. Haug *et al.* point out that their results might be explained alternatively in context with Coulomb charging effects.

3.1 Magnetoconductance

At low magnetic fields B the magnetoconductance of a 2DEG with a superlattice potential oscillates as function of $1/B$ with several Shubnikov–de Haas periods simultaneously, corresponding to different orbits at the Fermi surface in different minibands (Cole *et al.* 1977, Matheson and Higgins 1982, Evelbauer *et al.* 1986). Such oscillations have been observed in 2DEGs on vicinal planes and it was found that the magnetic forces dominate the superlattice potential at sufficiently high magnetic fields so that the orbits of the homogeneous system are recovered. This magnetic breakdown of the minigaps occurs roughly at fields according to $\hbar\omega_c > E_g^2/E_F$ (Blount 1962). Here $\omega_c = eB/m^*$ is the cyclotron frequency, E_g is the width of the minigap and E_F is the Fermi energy of the electrons.

Additional unexpected oscillations of the diagonal components of the magnetoresistivity tensor were found in 1D lateral superlattices with periods of several 100 nm, at magnetic fields well above the breakdown condition but smaller than the field for the onset of Shubnikov–de Haas oscillations (Weiss *et al.* 1988, Winkler *et al.* 1989, Gerhardts *et al.* 1989). They are periodic in $1/B$, akin to Shubnikov–de Haas oscillations, but the dependence of the amplitude of the oscillations on temperature is much weaker (Weiss *et al.* 1988). Thus they are much more easily observed than phenomena originating from the existence of minigaps in the single-particle dispersion. The oscillations have not only been observed in d.c. transport experiments but also in microwave transmission and magnetocapacitance measurements. As an example, the derivative of the microwave transmission with respect to gate voltage in a 2DEG with a grating-type superlattice potential of period $a = 300\,\mathrm{nm}$ is depicted in Figure 5 together with the d.c. resistance of the same sample (Brinkop *et al.* 1990). The sample is a GaAs-$\mathrm{Al}_x\mathrm{Ga}_{1-x}\mathrm{As}$ heterojunction with a front gate electrode whose insulator is modulated, as depicted in Figure 3b. The superlattice potential is generated by aplication of a gate voltage with respect to the electron system. The sample is brought into a microwave waveguide in

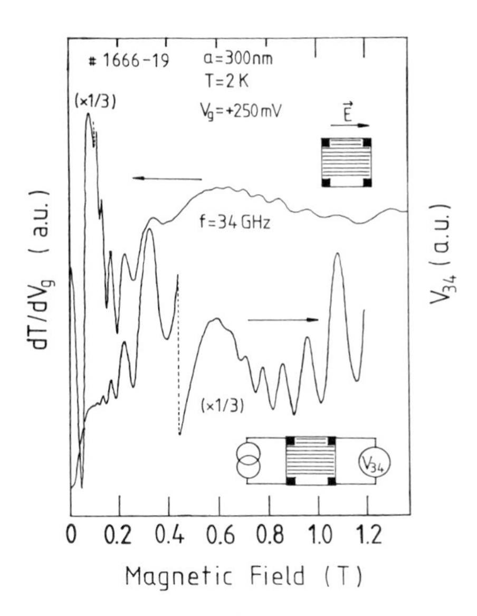

Figure 5: *Magnetoresistance (lower trace) and derivative of the microwave transmission with respect to gate voltage (upper trace) of a* GaAs-Al$_x$Ga$_{1-x}$As *heterojunction with a superlattice potential of period* 300 nm *(Brinkop et al. 1990).*

the centre of a magnet and cooled to a temperature of $T = 2\,\mathrm{K}$. The microwave transmission directly reflects the behaviour of the diagonal component of the conductivity matrix in the direction of the polarisation of the electric field. In Figure 5 the electric field of the microwave radiation of frequency $\nu = 34\,\mathrm{GHz}$ is polarised along the grating stripes, *i.e.* in the direction of constant potential contours which is taken to be the y-direction in the following. Shubnikov–de Haas oscillations are seen at high magnetic fields ($B > 0.6\,\mathrm{T}$). The mobility of the electrons is high ($\mu = 830\,000\,\mathrm{cm^2\,V^{-1}\,s^{-1}}$), so that the scattering rate $1/\tau$ is smaller than the frequency of the microwave probe, and the strong structure in the microwave signal at $B \approx 0.1\,\mathrm{T}$ arises from cyclotron resonance. Additional oscillations are seen at intermediate fields in the microwave transmission as well as in the component ρ_{xx} of the d.c. resistance. These oscillations are much weaker if the other polarisation is used in the microwave experiment so that σ_{xx} is probed.

The origin of the oscillations can be understood both in a semiclassical and in a fully quantum mechanical description as arising from commensurability between the cyclotron radius $R_c = \hbar k_F/eB$ and the period a of the superlattice. A 1D superlattice potential in the x-direction increases the bandwidth of the impurity-broadened Landau bands and thus the conductivity σ_{yy} perpendicular to the potential gradients. The amount to which the bandwidth increases depends on the Landau index because the number of superlattice periods sensed by electrons of different Landau states depends on the cyclotron radii (Winkler *et al.* 1989, Zhang and Gerhardts 1990). In a purely classical picture the modulation of the conductivity along the grating can be understood as a result of the drift velocity, $(1/B^2)(\mathbf{E} \times \mathbf{B})$, averaged along the orbit of the electrons through the superlattice potential (Beenakker 1989). To be more precise, both models result in oscillations of the component σ_{yy} of the conductivity with minima whenever $2R_c = a(m - \frac{1}{4})$, where m is an integer. Thus the oscillations in conductivity are predicted to be periodic in R_c/a, which is directly confirmed by the microwave experiment. Weiss *et al.* (1988) and Winkler *et al.* (1989) used the Hall-bar geometry in their measurements of d.c. resistance and thus the components of the magnetoresistance tensor were probed. The commensurability oscillations occur at magnetic fields where $\omega_c\tau > 1$ and the Shubnikov–de Haas oscillations are still small; thus approximately $|\sigma_{xy}| > |\sigma_{xx}| \approx |\sigma_{yy}|$ and $\rho_{xx} \propto \sigma_{yy}$. The strongest oscillations in resistance are thus observed in Hall bars where the current is directed perpendicular to the grating.

As will be explained in more detail later, the thermodynamic density of states is probed in measurements of capacitance. Shubnikov–de Haas oscillations are observed in the capacitance of the sample, corresponding to the strong modulation of the thermodynamic density of states in a magnetic field, when the Fermi energy is swept through the Landau levels. In magnetocapacitance measurements on 2DEGs with a superlattice potential it is observed that the the amplitudes of the Shubnikov–de Haas oscillations in capacitance are modulated with a period that corresponds to the commensurability oscillations in d.c. resistance (Weiss *et al.* 1989). This directly reflects the modulation of the thermodynamic density of states as a result of the modulation of the width of the Landau levels induced by the potential of the superlattice (Vasilopoulos and Peeters 1989, Gerhardts and Zhang 1990). Weak antiphase oscillations in ρ_{yy} have also been observed in transport experiments besides the strong commensurability oscillations in the resistivity component ρ_{xx}. These can be explained in a quantum mechanical model including the self-consistent Born approximation for the calculation of the magnetotransport coefficients (Zhang and Gerhardts 1990).

More recently, commensurability oscillations have also been observed on 2D superlattices (Alves *et al.* 1989, Gerhardts *et al.* 1991, Lorke *et al.* 1991b). Oscillations with extrema at magnetic fields corresponding to $2R_c/a = (m + \phi)$ are found in 2D superlattices, very similar to the 1D case. However, different phases ϕ are reported. Numerical simulations by Lorke *et al.* (1991b), integrating the classical equation of motion, indicate that the phase ϕ is controlled by the form of the 2D superlattice potential. If it is of a simple cosine form, $V(x,y) \propto (\cos kx + \cos ky + 2)$, a phase $\phi = \frac{1}{4}$ for the positions of the minima is determined from the results of the simulation. Gerhardts *et al.* (1991) find a phase $\phi = \frac{1}{4}$ in charge density superlatices generated by band-gap illumination. If, however, a cross-term is added according to $V(x,y) \propto (\cos kx + \cos ky + \cos kx \cos ky + 3)$, the phase for the minima changes to $\phi = \frac{1}{2}$ in correspondence with the experimental finding in lithographically defined superlattices (Lorke *et al.* 1991b). The dependence of the density of states in a 2D superlattice potential on magnetic field is very distinct from the density of states in a 1D potential. 1D potentials cause merely a broadening of the Landau levels, so that Landau bands are formed with widths governed by the ratio $2R_c/a$. In contrast, 2D superlattice potentials cause a broadening and a splitting of the Landau bands such that a number p of subbands are formed, where p/q is a rational number describing the number of flux quanta enclosed in a unit cell of the superlattice (Langbein 1969, MacDonald 1983). So far, there is very little experimental evidence for the observation of this peculiar commensurability property (Ismail *et al.* 1989b, Gerhardts *et al.* 1991).

The potential modulation of a periodically patterned gate electrode induces a charge carrier density modulation in the 2DEG which effectively screens the potential superlattice of the gate in the plane of the electron system. The following briefly describes a method which has been used to approximate the screened potential superlattice (Kotthaus and Heitmann 1982, Wulf 1987). The effective superlattice potential V_{eff} in the plane of the 2DEG may be decomposed into three parts. The contribution of the periodic gate electrode can be approximated by a Fourier expansion of the potential in a plane ($z = 0$) parallel to the gate at the sample surface: $V_g(x, z = 0) = \Sigma_\nu V_g(\nu q)\cos(\nu q x)$. Here $q = 2\pi/a$ is the reciprocal lattice vector defined by the period a of the gate pattern. In the plane of the 2DEG, $z = d$, this results in a potential

$$V_g(x,z) = \sum_{\nu=0}^{\infty} \frac{1}{a_+} V_g(\nu q)\cos(\nu q x)\exp(-\nu q z),$$

where $a_\pm = (\epsilon_s/\epsilon_i)\sinh(\nu q d) \pm \cosh(\nu q d)$, and ϵ_s and ϵ_i are the dielectric constants of the semiconductor and the insulator between the 2DEG and the gate respectively. The higher Fourier components of this potential are strongly weakened if $qd \geq 1$, so that it is often a good approximation to keep only the lowest one. Because of the strong confinement of the electron system in the z-direction it is furthermore often safe to neglect the thickness of the 2DEG: $n_s(x,z) = n_s(x)\,\delta(z-d)$. If the charge density modulation $n_s(x)$ is expanded in a Fourier series with the same periodicity as the gate potential, $n_s(x) = \sum_\nu n_s(\nu q)\cos(\nu q x)$, the corresponding potential V_{2DEG} induced by the charge has Fourier components

$$V_{\text{2DEG}}(\nu q) = \frac{e^2}{2\epsilon_s\epsilon_0 \nu q}\, n_s(\nu q).$$

The third contribution to the effective potential is produced by the image charges that take account for the discontinuities of the dielectric constants at the interfaces:

$$V_{\text{IC}}(\nu q) = \frac{a_-}{a_+}\, V_{\text{2DEG}}.$$

Thus the effective potential of the superlattice is given by the sum

$$V_{\text{eff}}(\nu q) = \frac{1}{a_+} V_g(\nu q)\exp(-\nu q z) + V_{\text{2DEG}}(\nu q) + V_{\text{IC}}(\nu q).$$

If $qd > 1$, a_-/a_+ reduces to $(\epsilon_s - \epsilon_i)/(\epsilon_s + \epsilon_i)$. The effect of the image charges is well approximated by omitting the third term from the sum and replacing the semiconductor's dielectric constant ϵ_s by an average dielectric constant $\epsilon = 0.5(\epsilon_s + \epsilon_i)$ in V_{2DEG}. If the Fermi wavelength λ_F in the 2DEG is much smaller than the superlattice period then, in absence of a magnetic field, the linear Thomas-Fermi model can be applied to relate the effective potential and the charge density modulation: $n_s(\nu q) = (g_v m^*/\pi\hbar^2)V_{\text{eff}}(\nu q)$. Here we suppose that $\nu \geq 1$ and $g_v m^*/\pi\hbar^2 = D_{2D}(E)$ is the density of states of the 2DEG at zero magnetic field, including the valley degeneracy g_v. At $\nu = 0$ charges in the depletion zone have to be taken into account (Wulf 1987). Screening can become strongly non-linear at high magnetic fields because of the strongly modified density of states (Wulf and Gerhardts 1988). The screened potential V_{eff} is thus described in the Thomas-Fermi approximation, with not too small a separation between gate and electron system ($qd > 1$), by

$$V_{\text{eff}}(\nu q) = \frac{1}{a_+} V_g(\nu q)\exp(-q\nu d)\left(1 + \frac{a}{\pi\nu a^*}\right)^{-1}. \tag{1}$$

Here $a^* = 4\pi\epsilon\epsilon_0\hbar^2/m^*e^2$ is the effective Bohr radius. For instance in GaAs the effective Bohr radius ($a^* = 9.7\,\mathrm{nm}$) is much smaller than the periods of present lateral superlattices. The thickness of the partially doped AlGaAs layer that separates the 2DEG from the gate in GaAs-$Al_xGa_{1-x}As$ heterojunctions is of the order of $d = 50\,\mathrm{nm}$. Equation (1) demonstrates that in present GaAs-$Al_xGa_{1-x}As$ heterojunctions the gate potential is effectively screened, either by the exponentials in the first bracket if the period is small, or by the second bracket if the period is large. Furthermore, higher Fourier components ($\nu > 1$) are effectively damped. The reduction of the potential is even larger in Si devices, due to the larger mass, the reduced dielectric constant and the larger valley degeneracy. On the other hand, screening is considerably less effective in InSb.

3.2 Plasmons

The superlattice of charge density induced by the periodic gate electrode results in a laterally modulated sheet conductivity. Plasmons can be excited when interacting with far-infrared radiation (FIR), because the periodically modulated lateral conductivity results in longitudinal components of the effective radiation field with wave-vector in the plane of the 2DEG. This is in contrast to plasmons in homogeneous 2DEGs, where the transverse radiation field does not interact with longitudinal collective excitations in general. Therefore a grating coupler on the surface of the sample has to be used to observe plasmons in FIR transmission spectroscopy on homogeneous 2DEGs. It partially transforms the incoming transverse radiation field into evanescent longitudinal components that interact with the plasmon excitations of the 2DEG. This very powerful technique for experimental investigation of plasmons on semiconductor surfaces is described in more detail in the chapter by Heitmann. In a density superlattice the plasmon dispersion is backfolded into the first minizone, and plasmons can be excited at the zone centre without the aid of a grating coupler. However, the single-particle spectrum is backfolded as well, so that the collective excitations intersect the single-particle spectrum within the first minizone on the entire energy range. Thus it could be expected that collective excitations in a density superlattice are Landau damped. In fact the notion that distinguishes 'single-particle' from 'collective' excitations becomes meaningless in a strong charge-density superlattice, as pointed out by Dahl (1990).

The behaviour of plasmon excitations when a two-dimensional electron system is transformed into a charge density superlattice have been investigated in FIR transmission experiments on Si-MOS samples as well as GaAs-$Al_xGa_{1-x}As$ heterojunctions. As an example, far-infrared spectra of plasmon excitations in a GaAs-$Al_xGa_{1-x}As$ heterojunction field-effect device are displayed in the lower part of Figure 6. In this device the distance between the gate and the electron system is periodically modulated by a holographically defined photo-resist grating with period $a = 500\,\mathrm{nm}$, as described in Figure 3b. The lowest spectrum is obtained at zero gate voltage where the electron system is essentially homogeneous. The corrugated gate acts like a grating coupler so that a plasmon is excited at wave vector $q = 2\pi/a$. The plasmon resonance position at $V_g = 0$ in Figure 6 is well described by the classical dispersion (Batke *et al.* 1986)

$$\omega_p(q) = \frac{e^2 n_s q}{2\bar{\epsilon}\epsilon_0 m^*}, \tag{2}$$

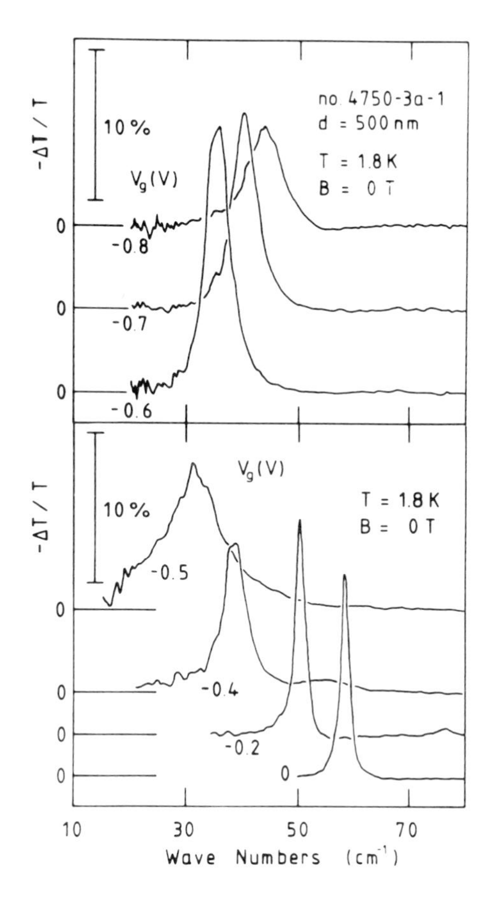

Figure 6: *Far-infrared transmission spectra of a* GaAs-Al$_x$Ga$_{1-x}$As *heterojunction with periodic gate as depicted in Figure 3b. The radiation is polarised perpendicular to the stripes of the corrugated gate (Kotthaus et al. 1988).*

where the carrier density $n_s = 6{\times}10^{11}$ cm^{-2} is determined by magnetotransport measurement, m^* is the effective mass in GaAs and $\bar{\epsilon} = 13.9$ is an effective dielectric constant that accounts for the polarisation of the medium as well as for the screening of the gate electrode. In the sample of Figure 6 higher-order plasmon excitations at $q = \nu(2\pi/a)$, with $\nu > 1$ are significantly weaker than the $\nu = 1$ resonance. The aspect ratio between the photo-resist stripes that define the grating coupler and the period is close to $\frac{1}{2}$ so that it is not surprising if the content of higher harmonics in the component of the longitudinal field induced by the grating coupler is low.

At negative gate voltages the electron system is preferentially depleted in those areas which are close to the gate electrode, so that a grating-type charge-density superlattice is induced. With decreasing gate voltage the plasmon resonance position decreases, the resonance broadens and the oscillator strength increases. In Figure 7 the squared plasmon resonance positions are plotted versus the gate voltage. The dashed line demonstrates the linear dependence of the resonance position at gate voltages close

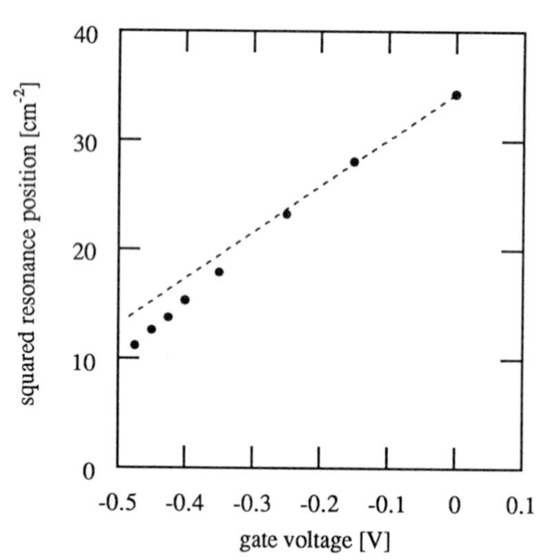

Figure 7: *Positions of the plasmon resonance versus gate voltage of the sample in Figure 6. The dashed line extrapolates the linear dependence of the resonance position at gate voltages close to zero.*

to $V_g = 0$. Here the density is weakly modulated and the positions of the resonance are well described by Equation (2) with an average density extracted from the integrated capacitance measured on the sample. This behaviour has also been observed in Si-MOS devices, where a relatively weak density modulation is induced by a laterally modulated oxide thickness (Mackens *et al.* 1984). However, the higher harmonics of the grating coupler field and the induced charge density modulation play an important role in the devices of Mackens *et al.* It is found that the back-folded plasmon dispersions split at points where they intersect each other in the first minizone, in close analogy to the opening of gaps in the single-particle dispersion. The minigaps of the plasmon dispersion have been described in a perturbative approach by Krasheninnikov and Chaplik (1981). The resonance positions of the heterojunction device in Figure 6 depart from linear behaviour to lower frequency at gate voltages where the density modulation becomes strong. Actually, this has been predicted by classical models that describe the plasmon excitations beyond the case of weak modulation (Eliasson *et al.* 1986, Cataudella and Ramaglia 1988). Of particular interest is the behaviour of the resonances in the limit in which the low density regions of the density superlattice become depleted completely, so that an array of electron stripes is formed.

The gate voltage at which this transition takes place can be unequivocally determined from the behaviour of the FIR excitations, magnetotransport as well as capacitance to be $V_g = -0.5\,\text{V}$. As an example the capacitance of the device in Figure 6 measured as function of the gate voltage is displayed in Figure 8. For $0 > V_g > -0.5\,\text{V}$ the electron system forms a homogeneous electrode and the capacitance is essentially constant. At $V_g = -0.5\,\text{V}$ the capacitance abruptly decreases, reflecting the reduction of the effective area when the electron system is contracted into isolated channels beneath the distant parts of the gate. In a model the corrugated gate may be approximated by a grating gate where the potential remains fixed to zero in the gaps between the

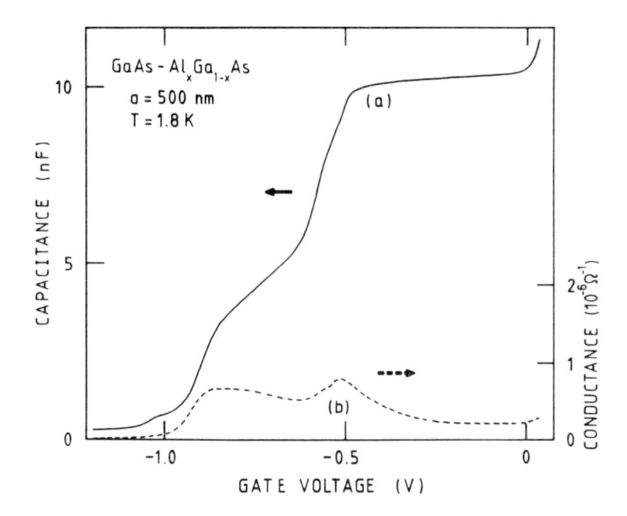

Figure 8: *Dependence of the capacitance on gate voltage measured on the sample of Figure 6 (Hansen 1988).*

grating stripes. If the ratio of the metal stripe width to the period is assumed to be $\frac{1}{2}$ in the model of Equation (1) a gate voltage of $V_g = -0.5\,\mathrm{V}$ is obtained for the voltage at which the effective potential modulation becomes equal to the Fermi energy. This is in surprisingly good agreement with the value found experimentally. The FIR resonances in the upper part of Figure 6 are measured at gate voltages below the transition point and thus reflect FIR resonances in an array of isolated electron channels. We note that below the transition point with decreasing gate voltage the resonance becomes narrow again, increases in frequency and decreases in oscillator strength.

The increase of oscillator strength in the resonance of a density modulated 2DEG in the gate voltage regime can be readily explained from the fact that the radiation field couples more effectively to longitudinal excitations with increasing spatial modulation of the conductivity in the charge-density superlattice. A dramatic broadening of the FIR resonances in the transisition regime is modeled in a quantum mechanical calculation of the excitation spectrum (Dahl 1990). Generally, the spectrum of the longitudinal excitations is determined by the zeros of the dielectric function. As a consequence, the spectrum of collective modes consists of the plasmon branch and also includes the so-called single-particle excitations. In a homogeneous 2DEG, however, the observed plasmon excitation has strongly predominating oscillator strength and is energetically well separated from the other modes. This is no longer the case in an electron system with a strongly modulated density. The spectra calculated by Dahl (1990) demonstrate that the oscillator strength is now distributed among several modes, so that a broad excitation spectrum results in the transition regime. In another picture the plasmon branch interferes with the back-folded single-particle excitations and is thus strongly Landau damped. Furthermore, the calculated spectra reflect the remarkable experimental finding that there is still a peak in absorption found at finite frequencies although the absorption spectrum is very broad. This result is also reproduced in more recent calculations by Wulf *et al.* (1990) and is in contrast to results of a classical model.

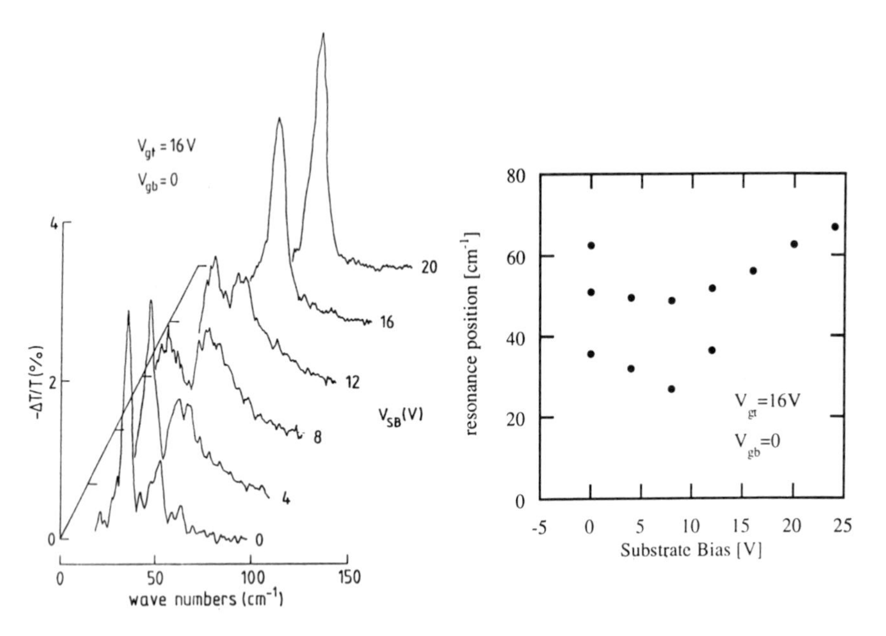

Figure 9: FIR *transmission spectra of a* Si-MOS *device with a stacked gate. The period of the grating-type bottom gate is* $a = 1\,\mu$m *and the opening between the grating stripes is* 200 nm. *The polarisation of the* FIR *radiation is perpendicular to the bottom gate grating. On the right-hand side the resonance positions are depicted as a function of the substrate bias (Alsmeier et al. 1990b).*

The transition from a strongly modulated 2DEG into an array of isolated electron channels has also been studied in so-called 'stacked-gate' Si-MOS devices. The gate configuration of a stacked-gate MOS device consists of a homogeneous top gate and a grating type bottom gate as depicted in Figure 4d. The transmission spectra displayed in Figure 9 have been recorded with different so-called substrate bias voltages, V_{SB} applied to the top gate. The amplitude of the superlattice potential can be controlled at fixed electron density averaged over one period of the superlattice in such devices by using a substrate bias. In order to do so the bottom gate voltage is chosen near the inversion threshold and a positive bias is applied at the homogeneous top gate, so that a relatively high electron density is induced beneath the openings of the bottom gate. The device is cooled down to liquid helium temperatures, so that minority carriers need to be generated by illumination with band gap radiation in order to establish a quasi-equilibrium of the inversion charges. The band gap radiation can be switched off after the inversion carriers have been injected and an additional gate voltage, the substrate bias, can be applied. This further modulates the potential landscape without changing the average electron density. With increasing substrate bias voltage the surface potential beneath the bottom gate openings is decreased and steeper potential minima are formed in which the electrons are bound more and more tightly.

The spectrum in Figure 9 recorded at $V_{SB} = 0$ thus exhibits the plasmon resonances of a 2DEG with strong density modulation. Three resonances are observed that are well described as the $\nu = 1$, 2 and 3-order plasmons of a charge-density superlattice with average density of $n_s = 7.5 \times 10^{11}$ cm^{-2}. Higher-order plasmons are observed here because the ratio of the grating stripe separation to the grating period is far from $\frac{1}{2}$. Again, the $\nu = 1$ plasmon initially decreases in frequency and broadens substantially with increasing modulation. The $\nu = 1$ and 2 modes merge into a rather broad but strong resonance out of which a narrow resonance emerges. In analogy to the behaviour of the resonances observed on the GaAs-Al$_x$Ga$_{1-x}$As heterojunction, the strong and narrow resonance indicates that the system has transformed into an array of isolated wires.

For comparison the FIR resonance positions of a GaAs-Al$_x$Ga$_{1-x}$As heterojunction with modulated gate are depicted for two different magnetic fields applied perpendicular to the sample surface in Figure 10. Here the system consists of isolated channels in the gate voltage regime at the left hand side of the dashed line, whereas at higher gate voltages it is a 2DEG with charge density modulation. At zero magnetic field the lowest order plasmon gradually transforms into the excitation of the isolated wires. In contrast, the transition of the resonances in the stacked gate Si-MOS device seems to be far more complex. In the Si device it may occur that the resonance of the wire array evolves out of the second order plasmon, due to the different ratio of channel width to period, but further detailed investigation is highly desirable to clarifiy the behaviour in the transition regime. The resonances in the gate voltage regime of 1D electron wires will be discussed in the next chapter.

In a magnetic field the cyclotron resonance is observed as a strong excitation besides the magnetoplasmon in the gate voltage regime $V_g > -0.5$ V. The triangles in Figure 10b denote the cyclotron resonance positions of the charge density superlattice at $B = 6.4$ T. We note that the position of the cyclotron resonance is drastically shifted to higher frequency with decreasing gate voltage. At low magnetic fields the cyclotron resonance position is hardly affected up to gate voltages very close to the transition point (Hansen 1988). Demel *et al.* (1990b) observed, on a sample similar to that of Figure 10, that oscillator strength is transferred from the cyclotron resonance to the plasmon at even higher magnetic fields ($B > 10$ T), indicating a crossover to the array of electron wires induced by the magnetic field. At a fixed gate voltage the cyclotron resonance is quenched and the magnetoplasmon becomes drastically stronger. These observations have been linked to the modified screening properties of the electron system in high magnetic fields (Wulf *et al.* 1988).

4 Quasi-one-dimensional electron channels in tight binding superlattices

The electron systems consist of arrays of isolated 1D electron wires in the GaAs-Al$_x$Ga$_{1-x}$As heterojunction of Figure 6 and 10 at gate voltage below $V_g = -0.5$ V, and in the Si-MOS device at substrate bias voltages above $V_{SB} = 16$ V. In both types of samples the FIR spectra of the wire array are very similar. There is only one single resonance of strong oscillator strength, which is significantly narrower than the plasmon excitations in the transition regime. Furthermore, the resonance increases in energy with increasing potential modulation. The FIR excitations of 1D wire arrays have also been studied intensively in other semiconductor systems such as deep and shallow

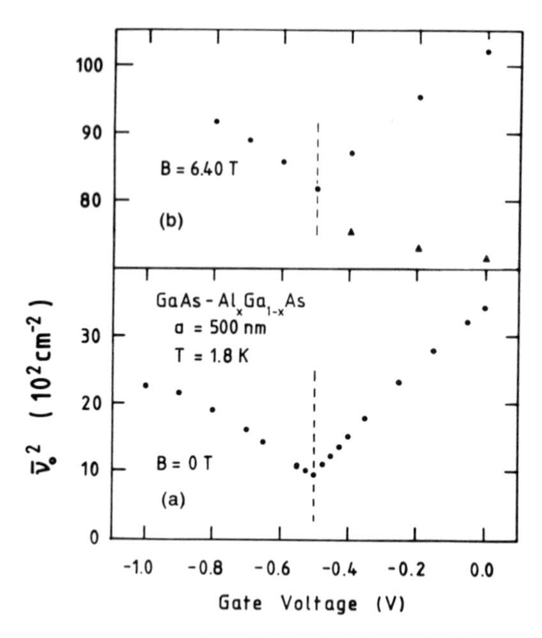

Figure 10: *Resonance positions of an* GaAs-Al$_x$Ga$_{1-x}$As *heterojunction with corrugated gate at (a) zero magnetic field and (b)* $B = 6\,\mathrm{T}$. *The dashed line indicates the gate voltage at which the electron system transforms from a* 2DEG *into an array of 1D electron channels.*

mesa etched GaAs-Al$_x$Ga$_{1-x}$As heterojunctions (Allen *et al.* 1985, Hansen *et al.* 1987a, Demel *et al.* 1988), InSb-MOS devices (Hansen *et al.* 1987b, Alsmeier *et al.* 1988), and deep-etched Al$_y$In$_{1-y}$As-Ga$_x$In$_{1-x}$As single quantum wells (Kern *et al.* 1990). Only one strong resonance is found in the transmission spectra of all of these systems. At zero magnetic field the resonance is observed only with the radiation polarised perpendicular to the wires. In a magnetic field the position ω_{res} of the resonance increases according to

$$\omega_{\mathrm{res}}^2 = \omega_0^2 + \omega_c^2 \,, \tag{3}$$

where ω_0 is the position of the resonance without a magnetic field and ω_c is the cyclotron resonance frequency. At finite magnetic field the resonance is also observed when the radiation is polarised along to the wires. In this polarisation the oscillator strength of the resonance increases with increasing magnetic field so that the oscillator strength of the resonance is essentially independent of polarisation at sufficiently high magnetic fields.

The behaviour of the FIR resonances in 1D electron wires investigated so far can be readily explained within a simple model based on the statement that the confinement is electrostatic and that the first-order approximation to the potential minimum is always parabolic. Numerical calculations of the electron density and the self-consistent potential distribution have been performed by Laux and Stern (1986) and Laux and Warren (1986) for stacked gate Si-MOS devices and for split-gate GaAs-Al$_x$Ga$_{1-x}$As heterojunctions (Laux *et al.* 1988). Laux *et al.* find that the electrostatic potential is nearly parabolic as long as the electron density is sufficiently low so that the corrections to the effective confinement potential from the electron charges within the wire are negligible. Also, the conclusions drawn from Equation (1) in the last chapter suggest that the confinement potential generated by remote charges is very smooth, and thus a harmonic *Ansatz* for the potential of these charges should be very good.

In the following the electrostatic potential, which is created by all charges excluding the electrons within the channel, will be called the bare potential in order to clearly distinguish it from the effective potential which includes the mutual interaction of the electrons in the channel. Actually, the bare potential includes the image charges and should, therefore, also be calculated self-consistently. With a harmonic bare potential in the x-direction, $V_{\text{bare}}(x) = \frac{1}{2}m^*\Omega^2x^2$, the Hamiltonian of the electron system in an isolated electron wire can be expressed as

$$H = \sum_{i=1}^{N} \frac{1}{2m^*}\mathbf{p}_i^2 + \sum_{i=1}^{N} \frac{m^*}{2}\Omega^2 x_i^2 + \sum_{i<j} U(\mathbf{r}_i - \mathbf{r}_j) + E_0\,,$$

where the system is assumed to be strongly bound at the semiconductor interface so that only the lowest mode with energy E_0 is occupied with respect to motion in z-direction; $\mathbf{p}_i$ and $\mathbf{r}_i$ are the momentum and position operators of the individual electrons in the x-y plane, respectively; x_i is the x-component of the position operator, and N is the total number of electrons in the channel. The potential $U(\mathbf{r}_i - \mathbf{r}_j)$ describes the mutual interaction of the electrons within the channel and is assumed to depend only on the distance $|\mathbf{r}_i - \mathbf{r}_j|$ between them. In close analogy to Kohn's theorem (Kohn 1961) for the cyclotron resonance in a homogeneous electron system, the interaction with a homogeneous radiation field polarised in the x-direction can be derived from the above Hamiltonian without any further assumptions (Brey *et al.* 1989, Maksym and Chakraborty 1990). The result is that a single resonance is excited at the characteristic frequency Ω of the *bare* potential. Thus the energy of the resonance is independent of the electron-electron interaction $U(\mathbf{r}_i - \mathbf{r}_j)$ and, in particular, independent of the number of electrons contained within the channel. To understand this point further, the above Hamiltonian can be expressed as function of a centre-of-mass coordinate and $N - 1$ further coordinates that describe the relative motion of the particles (Maksym and Chakraborty 1990). The Hamiltonian has the special property that it separates into a part describing the motion of the centre of mass and another which is a function only of the relative coordinates. The latter contains the electron-electron interaction, while the first describes a particle with charge Ne and mass Nm^* at the centre of mass in the potential $V_{\text{bare}}(x)$. A homogeneous radiation field couples only to the centre-of-mass coordinate, so that the whole electron system oscillates rigidly. The frequency at which energy is absorbed is still Ω since a particle with N-fold mass and N-fold charge oscillates with the same frequency, just with a smaller amplitude.

The model of Equation (1), which has been applied in the last section to estimate the gate voltage of the transition point in the GaAs-$Al_xGa_{1-x}As$ heterojunction with a corrugated gate, may also be applied to estimate the parabolic contribution to the bare potential. As before, the corrugated gate is approximated by a grating gate with zero potential in the gaps between the grating stripes and the ratio of the metal stripe width to the period is assumed to be $\frac{1}{2}$. The contribution of the electrons to the effective potential is mainly contained within the second bracket in Equation (1). If the second bracket is omitted for an estimate of the bare potential, the parabolic potential that fits the minima of the first ($\nu = 1$) harmonic has a characteristic frequency given by $\Omega/2\pi c = 32\,\mathrm{cm}^{-1}$ at the gate voltage $V_g = -0.5\,\mathrm{V}$, where the charge density superlattice decomposes into an electron wire array. This, again, is surprisingly close to the position of the resonance determined from the experiment (see Figure 6). There is an alternative approach to the bare potential from a model that has been proposed by Davies (1988) to calculate the electrostatic confinement potential in split gate structures close to cut-off. According to this model the characteristic frequency of the harmonic potential is given by

$$\Omega = \left[\frac{16e(-V_g)}{\pi m^* W^2}\left(\frac{2d}{W}\right)\right]^{1/2}\left[1+\left(\frac{2d}{W}\right)^2\right]^{-1} \tag{4}$$

where W is the width of the gap in the split gate. With the parameters of the above sample, $d = 50\,\mathrm{nm}$, $W = 250\,\mathrm{nm}$ and $V_g = -0.5\,\mathrm{V}$, we obtain $\Omega/2\pi c = 29\,\mathrm{cm}^{-1}$. The increase of the frequency with decreasing gate voltage is somewhat larger than suggested by the above formula so that we have $\Omega/2\pi c = 39\,\mathrm{cm}^{-1}$ at $V_g = -0.9\,\mathrm{V}$.

The above model is readily extended to the case of a finite magnetic field applied in the z-direction. After a proper coordinate transformation the Hamiltonian still represents a harmonic oscillator with modified frequency according to Equation (3). Closely related to the recorded FIR spectra are the real parts of the dynamic conductivities, as described in more detail in, for example, the chapter by Heitmann. The real parts of the dynamic conductivity of a harmonic oscillator are

$$\sigma_{xx} = \sigma_o \frac{\omega^2(1+\omega^2\tau_e^2+\omega_c^2\tau_e^2)}{\omega^2(1+\Omega^2\tau_e^2+\omega_c^2\tau_e^2-\omega^2\tau_e^2)^2+(\Omega^2-2\omega^2)^2\tau_e^2} \tag{5}$$

for polarisation of the radiation field in the direction of the confinement, and

$$\sigma_{yy} = \sigma_o \frac{\Omega^2\tau_e^2(\Omega^2-2\omega^2)}{\omega^2(1+\Omega^2\tau_e^2+\omega_c^2\tau_e^2-\omega^2\tau_e^2)^2+(\Omega^2-2\omega^2)^2\tau_e^2} + \sigma_{xx} \tag{6}$$

for parallel light polarisation (Hansen *et al.* 1987a), where a phenomenological scattering time τ_e has been introduced. In the experimental spectra the oscillator strength of the resonance in perpendicular polarisation is well described by Equation (5) with a carrier density n_s that has been determined independently. With constant scattering time τ_e and increasing cyclotron frequency ω_c the conductivity at resonance for perpendicular polarisation is expected to decrease monotonically from the static conductivity $\sigma_0 = e^2 n_s \tau_e / m^*$ to $\sigma_0/2$. At the same time, the corresponding conductivity for parallel polarisation increases from zero to $\sigma_0/2$ provided the scattering time is sufficiently long ($\omega\tau_e \gg 1$). The experimental spectra of electron wire arrays in magnetic fields show, in general, a more complex behaviour than the above prediction. It describes the overall variation of the intensity only qualitatively. This may be caused by the etched heterojunction surface that modulates the radiation field similar to a grating coupler. The condition of an unaltered polarisation after transmission becomes non-trivial in a magnetic field, if for example a linearly polarising grating coupler is combined with a circularly polarising electron system (Zheng *et al.* 1990). Furthermore, the assumption of a constant and isotropic scattering time may be too simple.

In contrast to the excitations in narrow quantum confined channels, several higher-order modes of dimensional resonances have been observed in stacked gate Si-MOS where

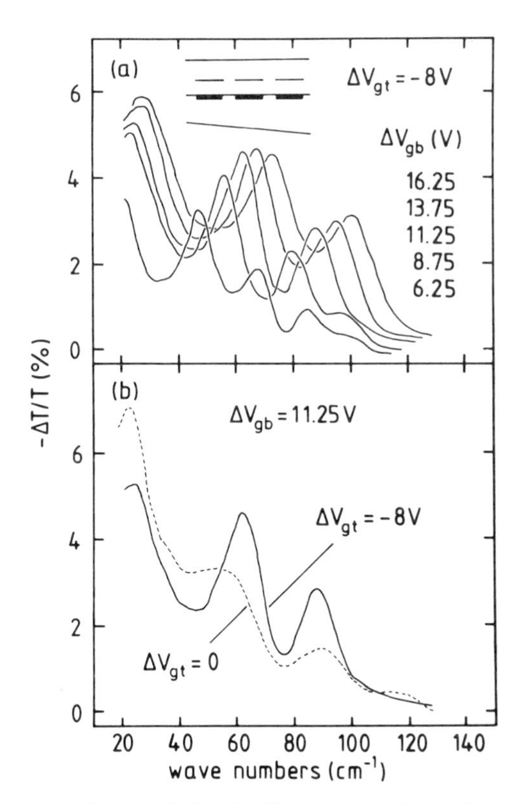

Figure 11: *FIR spectra of a stacked gate Si-MOS structure at zero magnetic field and temperature $T = 2$ K. The device is operated in the subgrating mode as sketched in the inset of (a). The top and bottom gate voltages are given with respect to the threshold voltage. (b) spectra recorded with two different top gate voltages (Alsmeier et al. 1989).*

the geometry of the bottom gate has been chosen such that the above assumptions are violated (Alsmeier *et al.* 1989). Here the device is operated in the so-called subgrating inversion mode where electron channels are induced beneath the positively biased bottom gate. The bottom gate has stripes of width $W = 1.5\,\mu$m and the period is $a = 2\,\mu$m. The homogeneous top gate is biased to $\Delta V_{gt} = -8$ V below threshold voltage so that a high potential barrier is created between adjacent channels. The distance $d_1 = 50$ nm between the bottom gate electrode and the inversion channel is 30 times smaller than the width W. From these dimensions it is clear, according to Equation (1), that the bare potential will be poorly described by a parabola and, moreover, the radiation field also has considerable modulation. The fundamental mode and up to 4 higher-order modes are observed as shown by the FIR spectra in Figure 11(a). The squared resonance frequencies plotted in Figure 12 versus index n are readily described as higher-order modes of a classical collective resonance that arises from the charge density polarisation induced by the radiation field at the system boundary (Allen *et al.* 1985, Eliasson *et al.* 1986, Cataudella and Ramaglia 1988, Alsmeier *et al.* 1989). In this model the resonance frequency is described by the channel width W, the two-dimensional electron density,

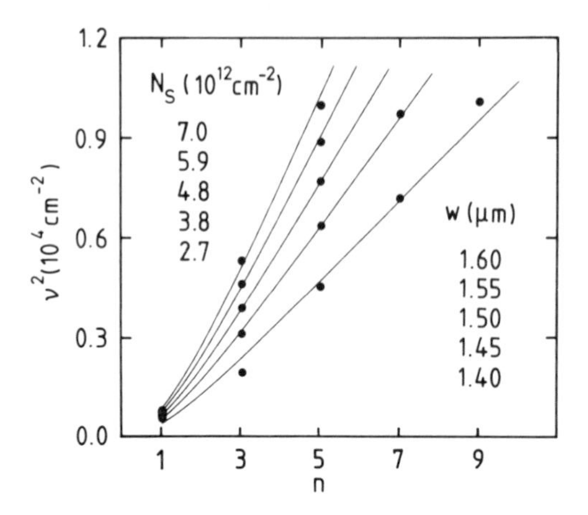

Figure 12: *Squared resonance positions of the sample of Figure 11 plotted versus mode index n. The different areal densities N_s are adjusted with the bottom gate voltage (Alsmeier et al. 1989).*

and an effective dielectric constant $\bar{\epsilon}(n)$:

$$\omega_{\mathrm{res}}^2 = \frac{e^2 n_s}{\epsilon_0 \bar{\epsilon}(n) m^* W} n. \tag{7}$$

Here $\bar{\epsilon}(n)$ may depend on the order n of the excitation (Chaplik 1972, Alsmeier *et al.* 1989). Figure 11(b) demonstrates that the strength of the higher-order modes decreases if the top gate voltage is chosen to be $V_g = 0$ and thus the confinement is less steep. In Figure 12 the positions of the resonance measured at low top gate voltage ($V_g = -8\,\mathrm{V}$) are plotted versus a mode index n. Resonances are observed at odd indices only, since modes with even index are associated with a symmetrical charge distribution and, consequently, are not optically active. The solid lines in Figure 12 are calculated according to Equation (7) with effective dielectric constants $\bar{\epsilon}(n)$ that depend on n and channel widths W that best fit the experimental data. The widths thus obtained are displayed for the respective densities.

From the preceding it is obvious that the FIR resonance positions of an electron system quantum confined in a *harmonic* bare potential contains information about this potential, rather than the self-consistent effective potential that determines how the electron states are quantised into 1D subbands. Electronic properties such as the static magnetoresistance or the capacitance can be probed to experimentally obtain independent information about the quantisation energies and the occupation of the 1D subbands. It is interesting to compare the energy separations of the 1D subbands thus obtained to the FIR resonance energies of the system, since the absorption of FIR radiation polarised perpendicular to the channels originates from inter-subband transitions of the electrons in the 1D channels. The Coulomb energy associated with the optically excited charge oscillation in the 1D channels can dominate the single-particle subband separations. In this case the FIR resonances, which have strong oscillator strength in the dipole approximation, will have considerably larger energies compared to the 1D subband separations. The resonances then are of dominantly collective or plasmon-

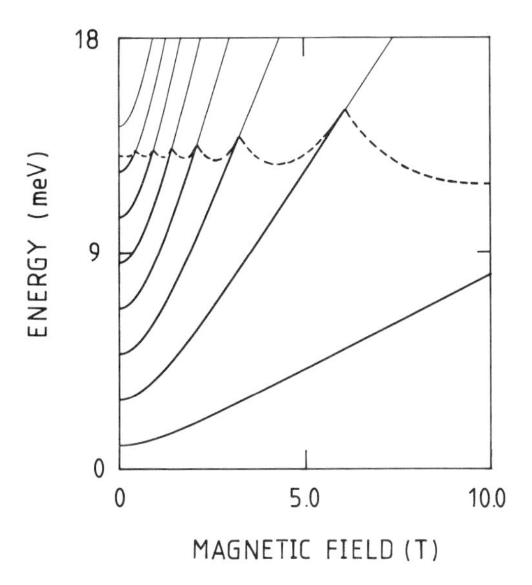

Figure 13: *The energies of the edges of the subbands of a 1D channel as function of a magnetic field applied perpendicular to the channel. A harmonic confinement has been assumed with level separation* $\hbar\Omega = 1.8\,\mathrm{meV}$. *The dashed line denotes the Fermi level assuming a constant 1D electron density* $n_1 = 4.8\times10^6\,\mathrm{cm}^{-1}$ *(Hansen 1988).*

type character reflecting the electrodynamic polarisation in the electron channel by the radiation field. Dimensional resonances of quantum wires are therefore often called inter-subband plasmons. If, on the other hand, the many-particle contributions are small, the resonance energies will be close to the subband separations. The theoretical aspects of the single-particle versus collective character and the excitation spectra of inter-subband plasmons have been discussed, mainly within the random phase approximation, in a number of articles (Zhu and Zhou 1988, Que and Kirczenow 1989, Chaplik 1989, Li and Das Sarma 1989, Gold and Ghazali 1990, Hu and O'Connell 1990, Haupt *et al.* 1991).

In magnetoresistance measurements the so-called magnetic depopulation of the 1D subbands with increasing magnetic field is observed (Berggren *et al.* 1988). Figure 13 demonstrates the behaviour of the 1D subband bottoms in a quantum wire as a function of a magnetic field applied perpendicular to the sample surface. The 1D subbands transform into hybrid bands which gradually become like Landau levels with increasing magnetic field. Note that there is a one-to-one correspondence between the 1D subbands arising from electric confinement at $B = 0$ and the Landau levels at very high magnetic fields, where the magnetic length $l_B = (\hbar/eB)^{1/2}$ is much smaller than the channel width W and thus the magnetic confinement predominates. For the sake of simplicity it is assumed in Figure 13 that the effective confinement potential is parabolic, since then the dependence of the 1D hybrid bands on magnetic field can be calculated analytically. However, the behaviour does not change qualitatively if the calculation is based on different model potentials.

The dashed line in Figure 13 denotes the position of the Fermi energy if the density

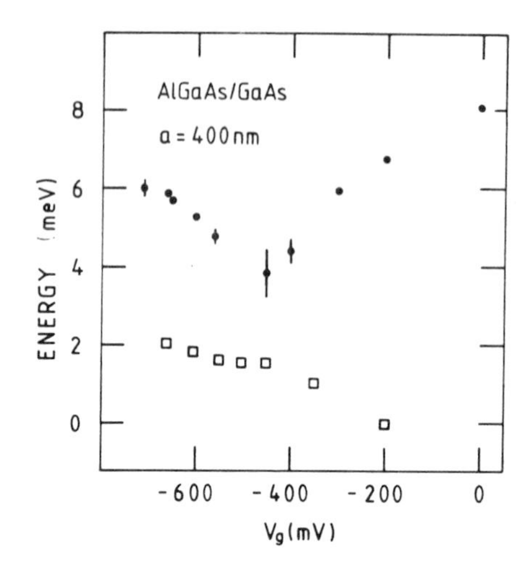

Figure 14: *Resonance energies of the* FIR *excitations (closed circles) at zero magnetic field in an* GaAs-Al_xGa_{1-x}As *heterojunction with corrugated gate and subband spacings (open squares) determined from quantum oscillations (Brinkop et al. 1988).*

of electrons in the 1D channel is constant. The Fermi energy oscillates with maxima whenever it intersects a hybrid band bottom, similar to oscillations of the Fermi energy between Landau levels in a 2D electron system. The density of states within a hybrid subband rises with increasing magnetic field and thus higher subbands are depopulated. The corresponding oscillations of the density of states at the Fermi energy are reflected in quantum oscillations of the channel resistance. The value of the Fermi energy can be determined from the magnetoresistance oscillations measured at high magnetic fields, because at high fields ($l_B \ll W$) the oscillations have the 2D Shubnikov–de Haas periodicity. At low magnetic fields, however, the quantum oscillations depart from the 2D behaviour; in particular, the number of resistance oscillations remains finite, *i.e.* smaller than or equal to the number of occupied subbands. The low field behaviour of the magnetoresistance can be analyzed within a model for the confinement potential to obtain information about the 1D subband energies.

1D subband energies obtained in this way are compared to inter-subband plasmon energies measured in the FIR transmission of the same sample in Figure 14. The wire array of Figure 14 is generated in an GaAs-Al_xGa_{1-x}As heterojunction with a corrugated gate (see Figure 3b) similar to the one of Figures 6 and 10. The 1D subband spacings determined from magnetotransport within a harmonic confinement model are marked by open squares and increase from about 1.5 meV at $V_g = -0.5$ V to 2.0 meV at $V_g = -0.66$ V. Simultaneously, the subband occupation decreases from 9 to 6. From a average of the oscillator lengths over subbands, effective channel widths are derived that decrease from 160 nm at gate voltage $V_g = -0.5$ V to 110 nm at $V_g = -0.66$ V. The FIR excitation energies are about a factor of 3 to 4 larger than the subband spacings. Thus the resonances are predominantly of collective character.

The data of Figure 15 demonstrate that the subband spacings can be considerably closer to the FIR resonance energies in quantum wires generated on InSb-MOS devices.

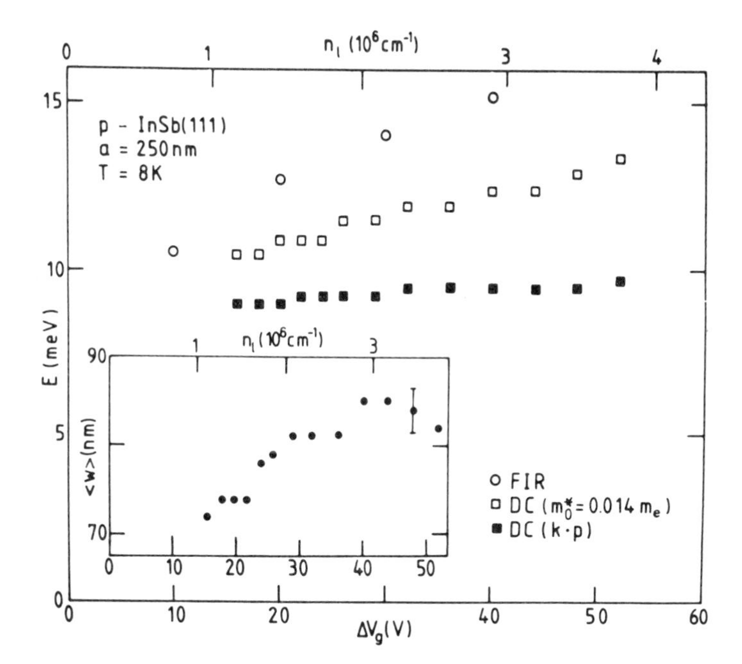

Figure 15: *Sub-band spacings and* FIR *resonance energies versus top gate voltage of 1D electron channels in an* InSb-MOS *device with gate configuration as depicted in Figure 3c. The top scales denote the linear electron density n_l deduced from magnetotransport. In the inset, a channel width derived from an average of the oscillator lengths in the occupied 1D subbands is plotted versus the top gate voltage (Merkt 1989).*

Open circles denote FIR resonance energies, while open squares denote subband spacings deduced from magnetotransport experiments assuming that the effective mass is constant. Filled squares are obtained with a mass that depends on the Fermi energy, which is more appropriate in InSb. Five subbands are occupied at the highest gate voltages. The FIR resonance energies decrease with decreasing electron density whereas the subband spacings deduced from the **k·p** approximation remain almost constant. Note that all three sets of data approach the same energy of $\hbar\Omega = 9\,\text{meV}$ if extrapolated to zero electron density. Thus collective contributions to the resonance energies become smaller than the quantisation energies and eventually vanish at vanishing electron density. The fact that the influence of the single-particle quantisation to the FIR excitations is more important in quantum-confined channels in InSb has been attributed to the smaller effective mass compared with GaAs, as well as to strong screening of the collective depolarisation by the gate configuration in close vicinity to the electron channels (Que and Kirczenow 1989, Merkt 1989).

Capacitance spectroscopy is an even more direct method to probe the variation of the density of states at the Fermi energy when the Fermi energy is swept through the bottom of a 1D subband. Samples for such capacitance measurements have a specially designed back contact from which carriers are injected into the 1D quantum wires. In GaAs-Al$_x$Ga$_{1-x}$As heterojunctions this is usually provided by n^+-doped GaAs that is separated by a tunnel barrier from the GaAs-$Al_xGa_{1-x}As$ interface to which the low-dimensional electron system is bound. The sample capacitance is measured between this back contact and a front gate.

The derivatives of the capacitance with respect to gate voltage, measured on GaAs-$Al_xGa_{1-x}As$ heterojunction devices with 1D quantum wires of different widths, are shown in Figure 16. The back contact in the MBE-grown heterojunctions of Figure 16 is provided by a Si-doped GaAs layer separated from the GaAs-$Al_xGa_{1-x}As$ interface by an 80 nm thick layer of undoped GaAs, which forms a shallow tunnelling barrier. This barrier sufficiently decouples the 1D electron channels from the back contact so that the influence of the contact on the 1D electron states is small although carriers can be exchanged between back contact and electron system on a time scale smaller than the inverse measurement frequency ($f = 20\,\mathrm{kHz}$). This technique of providing a back contact first proved useful in magnetocapacitance measurements on 2D systems, in which a back contact of low resistance and independent of magnetic field is essential (Smith *et al.* 1986). The way in which contact is made to the electron system becomes even more crucial if 0D electron systems are under investigation. Here, the weak coupling of the back contact to the 0D quantum dots is highly appreciated. Capacitance measurements employing a tunnel barrier for electron charging in combination with optical measurements where charge carrier exchange through contacts is not needed will be discussed in the next section.

The capacitance measured between the gate and the back contact consists essentially of two parts. One is determined by the geometries of the electrodes as well as the low-dimensional electron system in between. The second part describes the rise of the chemical potential μ with increasing occupation of the low-dimensional electron system and is proportional to the thermodynamic density of states $D_T(\mu)$:

$$D_T(\mu) = \frac{\partial}{\partial \mu} \int_0^\infty D(E)\, f\left(\frac{E-\mu}{k_B T}\right) dE = -\int_0^\infty D(E) \frac{\partial}{\partial E} f\left(\frac{E-\mu}{k_B T}\right) dE. \tag{8}$$

The contribution of the density of states in the 1D electron channels to the measured capacitance C_{tot} can in essence be described by a capacitance $C_e = e^2 D_T(\mu)$ in series with the geometric capacitance C_{geom} of the electrodes (Smith *et al.* 1985):

$$\frac{1}{C_{\mathrm{tot}}} = \frac{1}{C_{\mathrm{geom}}} + \frac{1}{C_e}. \tag{9}$$

The geometric contribution C_{geom} depends on the area of the electron system, the distances and geometries of front and back electrodes, the insulator material, and the extent of the electron system in the direction normal to the interface. At low temperatures the capacitance C_e is essentially proportional to the density of states at the Fermi energy $D(E_F)$. Each occupied 1D subband with energy E_i at the bottom adds a contribution proportional to $(E_F - E_i)^{-1/2}$ to the density of states, which therefore changes rapidly whenever a new subband is occupied.

This behaviour results in the oscillations of the derivative of the capacitance measured as function of the gate voltage *i.e.* channel occupation, which are observed in Figure 16. The technique of measuring the derivative is applied to enhance the fine structure of the capacitance with respect to the large background that stems from the

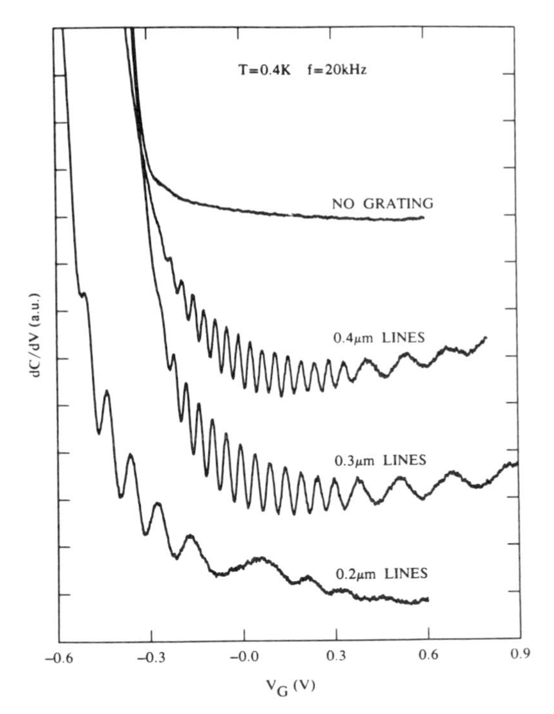

Figure 16: *The derivative of the capacitance recorded as a function of the gate voltage on 1D electron channels with different widths in an* GaAs-Al_xGa_{1-x}As *heterojunction with corrugated gate. The line widths denoted are the lithographic widths of the* GaAs *stripes defining the electron channels (Smith et al. 1987).*

geometric contributions. As depicted in cross-section in Figure 4c, the arrays of 1D electron channels in the samples of Figure 16 are defined by GaAs stripes fabricated from a formerly homogeneous GaAs cap layer by a selective etching process. After the etching process the heterostructure is covered with a homogeneous gold gate. Narrow electron channels reside beneath these stripes whereas beneath the gaps the heterojunction is depleted throughout the whole gate voltage regime recorded. The oscillations are not observed in an unstructured sample with a 2DEG, as demonstrated by the top trace. Only the lowest 2D subband is occupied within the range of gate voltage depicted. The periods of the capacitance oscillations recorded on structured samples increase with decreasing lithographic width of the GaAs stripes. This is expected since the spacing of the 1D subbands in energy increases with decreasing width of the confinement potential. Numerical calculations were performed to analyze the experiments, solving the Schrödinger and Poisson equations self-consistently in the Hartree approximation with boundary conditions derived from the geometry of the device (Laux *et al.* 1988, Smith *et al.* 1987). Calculated separations in gate voltage at which the Fermi level crosses subsequent subbands agree well with those of the measured oscillations capacitance. Calculations performed for low electron densities and geometrical widths between 0.2 μm and 0.4 μm result in typical level separations at the Fermi energy between 5 and

2 meV, and channel widths a factor two to three smaller than the geometrical line width of the GaAs stripes (Smith *et al.* 1987).

5 Electrons in quasi-zero-dimensional discs

Once the preparation technique is capable of fabricating 1D electron channels it can easily be applied to confine the electron systems in both lateral directions so that quasi-0D electron systems are generated. In 0D electron systems the energy levels are fully quantised and the degeneracy is essentially governed by the symmetry of the confinement potential. Thus the electron systems become similar to many-electron atoms or to impurity states in semiconductors. However, the number of electrons can be controlled in these man-made atoms and intriguing new properties can be expected because of different scales and confinement potentials that may drastically deviate from the atomic Coulomb potential. Moreover, the interaction between neighboring discs can be controlled, since the distances of adjacent electron discs can be deliberately chosen. Finally, 0D electron systems are of technological interest, since they will represent the basis of fabrication for quantum devices. Some results of capacitance measurements and FIR spectroscopy on matrices of many 0D electron systems will be discussed in the following.

5.1 Magnetocapacitance spectroscopy in quantum dots

Measurements of capacitance on matrices of 0D electron systems have been performed with GaAs-$Al_xGa_{1-x}As$ heterojunction devices that are very similar to those for the measurements on 1D channels discussed in the previous section (Smith *et al.* 1988b, Hansen 1989). In the heterojunction sample of Figure 17 the matrix of 0D electron systems is generated beneath a matrix of GaAs dots that are etched out of a 30 nm cap layer. Scanning electron micrographs show that the dots have the form of squares with rounded corners. The width of the squares is 300 nm and the period of the matrix is 500 nm. The matrix contains about 10^5 such squares in order to get a large and noiseless capacitance signal. The derivative of the capacitance signal with respect to gate voltage is recorded as a function of the gate voltage at different magnetic fields applied perpendicular to the surface of the sample as indicated in the inset of Figure 17. The derivative of the capacitance signal is zero at gate voltages below an onset voltage of $V_g = -0.27$ V and rises sharply at this voltage, indicating that electrons start to occupy the 0D states of the pockets of potential generated at the heterojunction. Similar to the case of the 1D quantum wires, the separation of the electron dots is so large that they are isolated in the whole gate voltage regime recorded, in the sense that the exchange of charge can take place only by tunnelling processes through the back contact. Oscillations of the capacitance signal are observed at gate voltages above the onset that are very similar to the oscillations observed on 1D channels. Corresponding behaviour is found in devices with dots of 200 nm and 400 nm width and 400 nm and 600 nm period, respectively. Comparison of measurements performed on devices with different diameters of dots reveals a systematic behaviour of the onset voltage and period of oscillation similar to that found in 1D channels. The onset voltage increases,

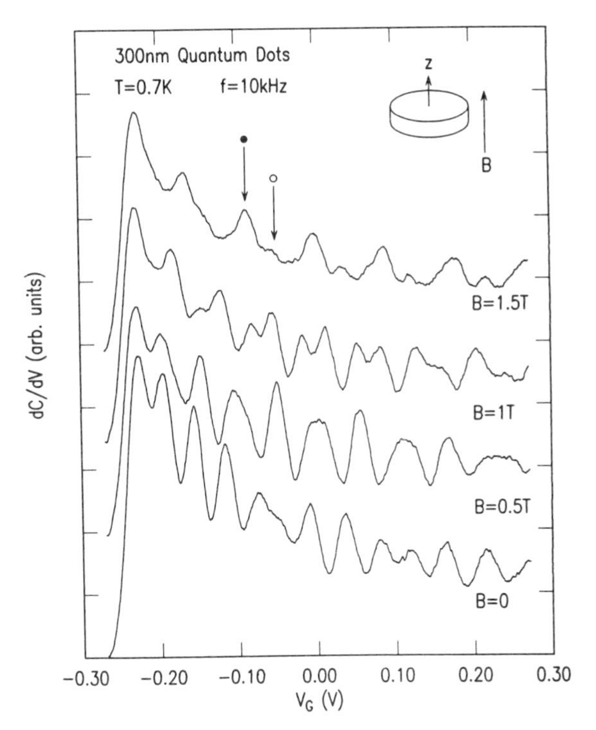

Figure 17: *Derivative of the capacitance with respect to gate voltage measured on an* GaAs-Al_xGa_{1-x}As *heterojunction with a corrugated gate. The configuration of the gate is sketched in Figure 4(b) and is periodically structured in both lateral directions with a matrix of* 300 nm *wide squares. The direction of the magnetic field with respect to the dot geometry is indicated in the inset (Hansen et al. 1989).*

and the oscillation period decreases, with decreasing dot diameter. An increase of the threshold voltage with decreasing dot size is expected, since fringing fields lift the potential minima beneath the dots of the GaAs cap.

The positions of the oscillation maxima in gate voltage change rapidly if a small magnetic field is applied, as demonstrated with the different traces of Figure 17. At high magnetic fields ($B > 2\,\mathrm{T}$) the positions in gate voltage increase linearly with the magnetic field similar to the Shubnikov–de Haas oscillations measured on the homogeneous 2D electron systems, where the areal density changes linearly with the gate voltage. At magnetic fields above 4 T an additional splitting similar to the spin splitting of Landau levels is observed. In Figure 18 the positions of the oscillation maxima in gate voltage are plotted versus the magnetic field.

Kumar *et al.* (1990) have performed numerical calculations within the Hartree approximation to model the states of the quantum dot in the capacitance devices. These calculations are far more elaborate than those for the 1D channels. Potentials, wave functions, and charge density distributions now vary in all three spatial dimensions. The level structure strongly alters whenever an additional electron is injected into the quantum dot, because the extension of the effective confinement potential increases with

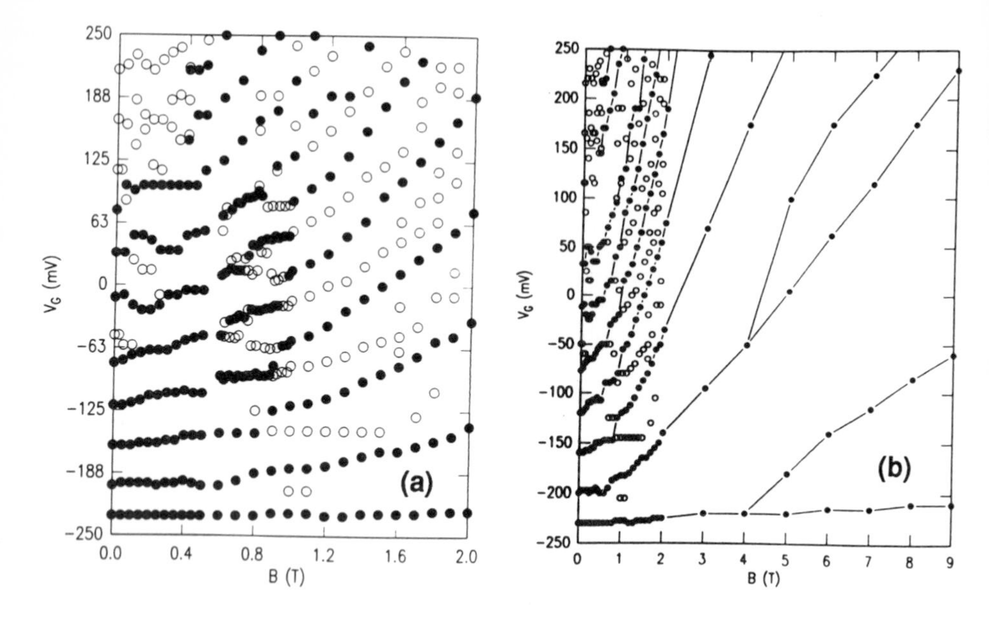

Figure 18: *Dependence on magnetic field of the maxima in the capacitance spectra measured as function of the gate voltage on the device in Figure 17 (Hansen et al. 1989).*

increasing occupation. For an analysis of the dependence on gate voltage it is important to include the back contact in the calculation properly (Kumar *et al.* 1990, Silsbee and Ashoori 1990). Because the capacitance associated with a single 0D electron disc is so small, the Fermi energies of the 0D electron system and the contact are in general not aligned. Whenever an electron is injected into the 0D system the additional charge on the small capacitor will basically lift the energy levels by the so-called Coulomb gap energy, which may be written as e^2/C where C is the capacitance of the 0D electron system with respect to the electrodes. The value of this parameter is very sensitive to the width and the background doping of the nominally undoped GaAs layer. The results of Kumar *et al.* (1990) show that the size of the electron system is considerably smaller than the width of the etched GaAs pattern, as in the case of 1D channels. The shape of the effective confinement potential is found to be almost circular although the GaAs cap consists of squares with rounded corners. The energy levels have the same degeneracies as the ones in a single-particle harmonic oscillator, apart from small splittings arising from the imperfect circular symmetry. The calculated gate voltages at which individual electrons are injected into the lowest quantum levels have typical separations of about 10 mV at zero magnetic field. In contrast, the first four capacitance maxima above the turn-on voltage $V_g = -0.27$ V in Figure 17 have a separation of about 40 mV. This value is similar to the calculated separations between gate voltages at which almost degenerate energy levels start to be filled. This indicates that not all single-electron charging processes can yet be resolved in the measured capacitance oscillations.

Ashoori *et al.* (1991) have fabricated capacitance devices with an $Al_xGa_{1-x}As$ tunnel barrier between the contact and 0D electron system, in which parameters of the tunnel barrier to the back contact can be obtained by experiment. They present results in which the oscillations in capacitance are about a factor 100 smaller than the data discussed above. The observations are discussed within a single-particle charging model. However, to simulate their data the authors have to assume that the turn-on voltage of different electron dots in their device fluctuates on a scale that exceeds the separation of gate voltages at which the injection of a charge carrier into a single dot takes place.

Additional valuable information about the nature of the capacitance oscillations can be obtained from their dependence on magnetic field. As in the case of 1D electron systems, a magnetic field applied perpendicular to the surface of the sample changes the electron states so that the energy levels gradually approach Landau energies at high fields, where the magnetic length l_B is much smaller than the extension of the electron system. Kumar *et al.* have extended their self-consistent Hartree model to include the effect of a magnetic field. The calculations show that the results of the Hartree approximation are still similar to the level structure in a single-particle harmonic oscillator. The dependence of the single-particle energy levels in a 2D harmonic oscillator potential $V(r) = \frac{1}{2}m^*\Omega^2r^2$ on a magnetic field was given by Fock (1928):

$$E_{n,l} = \hbar\,(2n + |l| + 1)\left(\Omega^2 + \tfrac{1}{4}\omega_c^2\right)^{1/2} + \tfrac{1}{2}l\hbar\omega_c\,. \tag{10}$$

At zero magnetic field the energy levels are determined by the electrostatic confinement, described by the characteristic harmonic oscillator frequency Ω. At finite field the energy levels rearrange to form levels close to the Landau energies $[n + \frac{1}{2}(|l| + l) + \frac{1}{2}]\hbar\omega_c$ in the limit of high fields. Physically l is the angular momentum quantum number and $N = n + \frac{1}{2}(|l| + l)$ is the Landau level index.

The dependence of single-particle energy levels on magnetic field in a circular disc or a rectangular box can be found in an article by Robnik (1986). Generally, at low magnetic fields the quantised energy levels and their degeneracies are determined by the symmetries of the electrostatic confinement potential. A magnetic field partially lifts the degeneracies and the energy levels rearrange to form a structure like Landau levels in the limit of high fields. There is no one-to-one correspondence of the electron states at $B = 0$ to the Landau levels at high magnetic fields, in contrast to the behaviour of 1D subbands. Instead the crossover from electrostatic to magnetic confinement is always accompanied by pronounced level crossings at intermediate fields. This behaviour is qualitatively reflected in the dependence of the capacitance on magnetic field shown in Figure 18 (Hansen *et al.* 1989). However, comparison with level structures calculated in single-particle models also suggests that the data do not resolve all single-particle energy levels. More recently, closely related transport experiments have been performed by McEuen *et al.* (1991) on relatively large electron discs with lateral tunnel contacts. More sophisticated models will also have to include exchange and correlation. Bryant (1987) has shown that exchange and correlation have a large effect on the level spectrum for 0D electron systems in quantum-well boxes of sizes comparable to the present experimental systems. More recently, Maksym and Chakraborty (1990) have calculated the dependence on magnetic field of the states in a quantum dot with 3 or 4 interacting electrons. The results demonstrate that the contribution to the total energy from interactions contains step-like structure as a function of the total angular momentum in high magnetic fields. The contribution to the total energy from interactions decreases with increasing total angular momentum, and causes the states to be in a higher total

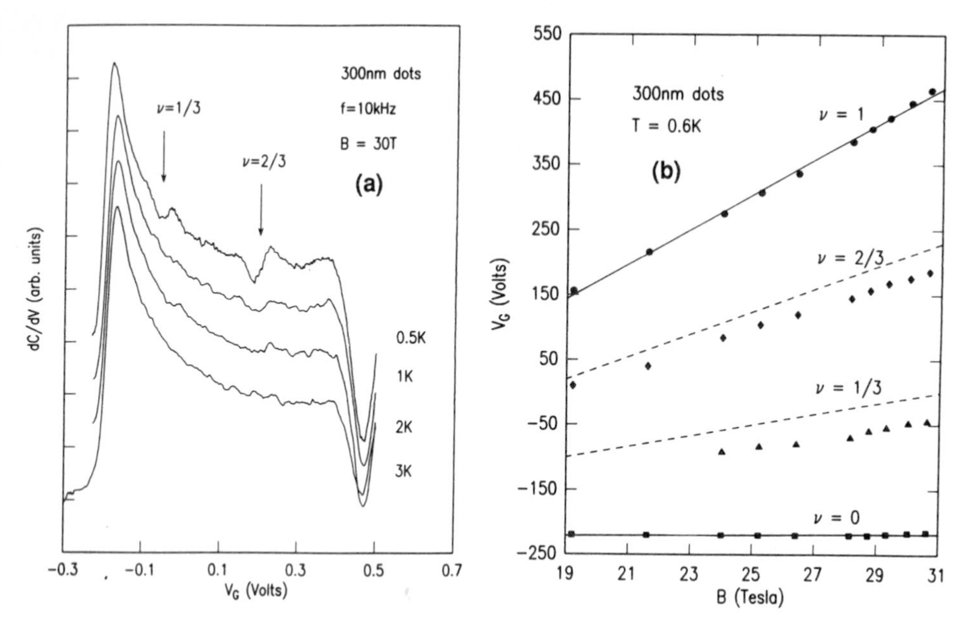

Figure 19: *(a) Derivative of the capacitance with respect to gate voltage recorded on a sample similar to that of Figure 17 in the magnetic quantum limit ($B = 30\,\mathrm{T}$) at four different temperatures. (b) Positions in gate voltage of the minima in the derivative of the capacitance measured at temperature $T = 0.6\,\mathrm{K}$ and different magnetic fields. The dashed lines indicate the interpolated gate voltages at which the filling factor is $\frac{1}{3}$ and $\frac{2}{3}$ (Hansen et al. 1990).*

angular momentum than in a single-particle model without interaction. The authors link the step-like structure in the dependence of the total energy on angular momentum to effective filling factors similar to the fractional filling factors in 2D electron systems.

It is intriguing to investigate whether energy gaps at fractional filling factors are still observable in quantum dots containing only a few electrons. Energy gaps at the Fermi energy arising at fractional Landau level filling factors cause additional plateaus in the Hall voltage in high mobility 2D electron systems (Chakraborty and Pietiläinen 1988). Fractional states have also been observed in capacitance measurements in 2D as well as 1D systems (Smith *et al.* 1986, 1988a). Figure 19 shows traces of the derivative of the capacitance recorded on a matrix of dots with 300 nm squares at a high magnetic field of $B = 30\,\mathrm{T}$ at different temperatures (Hansen *et al.* 1990). The turn-on voltage at which carriers start to be trapped in the 0D dots is close to the value at zero field, $V_g = -0.27\,\mathrm{V}$. The capacitance traces are recorded up to gate voltages at which a strong structure in the capacitance signal arises from occupation of the first spin-split Landau level. Additional structures are observed at lower filling factors that are marked with arrows in Figure 19. The structures become more pronounced with increasing magnetic field and decreasing temperature. As shown in Figure 19a they vanish if the temperature is increased up to $T = 3\,\mathrm{K}$ at $B = 30\,\mathrm{T}$. Figure 19b shows the dependence on magnetic field of the gate voltages at which the additional structures are observed.

The gate voltages increase linearly with the magnetic field, indicating that the filling factor at which they are observed is constant. For comparison, the dashed lines mark the positions of filling factors $\frac{1}{3}$ and $\frac{2}{3}$ estimated with the positions in gate voltage of the two lowest spin-split Landau levels and assuming a constant capacitance. The dependence on magnetic field, temperature and gate voltage indicate that the observed additional structures are indeed caused by fractional states in the 0D electron systems. These results demonstrate that numerical few-particle calculations of fractional quantisation may be directly compared with experiments in the future. At filling factor $\frac{1}{3}$ and magnetic field $B = 24\,\mathrm{T}$ only about 22 electrons are contained within a dot of diameter 120 nm.

5.2 Far-infrared absorption of quasi-zero-dimensional electron discs

The interaction of FIR with the electrons in a 0D electron system results in a dynamic polarisation and thus dimensional resonances, as in the case of 1D electron channels. Many properties of the dimensional resonances in 0D electron systems can be understood in the harmonic oscillator model for the bare potential, since the same arguments discussed above for 1D channels apply for 0D systems as well if the 1D harmonic oscillator potential is replaced by one with circular symmetry. Again, a single resonance is observed at zero magnetic field that describes the characteristic frequency Ω of the bare potential $V(r) = \frac{1}{2}m^*\Omega^2 r^2$. An important difference arises if a magnetic field is applied perpendicular to the sample surface: the zero-field resonance splits into two in 0D systems. Within the generalised Kohn theorem at sufficiently low resonance linewidths the dispersion of the branches is described by

$$\omega_\pm = \sqrt{\Omega^2 + \left(\frac{\omega_c}{2}\right)^2} \pm \frac{\omega_c}{2}. \tag{11}$$

The high-frequency branch approaches the cyclotron frequency $\omega_c = eB/m^*$ at high magnetic fields, whereas the second branch exhibits a negative dispersion in magnetic field and approaches zero frequency.

A negative dispersion in magnetic field is also seen for edge magnetoplasmons as discussed, for example, by Fetter (1986) to describe excitations in the radio frequency regime observed in macroscopic electron discs on liquid helium (Mast *et al.* 1985, Glattli *et al.* 1985). Actually, the high-frequency and low-frequency branches in high magnetic fields can be understood as a volume-type and a perimeter-type mode respectively. In a simple model the frequencies of the modes at zero field are determined by $\Omega = \omega_p\sqrt{2/3}$ where ω_p is the plasmon frequency according to Equation (2) with wave vectors $q = \nu/R$, with $\nu = 1, 2, \ldots$, that are a multiple of the inverse of the radius of the dot, R (Demel *et al.* 1990a). The dispersion in magnetic field of the lowest two branches is similar to Equation (11) with factors of order unity that multiply the elements of the right-hand side of Equation (11) to account for different charge distributions at the edge of the discs.

As an example the FIR spectra of quantum dots in dual stacked-gate Si-MOS devices are depicted in Figure 20. The gate configuration of the device is similar to that of the

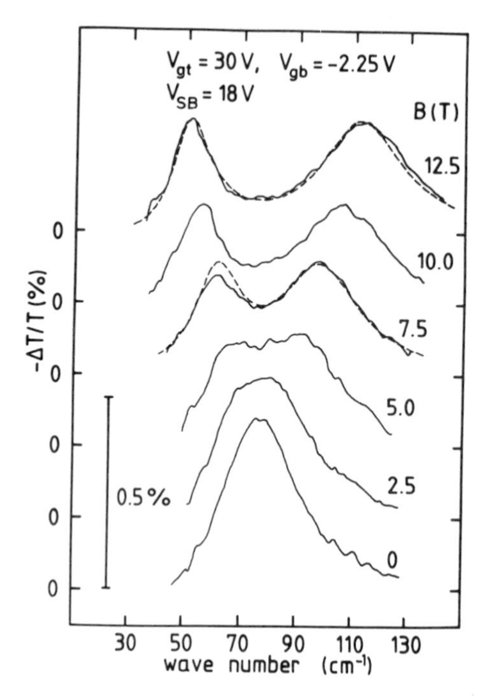

Figure 20: FIR-*transmittance spectra of a matrix of 0D electron systems on a stacked gate* Si-MOS *device in different magnetic fields applied perpendicular to the sample surface. The dotted lines are calculated from an approach to the dynamic conductivity using a harmonic oscillator (Alsmeier et al. 1990a).*

device with 1D channels in Figure 9, but the bottom gate consists of a grid with 400 nm period and 150 nm openings here. With gate voltages $V_{gb} = -2.25$ V and $V_{gt} = 30$ V applied on bottom gate and top gate respectively, a matrix of 0D electron systems is induced beneath the openings of the bottom gate. In Figure 20 a large substrate bias voltage $V_{\mathrm{SB}} = 18$ V has been applied on the top gate to increase the steepness of the confinement potential. At zero magnetic field a single resonance of energy 9.5 meV is observed. The resonance splits into two modes in non-zero magnetic fields with positions that are well described by Equation (11). Deviations from this behaviour occur at a lower bottom gate voltage $V_{gb} = -1.5$ V and substrate bias $V_{\mathrm{SB}} = 0$, which have been attributed to a softening of the walls of the confinement potential (Alsmeier *et al.* 1990). The number of electrons contained within each dot of the matrix can be derived from a fit of the oscillator strengths at zero magnetic field to the oscillator strengths calculated from the dynamic conductivity of a harmonic oscillator. An number $N = 140$ of electrons and a diameter of 100 nm for each dot are determined for the data in Figure 20. An electron number of 20 and a diameter as low as 40 nm are obtained from the resonances observed at 15 V top gate voltage. Quantum dots containing only 3 electrons have been realised on InSb-MOS devices and resonance energies of typically 8 meV are reported (Sikorski and Merkt 1989).

The dashed fits to the line-shapes of the resonance in Figure 20 are calculated with

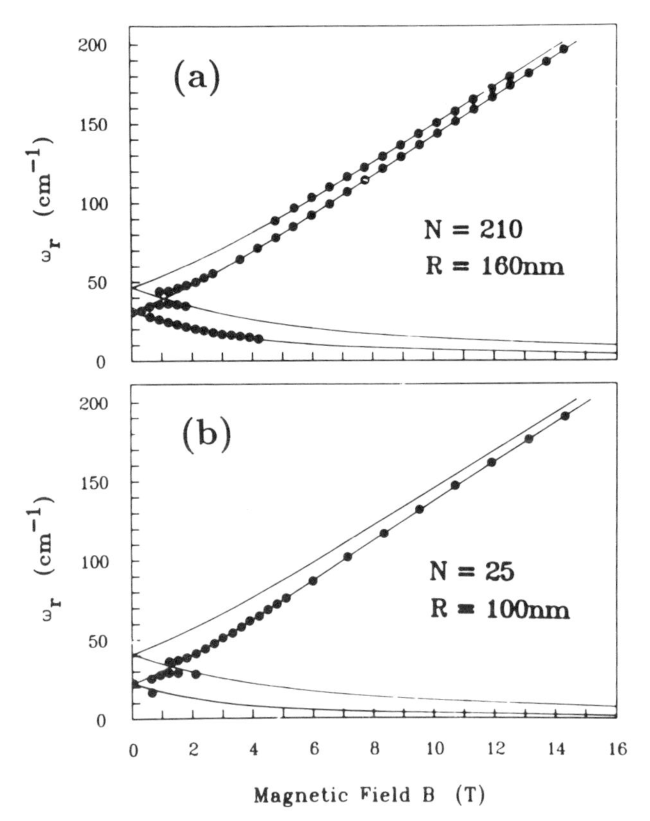

Figure 21: *Experimental dispersion of* FIR *resonances with magnetic field observed on quantum dots in deep mesa etched* GaAs-Al$_x$Ga$_{1-x}$As *heterojunctions. The number of electrons contained in one dot, N, and the radius of the dot, R, have been determined from the oscillator strengths and positions of the resonance (Demel et al. 1990a).*

the conductivity of a harmonic oscillator assuming a constant phenomenological scattering time. In this model the two resonant modes are expected to have equal heights and the oscillator strengths of the high-frequency mode S_+ and the low-frequency mode S_- are predicted to take the ratio $S_+/S_- = \omega_+/\omega_-$. Although, in general, the harmonic oscillator model describes the positions of the resonance rather well, the oscillator strengths and amplitudes measured on quantum dot devices deviate significantly. This is similar to the behaviour of the polarisation-dependent absorption strengths in 1D channels (Sikorski and Merkt 1989, Demel *et al.* 1990a, Kern *et al.* 1991a). The deviations have been interpreted as an indication of quantum confinement (Merkt 1989, Alsmeier *et al.* 1990a).

Demel *et al.* (1990a) investigated FIR excitations on quantum dots in deep mesa etched GaAs-Al$_x$Ga$_{1-x}$As heterojunction devices. The positions of resonances determined on such samples are plotted in Figure 21 versus the magnetic field. The lines in Figure 21 mark the positions of the resonance calculated according to Equation (11) with frequencies at zero field that best fit the experimental data. The dominant resonances are well described by the two modes emerging with increasing magnetic field from the lower frequency at zero field. However, significant deviations from the predictions of the harmonic oscillator model are observed as an anti-crossing behaviour at a magnetic field of about $B = 1.5\,\mathrm{T}$. Furthermore, a second high-frequency mode is resolved as a weak shoulder at higher magnetic fields $B > 4\,\mathrm{T}$ on the matrix with larger quantum dots.

It is remarkable that strong higher-order modes but no anti-crossing of the modes are observed in the dynamic conductivity of macroscopic discs realised on the surface of liquid helium (Mast *et al.* 1985, Glattli *et al.* 1985). There the experiments are performed in the radio-frequency regime with external electrodes so that higher-order modes can readily be excited, as well as the dipole mode. If the second high-frequency branch observed in Figure 21 originates from the second order perimeter-type mode with a wavelength of half the circumference of the dot, the zero-field frequencies of the fits should be in the ratio $\sqrt{2} : 1$ (Demel *et al.* 1990a). Shikin *et al.* (1991) calculate higher-order modes in a classical model basing on a parabolic bare potential and quote the ratio 1.5 for the frequencies of the first two modes, which is very close to the ratio of the frequencies at zero field in the fits in Figure 21. The branches of the higher-order modes with negative dispersion in magnetic field intersect the branch of the first order-mode that increases with magnetic field. An anti-crossing of the branches similar to that in Figure 21 is predicted if non-harmonic contributions are added to the parabolic confinement potential and the circular symmetry is broken (Shikin *et al.* 1991). The first electron discs on GaAs-$\mathrm{Al}_x\mathrm{Ga}_{1-x}$As heterojunction devices whose FIR response has been investigated had a diameter of $1.5\,\mu$m, rather macroscopic compared to the sizes made since (Allen *et al.* 1983). Nevertheless, the FIR resonances observed on these samples could well be explained within the harmonic oscillator model: no higher modes are excited and no anti-crossing behaviour is observed. From this fact it may be inferred that the observation of higher-order modes and an anti-crossing in small quantum dots indicates a quantum correction to the generalised Kohn theorem similar to the coupling in 2DEGs of the second harmonic cyclotron resonance to plasmons of small wave-vectors, known as the so-called non-local interaction (Demel *et al.* 1990a). More recently, the dynamic response of 0D electron systems with anharmonic contributions to the parabolic bare potential has also been investigated in a quantum mechanical model (Gudmundsson and Gerhardts 1991). It has been shown that the random phase approximation reproduces the predictions of the generalised Kohn theorem if the ground state is calculated self-consistently from a purely harmonic bare potential (Broido *et al.* 1990, Gudmundsson and Gerhardts 1991). With a circularly symmetric anharmonic contribution which makes the potential steeper, the appearance of a weak higher order mode at high magnetic fields can be simulated. Again, the anti-crossing is explained at least qualitatively by deviations from circular symmetry.

6 Coupled electron discs and antidots

In the preceding sections, the interaction of adjacent quantum dots has been neglected in the explanation of the experimental observations on matrices of electron dots. As has been shown by Bakshi (1990) this is well justified in most of the present systems. Intriguing behaviour is predicted if the Coulomb interaction between adjacent dots be-

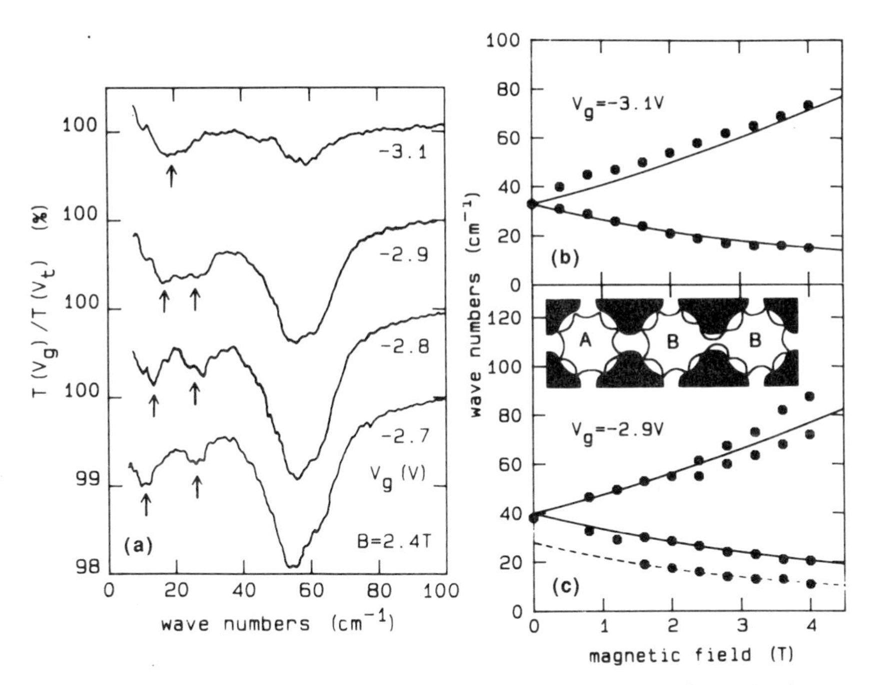

Figure 22: *(a) Far-infrared transmission spectra of an* GaAs-Al$_x$Ga$_{1-x}$As *heterojunction with a front gate that is periodically modulated in both lateral directions. A magnetic field* $B = 2.4\,\mathrm{T}$ *is applied perpendicular to the sample surface. Positions of the resonance as function of the magnetic field are depicted in (b) for isolated and in (c) for coupled electron dots (Lorke et al. 1990).*

comes important. A phase transition into polarised dots is predicted for closely spaced dots that contain a sufficiently large number of electrons (Kempa *et al.* 1991). Experiments reported so far have focused on a different regime, where the constituents of the electron dot matrix become electrically connected. In this regime the matrix of electron dots gradually transforms into an electron mesh. The classification of the constituents of the electron mesh into antidots or coupled dots seems somewhat arbitrary. In the following an electron mesh with strongly coupled electron dots will be called an antidot superlattice. If charge exchange between adjacent dots is inhibited, *e.g.* by a tunnelling barrier, the constituents of the mesh will be called coupled dots.

The transformation of a matrix of isolated quantum dots into a mesh of coupled dots has been investigated by Lorke *et al.* (1990) in FIR transmission experiments with GaAs-Al$_x$Ga$_{1-x}$As heterojunction devices in which the electron dots are defined by a corrugated gate as depicted in Figure 3b. The FIR spectra of a device with period $a = 450\,\mathrm{nm}$ are presented in Figure 22. At gate voltage $V_g = -3.1\,\mathrm{V}$ the electron system consists of isolated quantum dots. Correspondingly, two resonances are observed in a magnetic field applied perpendicular to the sample surface with dispersion given by Equation (11), as demonstrated in Figure 22b. With increasing gate voltages the diameter of the dot becomes larger at a fixed period of the matrix so that the dots become electrically coupled at $V_g = -2.9\,\mathrm{V}$. The strong interaction is reflected at finite magnetic fields in the FIR spectra taken at gate votages $V_g \geq -2.9\,\mathrm{V}$. The low-frequency branch of the FIR spectra splits into two resonances and, furthermore, a weak shoulder appears on the high energy side of the other branch.

From the behaviour at higher gate voltages Lorke *et al.* (1990) identify the shoulder at the high-frequency branch as a plasmon that propagates through the electron mesh with wavelength $\lambda = a$. The FIR couples to the plasmon via a modulation of the charge density and the modulated gate conductivity. The prominent alteration of the excitation spectrum is the splitting of the low-frequency branch. This behaviour is explained with the classical trajectories depicted in the inset of Figure 22. If the low-frequency excitation is understood as a perimeter-type mode, virtually two different trajectories become possible once adjacent dots become electrically coupled. If the coupling between adjacent dots is weak, another mode that oscillates between two adjacent dots is most likely apart from the mode orbiting around a single dot. The dog-bone shaped trajectory of this mode has roughly twice the length of the effective single dot circumference. Actually, the observed resonance frequencies form a ratio that augments this interpretation. The full line in Figure 22c indicates the dependence on magnetic field of the resonance positions calculated with Equation (11) and a frequency Ω at zero field that best fits the experimental data at gate voltage $V_g = -2.9\,\mathrm{V}$. The dashed line in Figure 22c is also calculated with Equation (11) and a frequency at zero field that is a factor $1/\sqrt{2}$ smaller than the previous one, indicating that the circumference that this mode senses is twice as large.

An antidot superlattice is formed if the insulator of the corrugated front gate consists of a layer with a periodic matrix of holes rather than a matrix of dots (Lorke *et al.* 1991a). At negative gate voltages a regular array of potential maxima is generated. Screening breaks down once the regions beneath the holes in the insulator are completely depleted of electrons, and a regular array of scattering centers is formed. At low magnetic fields, where the cyclotron orbit is comparable to the period of the antidot superlattice, strong resistance maxima reflect the interaction of the electron system with the regular array of scatterers. Surprisingly, the observations are well explained within classical models that consider classical particle trajectories of ballistic electrons in a 2D periodic potential (Lorke *et al.* 1991a, Weiss *et al.* 1991). In an ideal antidot superlattice, scattering of the electrons predominantly originates from the interaction with the superlattice potential rather than from intrinsic scattering processes such as phonon scattering or scattering of randomly distributed impurities. The trajectories of the electron motion, which are quasi-ballistic in this sense, can be classified into localised and delocalised orbits (Geisel *et al.* 1990). The localised orbits are stable, quasi-periodic trajectories that encircle a specific number of antidots. The phase space occupied by localised trajectories depends on the magnetic field. Since localised trajectories do not contribute to the transport process, the magnetoresistance increases if the phase space occupied by localised orbits is large. The magnetoresistance depicted in Figure 23 was recorded on an antidot superlattice with period 450 nm (Lorke *et al.* 1991a). A strong maximum in resistance is found at a magnetic field at which the diameter of the cyclotron orbit is equal to the period of the superlattice potential. More recently, several resistance maxima have been observed by Weiss *et al.* (1991) on an antidot superlattice defined by holes etched into a heterostructure mesa. Each resistance maximum could be assigned to a stable orbit encircling a specific number of antidots. At high magnetic

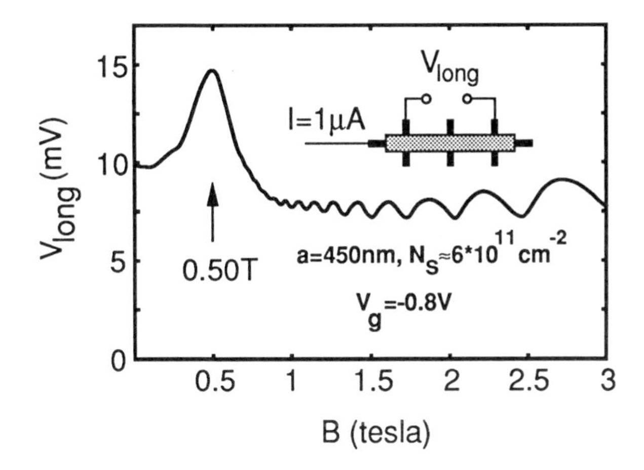

Figure 23: *Magnetoresistance of a superlattice of antidots generated with a corrugated gate at bias* $V_g = -0.8\,\mathrm{V}$ *in an* GaAs-$\mathrm{Al}_x\mathrm{Ga}_{1-x}$As *heterojunction (Lorke et al. 1991a).*

fields, where the cyclotron orbit becomes smaller than the spacings between adjacent antidots, the behaviour of an unpatterned 2DEG is recovered (Ensslin and Petroff 1990).

Classical pictures with single-particle trajectories have even been employed to explain the dynamic behaviour of antidot superlattices. FIR spectra of antidot superlattices were first recorded by Lorke *et al.* (1991a). The excitation spectra consist of a weak magnetoplasmon-type excitation and a strong resonance that is shifted from the cyclotron resonance positions according to Equation (11). Again the magnetoplasmon resonance is excited at wavelength $\lambda = a$ that is equal to the superlattice period, as in the matrix of coupled quantum dots. The shift of the cyclotron resonance has been attributed to virtual confinement of the cyclotron orbits within the superlattice potential. An additional intriguing low-frequency mode has been observed by Kern *et al.* (1991b) in an antidot superlattices fabricated on $\mathrm{Al}_y\mathrm{In}_{1-y}$As-$\mathrm{Ga}_x\mathrm{In}_{1-x}$As single quantum wells by deep mesa etching. The resonances are shifted to higher frequencies in these devices because of the lower mass of electrons in $\mathrm{Ga}_x\mathrm{In}_{1-x}$As. The low-frequency mode has probably eluded observation in the experiments on antidot superlattices in GaAs-$\mathrm{Al}_x\mathrm{Ga}_{1-x}$As heterojunctions because the frequencies of the resonance have been below the regime that is accessible by the Fourier transform spectrometer.

In Figure 24 the resonance positions observed on the antidot superlattice in an $\mathrm{Al}_y\mathrm{In}_{1-y}$As-$\mathrm{Ga}_x\mathrm{In}_{1-x}$As single quantum well are depicted versus the magnetic field. Three different modes are observed. At low magnetic fields the low-frequency mode (ω_-) is strongest. With increasing magnetic field the oscillator strength of the low-frequency mode is gradually transferred to a high-frequency mode (ω_+). At intermediate fields a weak cyclotron-type resonance (CR) is observed between these two modes. Dashed-dotted lines in Figure 24 are calculated with Equation (11) and frequencies Ω at zero field that fit the positions of the high- and the low-frequency modes, respectively. At high magnetic fields the two dominant modes behave similarly to the high- and low-frequency branches of excitations in electron dots. Thus the high-frequency mode is identified as a volume-type excitation originating from electrons that move on

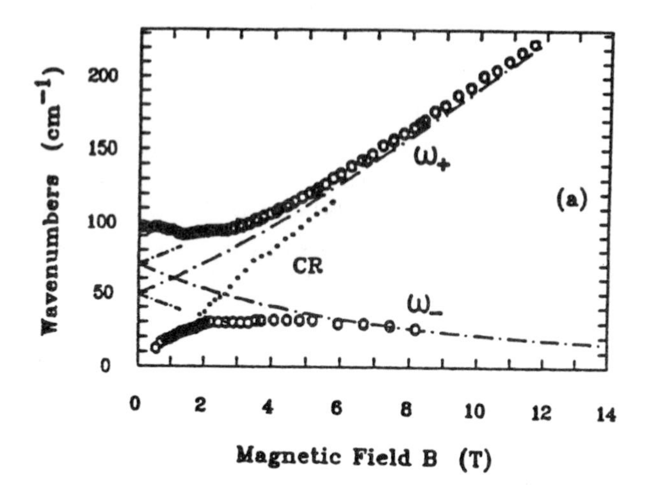

Figure 24: *Experimental dispersion in magnetic field of* FIR *resonances observed on deep mesa etched* $Al_yIn_{1-y}As$-$Ga_xIn_{1-x}As$ *single quantum wells. A matrix of* 200 nm *holes with period* $a = 300$ nm *has been punched into the device by the etching process. Dashed-dotted lines denote calculated dispersions fitted to either the* ω_+ *or the* ω_- *mode (Kern et al. 1991b).*

cyclotron orbits within the regions between the antidots. Again, the resonance frequencies are shifted above the cyclotron frequency because of the virtual confinement within the antidot superlattice. On the other hand, the low-frequency mode originates from perimeter-type trajectories that enclose individual antidots. Since the virtual confinement and the effective antidot circumference are independent in the antidot superlattice, it is not surprising that different frequencies fit the high- and low-frequency modes at zero field. Strong deviations from the behaviour of excitations in quantum dots emerge at low magnetic fields. The positions of the low-frequency branch now start to decrease with decreasing magnetic field and gradually approach the cyclotron resonance frequency. The radius of the cyclotron orbit is roughly equal to the radius of the antidots at the maximum position of the low-frequency branch. Thus the trajectories sense the volume of the electron mesh more and more at lower magnetic fields and, correspondingly, the excitation becomes a cyclotron resonance. On the other hand the high-frequency mode simultaneously becomes a magnetoplasmon propagating in the electron mesh. The position of this mode increases with decreasing magnetic field in contrast to the magnetic field dispersion of dimensional excitations in 0D dots [see Equation (11)]. This behaviour is reminiscent of the behaviour of plasmon excitations in 1D electron channels with finite wave-vectors parallel to the channel. Such excitations, again, exhibit a perimeter-type branch with negative dispersion in the field (Demel *et al.* 1991).

7 Summary

Lateral superlattices have been realised on various semiconductor systems such as Si- or InSb-MOS devices and GaAs-$Al_xGa_{1-x}As$ or $Al_yIn_{1-y}As$-$Ga_xIn_{1-x}As$ single quantum wells. The experimental systems investigated cover the whole spectrum of possible potential superlattices, that consist of charge density superlattices with weak potential modulation on the one end, and of arrays of isolated electron wires, electron meshes or matrices of dots on the other end. The experimental results of FIR spectroscopy, capacitance spectroscopy and magnetotransport measurements demonstrate their intriguing behaviour. Many of the experimental findings are quite well understood within surprisingly simple but illuminating models.

Acknowledgments

In parts of this article I have discussed the results of collaborations to which I did not directly contribute. I am therefore very grateful to those colleagues who generously allowed me to include their results in this article. It is also a pleasure to acknowledge particularly helpful discussions with F Brinkop, J Brum, C Dahl, D Heitmann, J P Kotthaus, A Lorke, U Merkt, T P Smith and R Winkler. Finally, I would like to thank H Drexler and Ch Lettau for a critical reading of the article.

References

Allen S J Jr, Störmer H L, and Hwang J C M, 1983. *Phys. Rev. B*, **28**, 4875.

Allen S J, DeRosa F, Dolan G J, and Tu C W, 1985. *Proc. 17th ICPS*, Eds Chadi J D and Harrison W A, New York, Springer, p. 313.

Alsmeier J, Sikorski Ch, and Merkt U, 1988. *Phys. Rev. B*, **37**, 4314.

Alsmeier J, Batke E, and Kotthaus J P, 1989. *Phys. Rev. B*, **40**, 12574.

Alsmeier J, Batke E, and Kotthaus J P, 1990a. *Phys. Rev. B*, **41**, 1699.

Alsmeier J, Kotthaus J P, Klapwijk T M, and Bakker S, 1990b. *Proc. 20th ICPS*, Eds. Anastassakis E M and Joannopoulos J D, Singapore, World Scientific, p. 2355.

Alves E S, Beton P H, Henini M, Eaves L, Main P C, Hughes O H, Toombs G A, Beaumont S P, and Wilkinson C D W, 1989. *J. Phys.: Condens. Matter*, **1**, 8257.

Ando T, Fowler A B, and Stern F, 1982. *Rev. Mod. Phys.*, **54**, 437.

Ashoori R C, Silsbee R H, Pfeiffer L N, and West K W, 1991. *Proc. Int. Conf. on Nanostructures and Mesoscopic Systems*, Academic Press, in press.

Bakshi P, Broido D A, and Kempa K, 1990. *Phys. Rev. B*, **42**, 7416.

Batke E, Heitmann D,, and Tu C W, 1986. *Phys. Rev. B*, **34**, 6951.

Beenakker C W J, 1989. *Phys. Rev. Lett.*, **62**, 2020.

Berggren K-F, Roos G, and van Houten H, 1988. *Phys. Rev. B*, **37**, 10118.

Bernstein G, and Ferry D K, 1986. *Superlattices and Microstructures*, **2**, 373.

Bernstein G, and Ferry D K, 1987. *Z. Phys. B*, **67**, 449.

Blount E I, 1962. *Phys. Rev.*, **126**, 1636.

Brey L, Johnson N F, and Halperin B I, 1989. *Phys. Rev. B*, **40**, 10647.

Brinkop F, Hansen W, Kotthaus J P, and Ploog K, 1988. *Phys. Rev. B*, **37**, 6547.

Brinkop F, Dahl C, Kotthaus J P, Weimann G, and Schalpp W, 1990. *Proc. Int. Conf. on Application of High Magnetic Fields in Semiconductor Physics*, Würzburg, Germany, in press.
Broido D A, Bakshi P, and Kempa K, 1990. *Solid State Commun.*, **76**, 613.
Brum J, 1991. *Phys. Rev. B*, **43**, 12082.
Bryant G W, 1987. *Phys. Rev. Lett*, **59**, 1140.
Cataudella V, and Ramaglia V M, 1988. *Phys. Rev. B*, **38**, 1828.
Chakraborty T, and Pietiläinen, 1988. *The Fractional Quantum Hall Effect*, in *Springer Series in Solid State Sci.*, **85**, New York, Springer.
Chaplik A V, 1972. *Sov. Phys. JETP*, **35**, 395 (*Zh. Eksp. Teor. Fiz.*, **62**, 746).
Chaplik A V, 1989. *Superlattices and Microstructures*, **6**, 329.
Cole T, Lakhani A A, and Stiles P J, 1977. *Phys. Rev. Lett.*, **38**, 722.
Dahl C, 1990. *Phys. Rev. B*, **41**, 5763.
Davies J H, 1988. *Semicond. Sci. Technol.*, **3**, 995.
Demel T, Heitmann D, Grambow P, and Ploog K, 1988. *Phys. Rev. B*, **38**, 12 732.
Demel T, Heitmann D, Grambow P, and Ploog K, 1990a. *Phys. Rev. Lett.*, **64**, 788.
Demel T, Heitmann D, and Ploog K, 1990b. *Proc. 20th ICPS*, Eds Anastassakis E M and Joannopoulos J D, Singapore, World Scientific, p 2403.
Demel T, Heitmann D, Grambow P, and Ploog K, 1991. *Phys. Rev. Lett.*, **66**, 2657.
Eliasson G, Wu J-W, Hawrylak P, and Quinn J J, 1986. *Solid State Commun.*, *60*, 41.
Ensslin K, and Petroff P M, 1990. *Phys. Rev. B*, **41**, 12307.
Esaki L, and Tsu R, 1970. *IBM J. Res. Dev.*, **14**, 61.
Esaki L, 1987. In *Physics and Applications of Quantum Wells and Superlattices*, Eds. Mendez E E and von Klitzing K, New York, Plenum.
Evelbauer T, Wixforth A, and Kotthaus J P, 1986. *Z. Phys. B*, **64**, 69.
Fetter A L, 1986. *Phys. Rev. B*, **32**, 7676; **33**, 3717.
Fock V, 1928. *Z. Phys.*, **47**, 446.
Geisel T, Wagenhuber J, Niebauer, and Obermair G, 1990. *Phys. Rev. Lett.*, **64**, 1581.
Gerhardts R R, Weiss D, and Klitzing K v, 1989. *Phys. Rev. Lett.*, **62**, 1173.
Gerhardts R R, and Chao Zhang, 1990. *Phys. Rev. Lett.*, **64**, 1473.
Gerhardts R R, Weiss D, and Wulf U, 1991. *Phys. Rev. B*, **43**, 5192.
Gershoni D, Weiner J S, Chu S N G, Baraff G A, Vandenberg J M, Pfeiffer L N, West K, Logan R A, and Tanbun-Ek T, 1991. *Phys. Rev. Lett.*, **65**, 1631.
Glattli D C, Andrei E Y, Deville G, Poitrenaud J, and Wiliams F I B, 1985. *Phys. Rev. Lett.*, **54**, 1710.
Gold A, and Ghazali A, 1990. *Phys. Rev. B*, **41**, 8318.
Gudmundson V, and Gerhardts R, 1991. *Phys. Rev. B*, **43**, 12098.
Hansen W, Kotthaus J P, Chaplik A, and Ploog K, 1987a. *High Magnetic Fields*, in *Semiconductor Physics*, Ed. Landwehr G, Springer Series in Solid State Sci., **71**, Heidelberg, Springer, p. 266.
Hansen W, Horst M, Kotthaus J P, Merkt U, Sikorski Ch, and Ploog K, 1987b. *Phys. Rev. Lett.*, **58**, 2586.
Hansen W, 1988. In *Festkörperprobleme (Advances in Solid State Phys.)* **28**, Ed. Rössler U, Braunschweig, Vieweg, p. 121.
Hansen W, Smith T P, Lee K Y, Brum J A, Knoedler C M, Hong J M and Kern D P, 1989. *Phys. Rev. Lett.*, **62**, 2168.
Hansen W, Smith T P, Lee K Y, Hong J M, and Knoedler C M, 1990. *Appl. Phys. Lett.*, **56**, 168.
Haug R J, Lee K Y, Smith T P, and Hong J M, 1990. *Proc 20th ICPS*, Eds. Anastassakis E M and Joannopoulos J D, Singapore, World Scientific, p. 2443.

Haupt R, Wendler L, and Pechstedt R, 1991. Preprint.
Hu G Y, and O'Connell R F, 1990. *J. Phys.: Condens. Matter*, **2**, 9381.
Ismail K, Chu W, Yen A, Antoniadis D A, and Smith H I, 1989a. *Appl. Phys. Lett.*, **54**, 460.
Ismail K, Smith T P, Masselink W T, and Smith H I, 1989b. *Appl. Phys. Lett.*, **55**, 2766.
Kamgar A, Sturge M D, and Tsui D C, 1980. *Phys. Rev. B*, **22**, 841.
Kapon E, Hwang D M, and Bhat R, 1989. *Phys. Rev. Lett.*, **63**, 430.
Kash K, Worlock J M, Sturge M D, Grabbe P, Harbison J P, Scherer A, and Lin P S D, 1988. *Appl. Phys. Lett.*, **53**, 782.
Kash K, Mahoney D D, and Cox H M, 1991. *Phys. Rev. Lett.*, **66**, 1374.
Kempa K, Broido D A, and Bakshi P, 1991. *Phys. Rev. B*, **43**, 9343.
Kern K, Demel T, Heitmann D, Grambow P, Ploog K, and Razeghi M, 1990. *Proc. 8th Int. Conf. on Electronic Properties of Two-dimensional Electron Systems*, Grenoble, France, *Surf. Sci.*, **229**, 256.
Kern K, Demel T, Heitmann D, Grambow P, Zhang Y H, and Ploog K, 1991a. *Superlattices and Microstructures*, **9**, 11.
Kern K, Heitmann D, Grambow P, Zhang Y H, and Ploog K, 1991b. *Phys. Rev. Lett.*, **66**, 1618.
Kohn W, 1961. *Phys. Rev.*, **123**, 1242.
Kotthaus J P, and Heitmann D, 1982. *Surf. Sci.*, **113**, 481.
Kotthaus J P, Hansen W, Pohlmann H, Wassermeier M, and Ploog K, 1988. *Surf. Sci.*, **196**, 600.
Kouwenhoven L P, Hekking F W J, van Wees B J, and Harmans C J P M, 1990. *Phys. Rev. Lett.*, **65**, 361.
Krasheninnikov M V, and Chaplik A V, 1981. *Fiz. Tekh. Poluprovodn.*, **15**, 32 (*Sov. Phys. Semicond.*, **15**, 19.)
Kumar A, Laux S E, and Stern F, 1990. *Phys. Rev. B*, **42**, 5166.
Langbein D, 1969. *Phys. Rev.*, **180**, 633.
Laux S E, and Stern F, 1986. *Appl. Phys. Lett.*, **49**, 91.
Laux S E, and Warren A C, 1986. International Electron Devices Meeting, Los Angeles, California, *IEDM Technical Digest*, 567.
Laux S E, Frank D J, and Stern F, 1988. *Surf. Sci.*, **196**, 101.
Li Q, and Das Sarma S, 1989. *Phys. Rev. B*, **40**, 5860.
Lorke A, Kotthaus J P, and Ploog K, 1990. *Phys. Rev. Lett.*, **64**, 2559.
Lorke A, Kotthaus J P, and Ploog K, 1991a. *Superlattices and Microstructures*, **9**, 103.
Lorke A, Kotthaus J P, and Ploog K, 1991b. *Phys. Rev. B*, **44**, 3447.
MacDonald A H, 1983. *Phys. Rev. B*, **28**, 6713.
Mackens U, Heitmann D, Prager L, Kotthaus J P, and Beinvogl W, 1984. *Phys. Rev. Lett.*, **53**, 1485.
Maksym P A, and Chakraborty T, 1990. *Phys. Rev. Lett.*, **65**, 108.
Mast D B, Dahm A J, and Fetter A L, 1985. *Phys. Rev. Lett.*, **54**, 1706.
Matheson T G, and Higgins R J, 1982. *Phys. Rev. B*, **25**, 2633.
McEuen P L, Foxman E B, Meirav U, Kastner M A, Meir Y, Wingreen N S, and Wind S J, 1991. *Phys. Rev. Lett.*, **66**, 1926.
Merkt U, Sikorski Ch, and Alsmeier J, 1989. In *Spectroscopy of Semiconductor Microstructures*, Eds Fasol G, Fasolino A and Lugli P, New York, Plenum Press, p. 89.
Merkt U, Huser J, and Wagner M, 1991. *Phys. Rev. B*, **43**, 7320.
Nieder J, Wieck A D, Grambow P, Lage H, Heitmann D, von Klitzing K, and Ploog K, 1990. *Appl. Phys. Lett.*, **57**, 2695.
Petroff P M, Gossard A C, and Wiegmann W, 1984. *Appl. Phys. Lett.*, **45**, 620.
Que W, and Kirczenow G, 1989. *Phys. Rev. B*, **39**, 5998.

Robnik M, 1986. *J. Phys. A*, **19**, 3619.
Scherer A, Roukes M L, Craighead H G, Ruthen R M, Beebe E D and Harbison J P, 1987. *Appl. Phys. Lett.*, **51**, 2133.
Sesselmann W, and Kotthaus J P, 1979. *Solid State Commun.*, **31**, 193.
Shikin V, Nazin S, Heitmann D, and Demel T, 1991. *Phys. Rev. B*, **43**, 11903.
Sikorski Ch, and Merkt U, 1989. *Phys. Rev. Lett.*, **62**, 2164.
Silsbee R H, and Ashoori R C, 1990. *Phys. Rev. Lett.*, **64**, 1991.
Smith T P, Goldberg B B, Stiles P J, and Heiblum M, 1985. *Phys. Rev. B*, **32**, 2696.
Smith T P, Wang W I, and Stiles P J, 1986. *Phys. Rev. B*, **34**, 2995.
Smith T P, Arnot H, Hong J M, Knoedler C M, Laux S E, and Schmid H, 1987. *Phys. Rev. Lett.*, **59**, 2802.
Smith T P, Lee K Y, Hong J M, Knoedler C M, Arnot C H, and Kern D P, 1988a. *Phys. Rev. B*, **38**, 1558.
Smith T P, Lee K Y, Knoedler C M, Hong J M, and Kern D P, 1988b. *Phys. Rev. B*, **38**, 2172.
Tanaka M, and Sakaki H, 1989. *Appl. Phys. Lett.*, **54**, 1326.
Tsubaki K, Sakaki H, Yoshino J, and Sekiguchi Y, 1984. *Appl. Phys. Lett.*, **45**, 663.
Ulloa S E, Castaño E, and Kirczenow G, 1990. *Phys. Rev. B*, **41**, 12350.
Vasilopoulos P, and Peeters F M, 1989. *Phys. Rev. Lett.*, **63**, 2120.
Warren A C, Antoniadis D A, Smith H I, and Melngailis J, 1985. *IEEE Electron Device Letters*, **6**, 294.
Weiss D, Klitzing K v, Ploog K, and Weimann G, 1988. In *High Magnetic Fields in Semiconductor Physics II*, Ed Landwehr G, Springer Series in Solid State Sci. **87**, Berlin, Springer, p. 357.
Weiss D, Zhang C, Gerhardts R R, von Klitzing K, and Weimann G, 1989. *Phys. Rev. B*, **39**, 13020.
Weiss D, Roukes M L, Menschig A, Grambow P, von Klitzing K, and Weimann G, 1991. *Phys. Rev. Lett*, **66**, 2790.
Winkler R W, Kotthaus J P, and Ploog K, 1989. *Phys. Rev. Lett.*, **62**, 1177.
Wulf U, 1987. *Phys. Rev. B*, **35**, 9754.
Wulf U, Gudmundsson V, and Gerhardts R R, 1988. *Phys. Rev. B*, **38**, 4218.
Wulf U, Zeeb E, Gies P, Gerhardts R R, and Hanke W, 1990. *Phys. Rev. B*, **42**, 7637.
Zhang Chao, and Gerhardts R R, 1990. *Phys. Rev. B*, **41**, 12850.
Zheng L, Schaich W L, and MacDonald A H, 1990. *Phys. Rev. B*, **41**, 8493.
Zhu, and Zhou Sh, 1988. *J. Phys. C: Solid State Phys.*, **21**, 3063.

Optical Properties of Type I and Type II GaAs-AlGaAs Nanostructures

K Brunner, F Hirler, G Abstreiter, G Böhm, G Tränkle, and G Weimann

Walter Schottky Institut
Garching, Germany

1 Introduction

In recent years a variety of techniques has been applied to fabricate semiconductor nanostructures revealing one- and zero-dimensional electronic and optical properties. Besides some direct growth methods (Petroff *et al.* 1982, Fukui and Saito 1987) lateral structuring of heterostructures grown by molecular beam epitaxy is used in order to achieve low-dimensional systems of high quality. Mask patterning by electron beam lithography or optical interference lithography are already well controlled. The transfer of the patterns to a persistent modulation of the band edge within the sample, however, often results in surface depletion, crystal damage, and low optical efficiency. This damage seems to originate from bombardment by high energetic particles in techniques like reactive ion etching (Demel *et al.* 1988) or ion beam induced inter-diffusion (Li *et al.* 1990). Here two soft methods are presented which result in a lateral type I and type II band edge modulation within buried GaAs-$Al_xGa_{1-x}As$ single quantum well structures. The lateral confinement of electrons and holes within the same or within adjacent regions, respectively, strongly affects the optical properties.

Systematic investigations of photoluminescence (PL) at liquid helium temperatures are applied to identify the basic mechanisms of recombination.

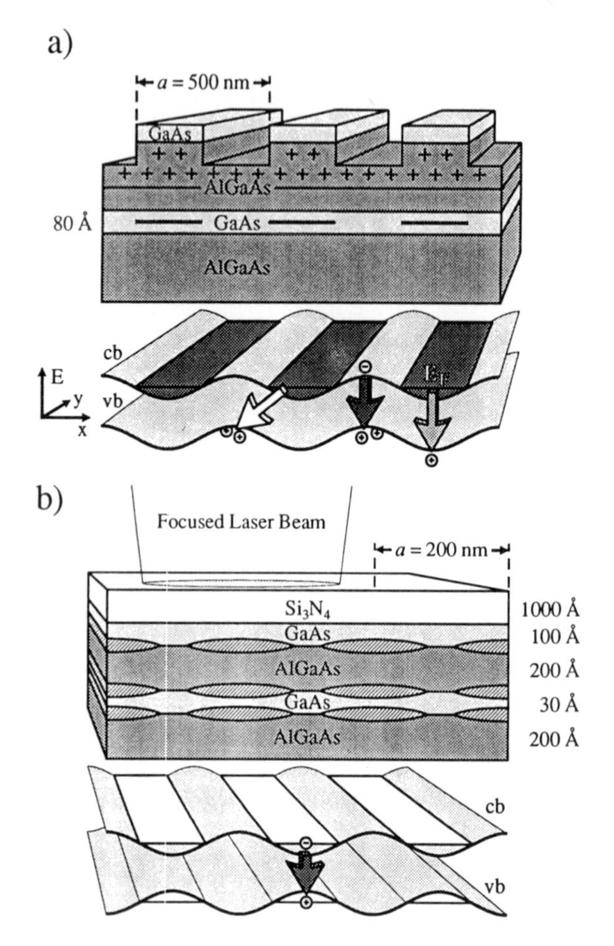

Figure 1: *Schematic view of the samples and lateral modulation of the band edge for (a) lateral type II superlattice, and (b) lateral type I superlattice.*

2 Sample preparation

A shallow wet chemical etching technique is applied to get a lateral type II modulation. A holographic photoresist grating of period $a = 500\,\mathrm{nm}$ on top of the remote doped 80 Å GaAs-$Al_xGa_{1-x}As$ quantum well serves as a mask. To minimise defects in the active well, we etch only into the doped AlGaAs barrier leaving the spacer untouched (Figure 1a). As a consequence of the modification of dopant and surface charges, the valence and conduction band within the quantum well are modulated laterally, and electrons and laser-induced holes are separated into adjacent wires. Depending on the overlap and occupation density of electron and hole states, spatially direct as well as spatially indirect recombinations are possible. The etch depth and the amplitude of potential modulation are controlled during etching via the anisotropic change of conductivity of the electron gas (Hirler *et al.* 1991).

In a lateral type I structure electrons and holes must be confined in the same spatial region by a modulation of the effective band gap. Besides confinement by strain gradients (Kash *et al.* 1990) only a lateral modulation of sample composition achieves this. Thus we have developed an optical technique to achieve local thermal inter-diffusion of quantum well structures with little damage. The layer inter-diffusion results in a lateral barrier for electrons and holes due to an increase of the effective energies of the band edge of the quantum well. A schematic view of a periodic wire structured sample and of the band edge modulation are shown in Figure 1b. The inter-diffusion is induced by a visible focused laser beam (FLB) with a spot diameter of about half a micrometre and a power of about 6 mW. The absorbed laser spot locally heats the undoped GaAs-$Al_xGa_{1-x}As$ quantum well structure ($L = 30$ Å). The exponential onset of inter-diffusion with local temperature causes a narrow lateral profile of Al-Ga inter-diffusion. Its total width is calculated to be about 200 nm (Brunner *et al.* 1991). Decomposition of the sample during processing with the laser is prevented by a film of Si_3N_4 on the surface. Various lateral patterns can be drawn by moving the sample on a high precision *xyz*-translation stage underneath the laser spot.

3 Lateral type II superlattices

Figure 2 shows a series of photoluminescence spectra of shallow etched type II superlattices after various processing steps. The probing spot size is 110 μm and the excitation wavelength is 710 nm. The unstructured reference sample (Figure 2a) exhibits the well-known line-shape of modulation-doped quantum wells, with a high energy tail and a cut off at the Fermi energy (Pinczuk *et al.* 1984). To study the influence of the etch process on the optical properties, an unstructured sample is etched homogeneously. It shows a narrow luminescence line (Figure 2d) with a full-width at half-maximum of 1.7 meV, very similar to high quality, undoped 80 Å GaAs quantum wells. This demonstrates that the etching procedure does not induce additional defects in the active region. The shift of the recombination energies with respect to the two-dimensional luminescence is due to renormalisation of the band gap (Tränkle *et al.* 1987).

Figures 2b and 2c show the PL spectra for quantum wire samples with different etch depths. The lineshape of the luminescence is completely inverted with respect to the 2 D reference even at a very shallow etch depth, similar to that reported by Weiner *et al.* (1989). The peak is shifted to the high energy side, but the onset of the luminescence at low energy remains roughly at the same energy. Even at larger etch depths, and for different laser excitation densities, its position and shape are almost the same. Therefore, we believe this luminescence to be *direct* in *real space*.

The deeper etched sample shows in addition an asymmetric and broad luminescence feature at energies below the fundamental gap, similar to that reported for bulk *nipi*-structures (Döhler *et al.* 1981). We interpret it as *spatially indirect* recombination of laterally separated electrons and holes. With increasing excitation intensity (Figure 2c), additional electrons and holes are generated, which are separated by the lateral field, and thereby flatten out the potential electrostatically. As expected, the indirect peak moves continuously towards the direct luminescence concomitant with a stronger increase of the intensity of the latter one.

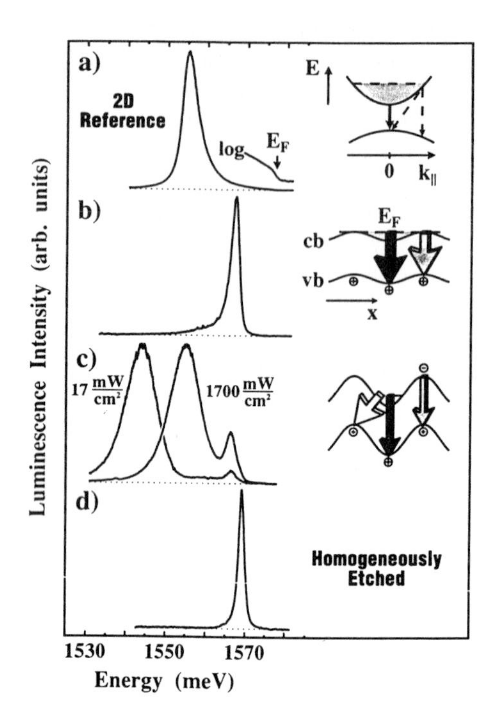

Figure 2: PL *spectra of samples with different etch depths: (a) unstructured sample, (b) sample with small etch depth, (c) deeper etched sample with two different illumination densities, and (d) unstructured, homogeneously etched sample.*

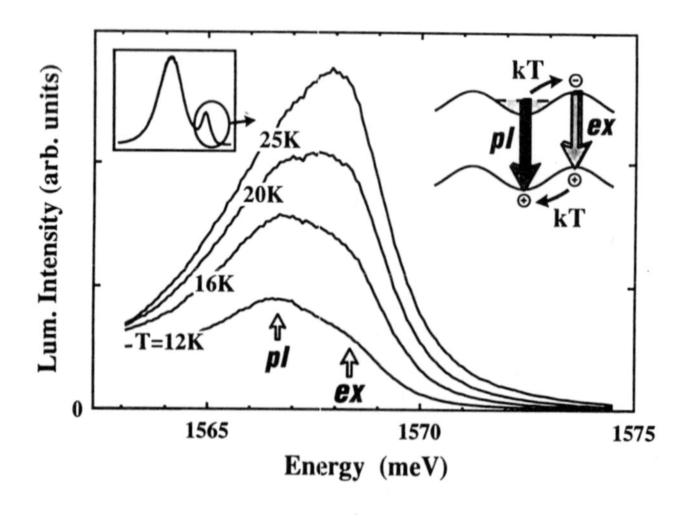

Figure 3: PL *spectra of the direct transitions for different temperatures.*

The nature of the direct luminescence is rather complex, as shown by the variation with temperature in Figure 3. The shoulder on the high energy side increases faster with temperature than the low energy peak. As indicated by the grey arrows, two different contributions from areas with high and areas with low carrier density are possible. At the potential maximum the carrier density is low and the luminescence will be more *exciton-like* ('*ex*' in figure 3), similar to an undoped quantum well. We can simulate this situation by etching a sample homogeneously (Figure 2d). Indeed, the narrow peak is very close in energy to the high energy shoulder. Furthermore the latter shows an activation energy approximately equal to the total potential modulation minus the Fermi energy. On the other hand, the luminescence from the area with the high electron density will be more *plasma-like* ('*pl*' in Figure 3), similar to the 2D reference. But the lineshape will be modified as a consequence of the reduced dimensionality, because conservation of momentum perpendicular to the wires is no longer valid. In addition, a maximum at the high energy side due to Fermi edge enhancement is expected. For these reasons, we believe the high-energy feature to be due to exciton-like direct transitions, and the low energy peak to represent the maximum of the plasma-like transitions.

For our type II superlattices we finally estimate the total amplitude of the potential modulation by the energy difference of the spatially direct, and indirect luminescence. Values up to 50 meV are achieved. The corresponding one-dimensional energy levels with a spacing of the order of 2 meV are not resolved in the luminescence spectra.

4 Lateral type I superlattices

Figure 4 shows PL spectra of type I wire structures with different periods a, which have been fabricated by FLB-induced inter-diffusion at constant laser processing parameters. While a type II band edge modulation affects PL spectra even at a period as large as 500 nm, the influence of a type I modulation depends strongly on its period. At periods larger than about 400 nm no significant change of PL is observed compared to the as grown quantum well. We attribute this to wide as-grown quantum well regions in between the FLB written lines, which represent the lateral barriers. PL only probes these as-grown regions of minimum energy, as schematically shown in the inset of Figure 4.

The PL spectrum is totally changed by decreasing the period to a value of 200 nm. Five dominant peaks are observed, which are separated in energy by about 8 meV. The peak lowest in energy is blue-shifted by about 5 meV. We interpret these peaks as spatially direct transitions of one-dimensional electron and hole states, which are localised within narrow, barely inter-diffused regions between neighbouring lines written with the FLB. At a period of 200 nm, which compares well to the proposed total width of the lateral inter-diffusion profile, the strongest modulation of the band edges and largest lateral confinement energies are expected. Assuming a sinusoidal modulation of the band gap with a height of 80 meV, energy separations of about 5 meV for electrons and comparably smaller values for holes are derived from detailed calculations of Bockelmann and Bastard (1991). Thus the observed PL shift and splitting corresponds roughly to the expected transition energies between one-dimensional electron and hole subbands. At a further reduced period length of $a = 150$ nm a large blue shift, and a decreased peak splitting indicate a more homogeneous inter-diffusion, which is due to

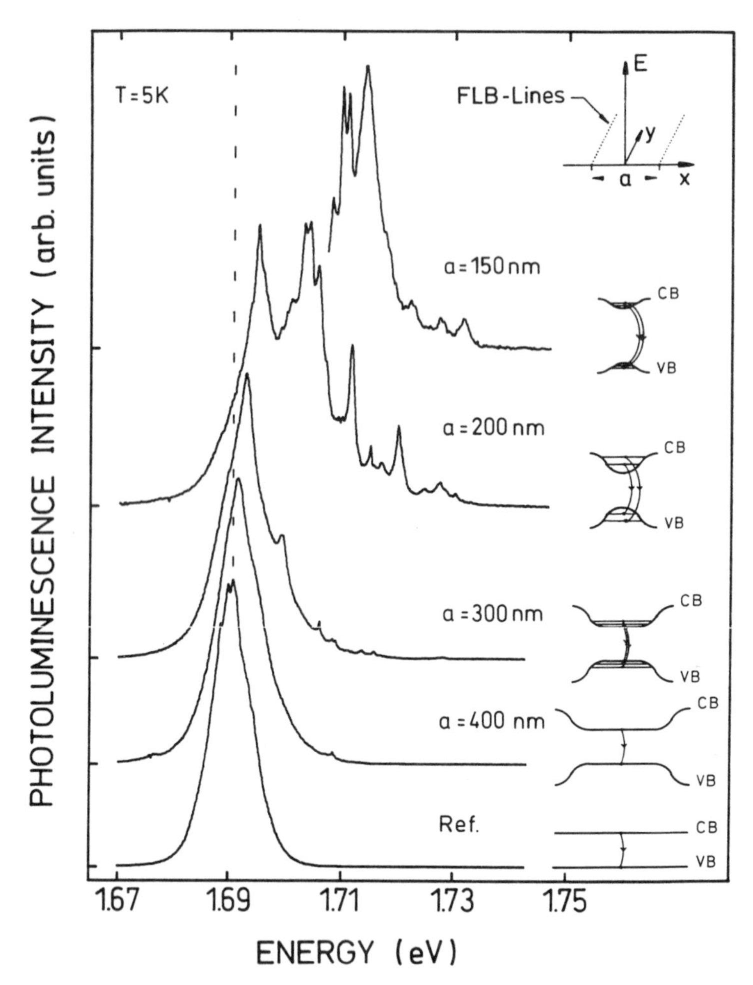

Figure 4: PL *spectra of wire structures with different periods a, written by* FLB*-induced inter-diffusion.* PL *excitation was* $E = 1.76\,\text{eV}$, $P = 10\,\mu\text{W}$, *with a spot size of* $1.5\,\mu\text{m}$. *The expected modulation of the band edge is shown schematically in the insets.*

overlapping profiles of inter-diffusion.

The energy positions of the dominant peaks and the spectral fine structure are changed by varying the position of the $1.5\,\mu\text{m}$ sized PL probe within the wire structure of period $a = 200\,\text{nm}$. This indicates that inhomogeneity in local sample composition causes the fine structure in between the main peaks. The energy separation of 8 meV between these dominant peaks, however, remains nearly constant. So the lateral quantisation energy, which is essentially determined by the period a and by the lateral profile of inter-diffusion, varies only slightly within the wire structure.

PL excitation spectra are shown in Figure 5. The as-grown quantum well reveals peaks due to the heavy-hole and light-hole excitons. For the wires with $a = 200\,\text{nm}$ the intensity of the low-energy PL peak has been detected. This excitation spectrum PLE1

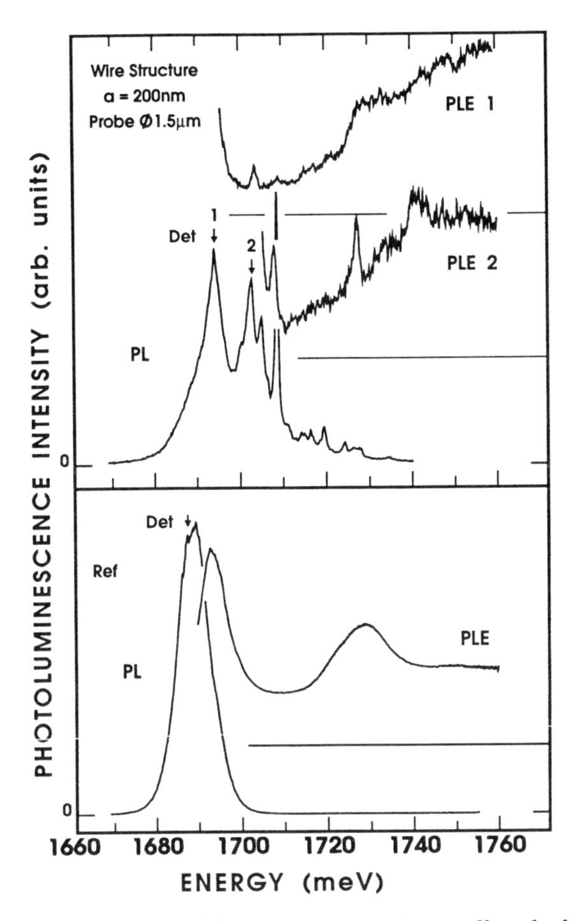

Figure 5: PL *excitation spectra of the as-grown quantum well and of the wire structure with period* $a = 200\,\mathrm{nm}$. *The detection energies are indicated by arrows.*

shows peaks at the positions of the neighbouring PL peaks. Detecting the second PL peak the third one gets much more pronounced (PLE2). The correspondence of the PL and PLE peaks confirms their intrinsic homogeneous nature due to lateral confinement. In PL of the wire structure the relative intensity of the high energy peaks increases with excitation density. This shows that filling of subbands and the competition of recombination with carrier relaxation is important. Shifts of PL peak energies with excitation density are not observed. As expected, flattening of the modulation by photo-generated carriers, which is a crucial feature of type II superlattices, is negligible in type I structures.

In conclusion, the optical properties of lateral type II superlattices, even in systems where quantisation effects are not important, are dominated by spatially direct and indirect transitions as well as momentum selection rules. In lateral type I superlattices PL spectroscopy with high spatial resolution reveals direct transitions between one-dimensional states and resolves inhomogeneities within the sample. Variation of excitation density does not affect the transition energies of direct recombinations in both types of superlattices. The indirect luminescence in type II structures, however, shifts strongly.

Acknowledgments

We would like to thank J Smoliner, and E Gornik for useful discussions, and for providing the holographic lithography setup. The work was financially supported by the Deutsche Forschungsgemeinschaft.

References

Bockelmann U, and Bastard G, 1991. *Phys. Rev. B*, in press.
Brunner K, Abstreiter G, Walther M, Böhm G, and Tränkle G, 1991. *Surf. Sci.*, in press.
Demel T, Heitmann D, Grambow P, and Ploog K, 1988. *Appl. Phys. Lett.*, **53**, 2176.
Döhler G H, Künzel H, Olego D, Ploog K, Ruden P, Stolz H J, and Abstreiter G, 1981. *Phys. Rev. Lett.*, **47**, 864.
Fukui T, and Saito H, 1987. *Appl. Phys. Lett.*, **50**, 824.
Hirler F, Küchler R, Strenz R, Abstreiter G, Böhm G, Smoliner J, Tränkle G, and Weimann G, 1991. *Surf. Sci.*, in press.
Kash K, Worlock J M, Gozdz A S, Van der Gaag B P, Harbison J P, Lin P S D, and Florez L T,1990. *Surf. Sci.*, **229**, 245.
Li Y J, Tsuchiya M, and Petroff P M, 1990. *Appl. Phys. Lett.*, **57**, 472.
Petroff P M, Gossard A C, Logan R A, and Wiegmann W, 1982. *Appl. Phys. Lett.*, **41**, 635.
Pinczuk A, Shah J, Störmer H L, Miller R C, Gossard A C, and Wiegmann W, 1984. *Surf. Sci.*, **142**, 492.
Tränkle G, Leier H, Forchel A, Haug H, Ell C, and Weimann G, 1987. *Phys. Rev. Lett.*, **58**, 419.
Weiner J S, Danan G, Pinczuk A, Valladares J, Pfeiffer L N, and West K, 1989. *Phys. Rev. Lett.*, **63**, 1641.

Planar Coupled Electron Waveguides

C C Eugster and J A del Alamo

Massachusetts Institute of Technology
Cambridge, Massachusetts

Using a split-gate scheme, we have fabricated a novel quantum-effect device that consists of two electron waveguides placed in very close proximity to each other. The field-effect action of the gates controls the degree of interaction between the waveguides as well as determines the number of occupied modes in each waveguide. For the simplest case in which only one waveguide is formed, we observe very strong oscillations in the tunnelling current leaking out through a thin sidewall into a two-dimensional electron gas (2DEG) as the electron density within the waveguide is modulated. We interpret these oscillations as the result of the one-dimensional (1D) density of states sweeping through the Fermi level. This experiment constitutes the first tunnelling spectroscopy of an electron waveguide.

The rapid scaling of modern device dimensions will eventually precipitate fundamental changes in how devices function. A growing area of research which might provide a bypass to the immediate scaling limitations of conventional devices lies within the field of quantum-effect electronics (Holden 1991). The attractive feature of quantum-effect devices is that scaling down the dimensions actually enhances their performance. In these devices the electrons behave more like waves than particles. This has resulted in the proposition of many novel device concepts that are analogous to electromagnetic wave devices (Kriman *et al.* 1988, Miller *et al.* 1989, Smith *et al.* 1989, Ismail *et al.* 1989, del Alamo and Eugster 1990, Tsukada *et al.* 1990, Dagli *et al.* 1991). We are interested in implementing one of these analogies, the electron directional coupler (del Alamo and Eugster 1990, Tsukada *et al.* 1990, Dagli *et al.* 1991). In this paper, we will show our approach towards the implementation of an electron directional coupler and we will report preliminary experimental results that indicate we have indeed achieved a coupled electron waveguide structure. This device has also been configured to implement a 'leaky' electron waveguide which has resulted in the first tunnelling spectroscopy of an electron waveguide (Eugster and del Alamo, 1991).

In our device, schematically shown in Figure 1, a two-dimensional electron gas (2DEG) at the interface of a high-mobility modulation-doped heterostructure is shaped

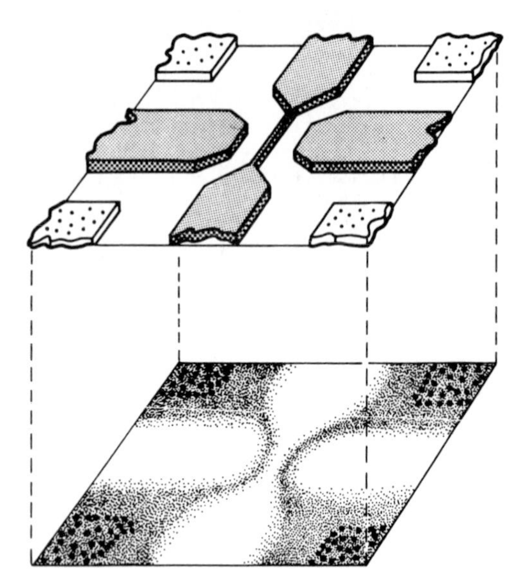

Figure 1: *Schematic illustration of coupled electron waveguide structure. Top plane shows the gates, defined by electron beam lithography, and ohmic contacts at the surface of semiconductor. The bottom plane shows the two closely spaced electron waveguides at the interface of the heterostructure resulting from the proper biasing condition of the gates.*

in the form of two one-dimensional electron waveguides that are in very close proximity to each other over a certain length. The waveguides are separated in the extrinsic region so that interaction between them only takes place in the intrinsic region. Four contacts allow access to the input and output of each waveguide. If electrons are injected into the first mode of one waveguide, then it should be possible to directionally couple them between the waveguides (del Alamo and Eugster 1990, Tsukada *et al.* 1990, Dagli *et al.* 1991). Such a coupled waveguide structure can be achieved using a split-gate scheme in which three metal gates on the surface are properly biased to create the two closely spaced waveguides. In this scheme, the degree of coupling between the two waveguides can be controlled through the field-effect action of a thin middle gate. In addition, the carrier density in each waveguide can be controlled through the fringing fields of the side gates. We can thereby control the degree of interaction between the waveguides as well as control the number of occupied modes (subbands) in each waveguide.

Fabrication of the devices began with MBE growth of a GaAs-$Al_xGa_{1-x}As$ MODFET structure. The mobility of the 2DEG is $170,000\,cm^2\,V^{-1}\,s^{-1}$ and the carrier density is $7\times10^{11}\,cm^{-2}$ at 4 K. The 2DEG is at a depth of 545 Å from the semiconductor surface. Four alloyed Au-Ge-Ni ohmic contacts, one at each corner, allow access to the two ends of each waveguide. The actual split-gate structure shown in Figure 2 was defined using electron-beam lithography at the National Nanofabrication Facility at Cornell University (Rooks *et al.* 1991). In order to minimise electron beam writing time, only the periphery of the confining gates were drawn (Eugster *et al.* 1990). We fabricated several devices with the separation, W, between the side gates and the middle gate

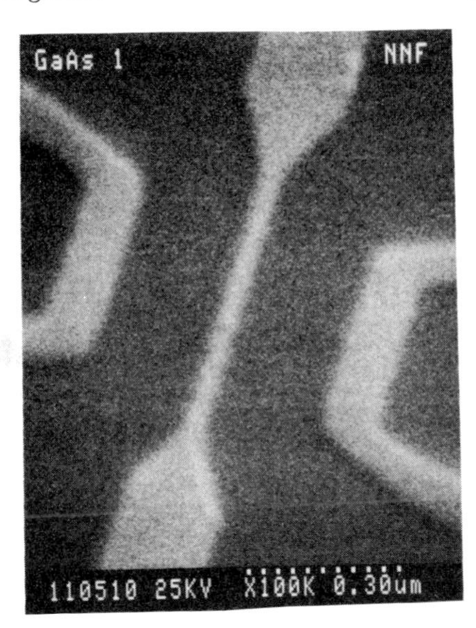

Figure 2: *Scanning electron micrographs of the gates defined by electron beam lithography. The right micrograph is an enlargement of the intrinsic region of the device. The middle gate which controls the degree of coupling is only* 30 nm *wide.*

being 0.2 μm or 0.25 μm and the length, L, of the intrinsic portion of the split gates ranging from 0.1 μm to 0.5 μm. The lengths were chosen to ensure ballistic transport through the channels at the temperature of measurement. The middle gate in the intrinsic region was 30 nm wide and was fabricated using a single-pass electron-beam lithography technique. The middle gate was widened in the extrinsic region to 0.5 μm to provide isolation in the outer regions through short-channel effects.

In order to confirm that this device indeed results in two closely located electron waveguides, we measured the conductance of each channel in a given device. A small a.c. drain-source bias of 100 μV was applied between the contacts of the measured waveguide. The voltages on the middle gate and the respective side gate were swept together with the current being monitored using a lock-in technique. The other side gate was biased at zero. Figure 3 shows the conductance of each waveguide for a 0.5 μm long, 0.25 μm wide device at 3 K. One of the waveguides shows near perfect quantisation of the conductance in steps of $2e^2/h$ (van Wees *et al.* 1988, Wharam *et al.* 1988, Timp *et al.* 1989). There are approximately eight modes in the waveguide when it is formed. These features wash out at temperatures approaching 10 K and for drain-source biases of 2–3 mV. The other waveguide shows somehow poorer quantisation, perhaps resulting from the presence of remote impurities in the vicinity of the waveguide (Nixon *et al.* 1991). This might be expected given the relatively low mobility of the material. For the implementation of the directional coupler where symmetry is crucial, the asymmetry seen in Figure 3 can be corrected for by using different biases on the side confining gates. However, even with this correction, directional coupling will not be observed since one waveguide lacks perfect quantisation.

We also studied the tunnelling current in the intrinsic portion of the device. To gain

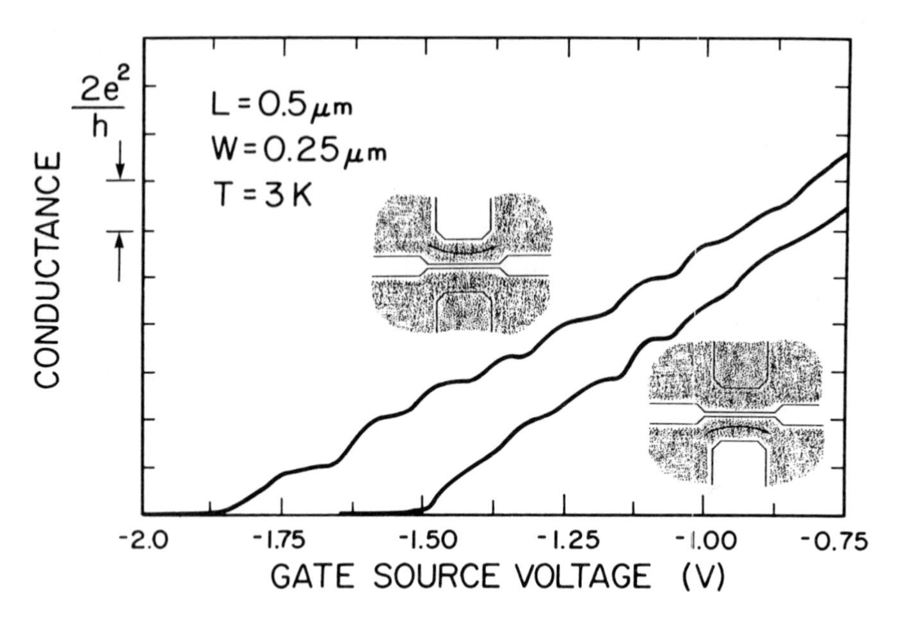

Figure 3: *The conductance of each waveguide in a* 0.5 μm *long device. The insets illustrate the bias configuration which was used for the two measurements. Shaded regions on the insets represent electron concentration at the heterointerface. Arrow represents current flow.*

physical understanding, we have studied in detail the simplest case in which only one waveguide is on and the 1D to 2D tunnelling current is measured. We refer to such a structure as a 'leaky' electron waveguide in which some of the current flowing through the waveguide leaks through the thin sidewall barrier created by the middle gate bias.

In order to implement such a structure, the middle gate bias V_{GM} was chosen to provide perfect isolation in the extrinsic parts of the device yet still allow tunnelling to take place in the intrinsic region. This is easily achieved since the 2DEG first turns off at $V_{GM} \approx -0.6$ V underneath the wider parts of the middle gate in the extrinsic regions, and then due to short-channel effects, turns off in the narrow intrinsic region at a more negative voltage $V_{GM} \approx -0.8$ V. In our experiments, the middle gate voltage was fixed at $V_{GM} = -0.85$ V which allowed a small tunnelling current out of the waveguide through the middle barrier in the intrinsic region. For the results reported in this measurement the side gates are referred to as 'top' and 'bottom' gates. The bottom gate bias V_{GB} was set to zero in order to maintain a 2DEG underneath that gate. The bias on the top confining gate, V_{GT}, was swept negative from zero and was used to modulate the electron density in the waveguide (see inset to Figure 4).

The biasing between the ohmic contacts was as follows. An a.c. voltage of 100 μV was applied across the waveguide with respect to ground. The contact to the side 2DEG was also grounded. Both grounds were actually the virtual grounds of two lock-in amplifiers working in the current mode that monitored the current flowing through the waveguide, I_{S1}, and the tunnelling current, I_{S2}. The measurements were carried out at 1.6 K using standard lock-in techniques with $f = 11$ Hz.

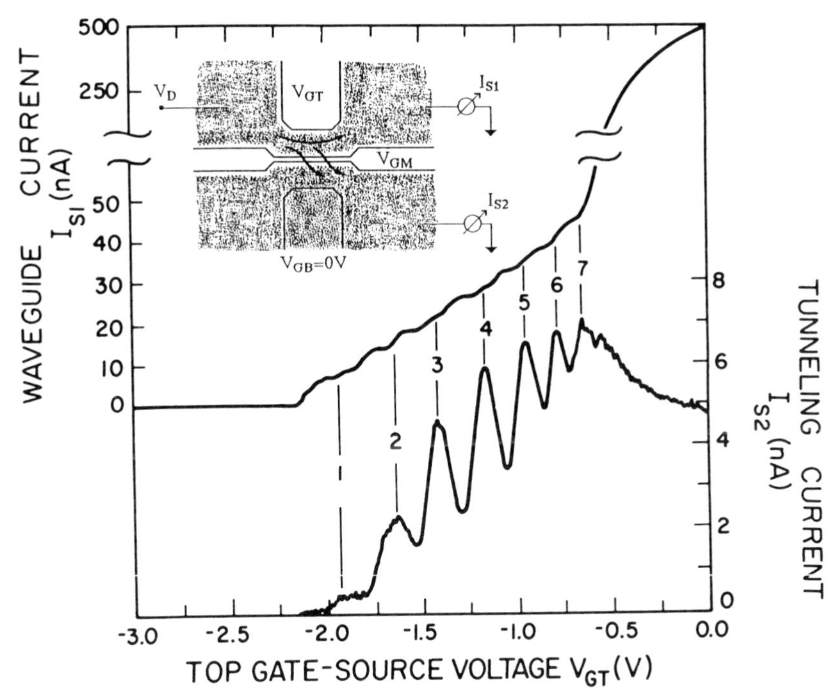

Figure 4: *The $I_S(V_{GT})$ characteristic of a leaky waveguide with $L = 0.1\,\mu$m and $W = 0.2\,\mu$m at $T = 1.6$ K. I_{S1} is the current flowing through the waveguide and I_{S2} is the tunnelling current leaking out the thin sidewall barrier. The biasing configuration is shown in the inset.*

Under such a biasing scheme and for a device with $L = 0.1\,\mu$m, $W = 0.2\,\mu$m, the waveguide current I_{S1} and the tunnelling current I_{S2} were measured independently at 1.6 K as the top gate bias, V_{GT}, was swept, as shown in Figure 4. The waveguide current I_{S1} decreases as V_{GT} is made more negative, until at around $V_{GT} = -0.6$ V, the 2DEG threshold voltage, a sudden change of slope occurs. At this point the 2DEG under the top gate is turned off and a 1D waveguide is formed. The ensuing long tail arises from the slow depletion of the electron concentration in the waveguide through the fringing field of the top gate. The waveguide turns off completely at $V_{GT} = -2.2$ V. The tunnelling current I_{S2} through the thin middle barrier, measured simultaneously with I_{S1}, is also plotted in Figure 4. Before the formation of the 1D waveguide the tunnelling current is featureless. At the onset of the formation of the top waveguide ($V_{GT} = -0.6$ V), strong oscillations are observed in the tunnelling current I_{S2} as the carrier density in the channel is modulated. When the waveguide is turned off, the oscillations also disappear.

When we examine in detail the features in the tail region of the upper current shown in Figure 4, we again observe the quantised current steps. The rise of each step in I_{S1} roughly lines up with the onset of a new oscillation in I_{S2}. There are approximately seven steps of magnitude $2e^2/h$ for I_{S1} consistent with the total number of oscillations observed in I_{S2}. The oscillations are reproducible and we have observed them in several different devices. They also are robust to temperature cycling all the way up to room temperature. The total number of oscillations is always consistent with the total number of steps in the tail region of the $I_{S1}(V_{GT})$ characteristics for the different devices. In fact for the wider devices ($W = 0.25\,\mu$m) eight oscillations are observed, resulting from the larger number of initial subbands occupied. In all cases, the oscillations in I_{S2} were found to wash out with increasing driving potential V_{DS} and with increasing temperatures. The oscillations were no longer visible for V_{DS} between 2 and 3 mV and temperatures around 10 K.

These experimental observations are consistent with a picture in which the oscillations in the tunnelling current arise from mapping out the 1D density of states (1DOS) in the waveguide. The quantised conductance steps of magnitude $2e^2/h$ observed in the waveguide confirm that near-ballistic transport is taking place (Landauer 1981). The onset of each step corresponds to the sweeping of a 1DOS subband through the Fermi level (van Wees *et al.* 1988, Wharam *et al.* 1988, Timp *et al.* 1989). The current flowing through the waveguide is proportional to the product of the 1DOS, g^{1D}, and the electron wavevector in the direction of the current flow, $k_{\|}$. The 1DOS is inversely proportional to the square root of the energy of the electrons (Petroff 1990) while $k_{\|}$ is proportional to it, thereby resulting in a constant product. This gives rise to the constant steps in the current observed in the tail region of I_{S1}.

For low enough temperatures and a narrow window of injected electrons around the Fermi level, we can approximate the tunnelling current I_{S2} to be

$$I_{S2} \approx \frac{e^2\hbar}{m^*}\sum_j k_{\perp j}\, T_j(E_F - E_b)\, g_j^{1D}(E_F - E_j)\,\Delta V, \tag{1}$$

where we have accounted for the contribution to the current from each occupied subband j. The voltage ΔV is the potential drop across the waveguide and differs from V_{DS} by the potential drops across the series and contact resistances. At low temperatures these resistances are negligible (approximately 200 Ω). E_j is the energy at the bottom of the jth subband, E_b is the height of the tunnelling barrier, $k_{\perp j}$ has a value determined by the confining potential and is a constant for each subband due to lateral quantisation. The transmission coefficient, $T_j(E_F - E_b)$, is the same for the different subbands to first order since the barrier height is fixed relative to the Fermi level.

As seen from Equation (1), the tunnelling current I_{S2} is proportional to $g^{1D}(E_F - E_j)$. Modulating V_{GT} sweeps the subbands through the Fermi level. As $k_{\perp j}$ and $T_j(E_F - E_b)$ are constant for a given subband, I_{S2} *reproduces* the 1D density of states g^{1D} for each subband. It is important to note that in sweeping V_{GT} the Fermi level is unaffected with respect to the tunnelling barrier E_b since V_{GM} is fixed at a constant value.

The effects of finite temperature and voltage present in the experiment result in a rounding of the features of I_{S2} from g^{1D}. By using a convolution technique (Bagwell and Orlando 1989), it has been shown that the effects of finite voltage ΔV and finite temperature T are equivalent and independent broadening phenomena. A finite voltage averages the current over an energy range $e\Delta V$ near the Fermi level, while finite temperature averages the current over a region $3.5\,k_BT$ around E_F. In our results both the finite voltage and temperature broadening indicate a characteristic energy of 2–3 meV at which the oscillations wash out. An explanation for this effect comes from the fact that as the window of injected electrons into the waveguide, determined by the bias and temperature, approaches the inter-subband spacing, the resolution of tunnelling from a single subband diminishes and Equation 1 is no longer valid. The 2–3 meV characteristic energy which we arrive at is in very good agreement with the subband spacings which have been calculated for similar split-gate wires (Snider *et al.* 1990).

In summary, we have fabricated and measured planar coupled 1D electron waveguide structures. We have observed quantised conductance steps in the current through each waveguide on a given device. We also observe strong oscillations in the tunnelling current out of the sidewall of a 'leaky' waveguide as the electron density in the waveguide is modulated. We interpret these results with a picture in which the tunnelling current is directly mapping out the 1DOS of the waveguide, thereby constituting the first tunnelling spectroscopy of an electron waveguide. The versatility of these coupled waveguide structures offers many exciting possibilities for novel devices as well as the potential for uncovering new physics.

Acknowledgments

We would like to acknowledge M J Rooks for electron beam lithography at NNF, Cornell University; Prof C G Fonstad and K Ismail for epitaxial sample growth; and Profs M A Kastner, H I Smith, P Lee, Q Hu and T P Orlando for stimulating discussions. This work has been funded by NSF contracts 87-19217-DMR and DMR-9022933. C C Eugster acknowledges an IBM Graduate Fellowship award.

References

Bagwell P F, and Orlando T P, 1989. *Phys. Rev. B*, **40**, 1456.
Dagli N, Snider G, Waldman J, and Hu E, 1991. *J. Appl. Phys.*, **69**, 1047.
del Alamo J A, and Eugster C C, 1990. *Appl. Phys. Lett.*, **56**, 78.
Eugster C C, and del Alamo J A, 1991. *Phys. Rev. Lett.*, **64**, 3586.
Eugster C C, del Alamo J A, and Rooks M J, 1990. *Jap. J. Appl. Phys.*, **29**, L2257.
Holden A J, 1991. *Inst. Phys. Conf. Ser.*, **112**, 1.
Ismail K, Antoniadis D A, and Smith H I, 1989. *Appl. Phys. Lett.*, **55**, 589.
Kriman A M, Bernstein G H, Haukness B S, and Ferry D K, 1988. *Proc. 4th Intl. Conf.*.
Landauer R, 1981. *Phys. Lett.*, **85A**, 91.
Miller D, Lake R, Datta S, Lundstrom M, Melloch M, and Reifenberger R, 1989. *Proc. Int. Symp. Nanostructure Physics and Fabrication*, Eds. Reed M, and Kirk W P, Boston, Academic Press, p. 165.
Nixon J A, Davies J H, and Baranger H U, 1991. *Phys. Rev. B*, **43**, 12638.
Petroff P M, 1990. *Physics of Quantum Electron Devices*, 353, Ed. F Capasso, New York, Springer-Verlag.
Rooks M J, Eugster C C, del Alamo J A, Snider G L, and Hu E L, 1991. To appear in *J. Vac. Sci. Tech. B*.
Smith C, Pepper M, Ahmed H, Frost J, Hasko D, Newbury R, Peacock D, Ritchie D, and Jones G, 1989. *J. Phys. Condens. Matter*, **1**, 9035.
Snider G L, Tan I-H, and Hu E L, 1990. *J. Appl. Phys.*, **68**, 5922.

Timp G, Behringer R, Sampere S, Cunningham J E, and Howard R E, 1989. *Proc. Int. Symp. Nanostructure Physics, and Fabrication*, Eds. Reed M, and Kirk W P, Boston, Academic Press, p. 331.

Tsukada N, Wieck A D, and Ploog K, 1990. *Appl. Phys. Lett.*, **56**, 2527.

van Wees B J, van Houten H, Beenakker C W J, Williamson J G, Kouwenhoven L P, van der Marel D, and Foxon C T, 1988. *Phys. Rev. Lett.*, **60**, 848.

Wharam D A, Thornton T J, Newbury R, Pepper M, Ahmed H, Frost J E F, Hasko D G, Peacock D C, Ritchie D A, and Jones G A C, 1991. *J. Phys. C*, **21**, L209.

Fabrication of Three-Dimensional Superlattices of Nanostructures

V Bogomolov, Y Kumzerov, and S Romanov

A F Ioffe Physico-Technical Institute
St Petersburg, Russia

The modern tendency in developing nanostructures is towards the decrease of their size for the best operating characteristics. Unfortunately, this process is accompanied by the decrease of the power parameters of nano-devices and increase of the thermal fluctuations. One possible way to overcome these problems is to use arrays of nano-structures. In this respect it is especially tempting to employ three-dimensional (3D) arrays because of their high density of elements.

In our opinion, it is necessary to satisfy the following requirements for successful application of such arrays:

i) identity of the nano-elements in the array;
ii) synchronous operation of the elements in the array.

In solving the first problem, we apply the method which uses a crystalline porous dielectric matrix with voids impregnated by the 'guest' material (Bogomolov 1978). As a matrix we use in this work the artificial regular packing of silica spheres (face-centered cubic symmetry), which is similar in structure to the precious gem opal (Sanders 1964). The empty cavities that exist between the adjacent spheres also form a network of high symmetry and their total volume is about 25% of the whole. It is easy to see from the scanning electron micrograph (Figure 1) that this packing preserves its regularity on the macroscopic scale. The desired material (one of the metals Bi, Ga, In, Pb, Sn, semiconductors InSb, Se, Te or alloys BiPb, In-InSb) was introduced into the cavities from the melt under high hydrostatic pressure. The resulting structure is shown in a transmission electron micrograph of a cut sample (Figure 2). In the upper right corner one can see the shadow of the silica sphere and, in the centre, the series of the grains of two types connected with each other by thin bridges. The characteristic size of the large grains was about $D_1 = 100\,\mathrm{nm}$, small grains $D_2 = 56\,\mathrm{nm}$ and minimum bridge size $d = 10$–$20\,\mathrm{nm}$ due to the matrix growth conditions. This matrix permits us to fabricate arrays of nano-particles with a density of about $10^{14}\,\mathrm{cm}^{-3}$. The advantage of this method is that the size, shape and position of the grains are strictly determined

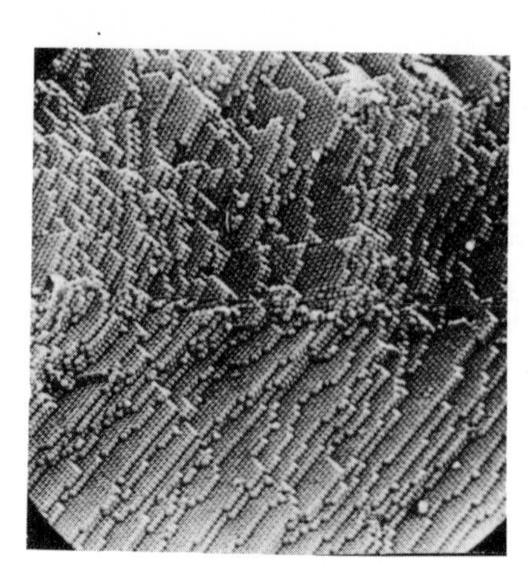

Figure 1: *Scanning electron micrograph of silica spheres.*

by the crystal-like matrix structure. Since the inequality in size of the matrix spheres is less than 5%, the inequality in size of the grain structure is of the same order. Since we use samples with sizes such as $3 \times 1 \times 0.2\,\mathrm{mm}^3$ in our experiments, their structure looks polycrystalline due to matrix packing defects.

As for the second demand, we believe that we need to provide tight binding between neighbouring grains so that there is a phase coherence between these grains. In this case the array may be considered as a superlattice of nano-elements. Moreover the distance between adjacent grains must not be more than the characteristic length of the physical effect intrinsic to this superlattice.

Thus the samples under investigation may be regarded as 'secondary crystals' with specific properties which are determined by (i) the type of material, (ii) the particle size and (iii) the arrangement of these particles in space.

What physical consequences result from this structure? Figure 3a shows a schematic diagram of the system of grains. Obviously, the properties of the grains are nearly the same as those of the bulk material, and the properties of bridges are strongly influenced

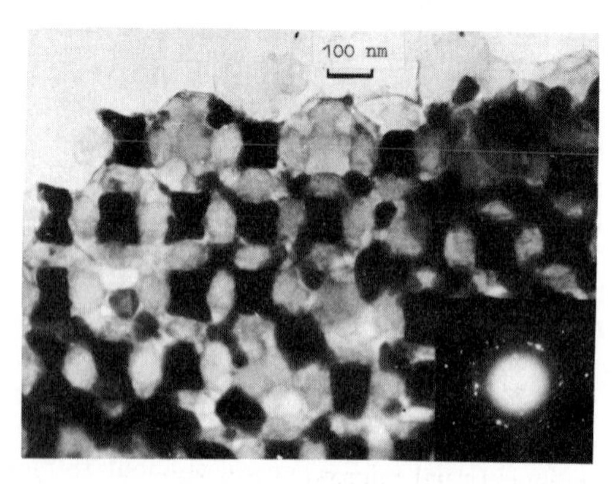

Figure 2: *Transmission electron micrograph of nanostructure superlattice.*

by classical and quantum size effects. Therefore it might be expected that transport properties would be controlled by the bridges and the optical properties by the grains. According to physical properties, we may divide the samples into the following three types.

1. **Semiconductor superstructures**

In the case of semiconductors, there are alternating changes in the width of the gap due to the quantum size splitting of the energy levels in the bridges. This splitting is proportional to $1/(m^* d^2)$ and may come to 75 meV for the light electrons in InSb. The energy diagram looks like the series of quantum wells shown in Figure 3b. Each section grain-bridge-grain may be considered as a single quantum device of the heterojunction type. The association of these single junctions into the superlattice is generated by the overlapping of the wave functions of donor levels in InSb, whose radius exceeds the inter-grain distance. When current passes through this system, electrons alternately find themselves in nonequilibrium conditions and a recombination process may take place. If the Fermi level is moved away from the gap by, for example, selective doping of the bridges, the emission of radiation may surpass the absorption and this system may be regarded as an active medium.

Preliminary experiments showed (a) conservation of the type of main carriers after introduction, (b) non-linear current-voltage characteristics, $I(V)$, in fields less than $3\,\mathrm{kV\,m^{-1}}$, and (c) very weak temperature dependence of resistance from 4 to 300 K. This data is in accordance with the expected influence of bridges on transport properties.

2. **Superconductor superstructure**

In the case of a superconducting system, the modulation of its properties in space arises due to the strong dependence of the order parameter of superconductivity on the current density. When the current flows through this system, the superconductive state is destroyed at first in the bridges and this system(Figure 3c) may therefore be considered as an array of weak links (Josephson junctions). A strong interaction is produced since the coherence length of superconductivity exceeds the inter-grain distance. So this system is a 3D Josephson superlattice. Its peculiar feature is that all junctions in the resistive state coherently emit microwave radiation. It should be mentioned that the problem of propagation of radiation in an active composite dielectric-metal-superconductor structure has not been solved.

We mainly pay attention to the study of superconducting properties of this system because of its high sensitivity to the structure near the point of phase transition. We have found the following:

1. An increase of the superconducting transition temperature (for In and Sn) and a widening of the transition region due to the reduced dimension of the bridges.

2. An oscillating dependence of the critical current on the magnetic field due to the regular disposition of the grains.

3. Anomalous hysteresis of $I(V)$ and interaction between adjacent samples, probably due to the (self-) influence of the emitted radiation.

4. A response to the external microwave radiation with a frequency-sensitive component.

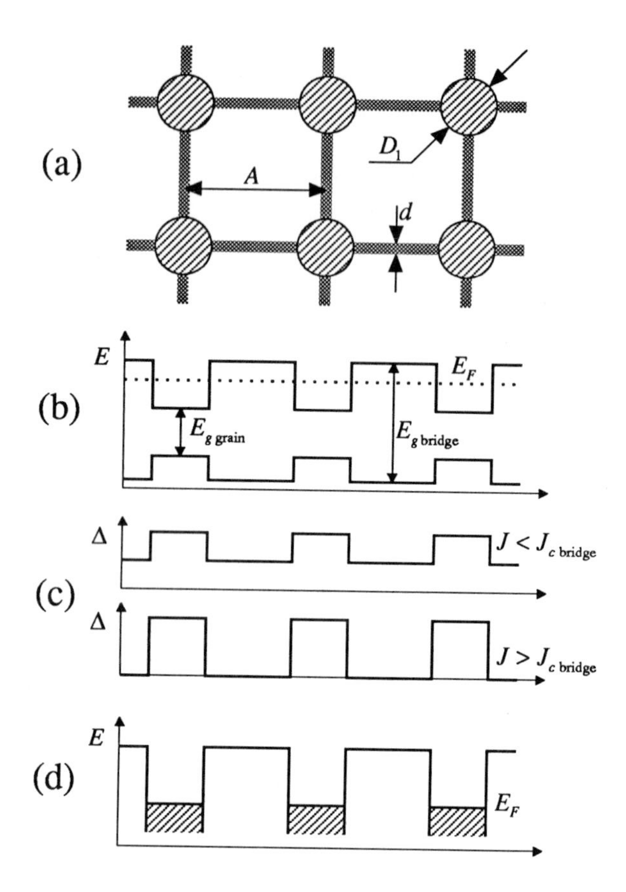

Figure 3: *Schematic diagram of grain system. (a) Grains and bridges showing dimensions. (b) Energy band diagram of semiconducting superstructure. (c) Order parameter in superconducting superstructure. (d) Energy barriers in a percolation system.*

3. Lattice of isolated grains

It is possible to break the bridges of the identical isolated metal particles by thermal treatment. Such a system may be very interesting for acoustical studies because of its very low phonon frequencies.

It is also possible to obtain percolation-type conductivity in a system of randomly connected grains. The resistance of this system is more than 10^7 times greater than the resistance of closed systems; it shows nonlinear $I(V)$ ($I \propto V^n$ with $n = 1, 2, 3$), weak dependence of resistance on temperature (temperature coefficient of resistance less than $10^{-4}\,\mathrm{K}^{-1}$ from 77 K up to 300 K) anda classical-type response to radio frequency signals. It needs to be kept in mind that the properties of the systems under consideration differ significantly from the properties of the one or two-dimensional arrays of nano-elements, especially in the case of active sytems. As an illustration of this unusual behaviour, we may use the 'anomalous' hysteresis in $I(V)$ (Figure 4). Usually the forward branch of the $I(V)$ lies above the backward branch, because of heating for example. For this

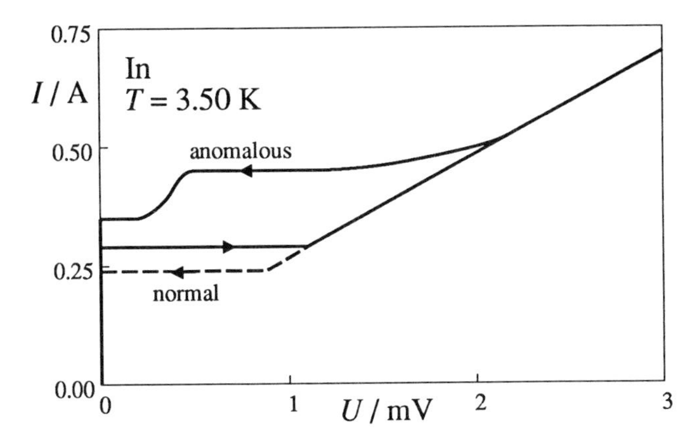

Figure 4: *Anomalous hysteresis in $I(V)$ of an indium sample at* 3.5 K.

sample, however, there is an inverse order of branches. The greater the current value through the sample, the greater the 'anomalous' difference (Bogomolov *et al.* 1982). On the contrary, if the thickness of the sample is reduced, the effect is decreased and is tranformed to a 'normal' one (if the thickness is less than 50 grain layers).

In conclusion we should like to emphasise that 3D superstructures demonstrate new and very interesting physics, and are highly attractive for fundamental study and possible applications.

References

Bogomolov V, 1978. *Usp. Fiz. Nauk*,**124** 171 (in Russian) .

Bogomolov V, Zhuravlev V, Zadorozhnii A, Kolla E and Kumzerov Y, 1982. *JEPT Lett.*, **36**, 298.

Sanders J V, 1964. *Nature*, **204** (number 4963), 35.

Contributed Papers

The titles of papers presented by participants at the School, either orally or as posters, are listed below.

F Aristone Non-parabolicity effects on the calculated current density of double barrier resonant tunnelling systems

P Balasingham Chemical potential oscillations induced by coherent transport in mesoscopic devices

V Bayot Disorder-induced dimensional transition in graphite

R Beresford Two-band model of inter-band tunnelling current

G Berthold Electrical transport properties of barrier-separated low-dimensional systems

U Bockelmann Inter-band transitions in quantum wires

F Bolton Some effects of electron confinement in quantum dots

H Bruus Can phase coherence in the quantum Hall effect be observed in mesoscopic rings?

A Chamarro Study of memory of luminescence polarisation in CdS_xSe_{1-x} doped glasses: evidence of Wurzite structure of CdS_xSe_{1-x} nano-crystals

C Chapelier The search for persistent mesoscopic currents

S Charbonneau Optical characterisation of FIB-implanted GaAs-InGaAs quantum well structures

R Cusco-Cornet Magneto-transport in a 1D lateral surface superlattice

M Dabbicco Linear and nonlinear absorbtion in ternary alloy AlGaAs-AlAs MQW structures

W Demmerie Electrical transport properties of barrier-separated low-dimensional systems

A Dmitriev Resonant tunnelling time

C N Garrido Diagrammatic approach of optical properties of 2D disordered systems

M Gerling Hydrostatic pressure dependence of the band offset of GaAs strained to InP

G Goldoni Surface magnetic states and coherent focussing in a two-dimensional hole gas

H Grossner The contact problem of a resonant tunnelling diode

O Gulseren Electric field effects on multiple quantum wells

G Gusev Properties of mesoscopic systems under microwave field irradiation

A Gustafsson Cathodo-luminescence studies of quantum wells and quantum well wire structures of InP-GaInAs

N Haider Atomistic simulation of large structures through MBE

R Heemskerk Non-equilibrium distribution of edge and bulk currents in a Si MOSFET under quantum Hall conditions

F Hirler Optical properties of type I and type II GaAs-AlGaAs nanostructures

X Y Hou A computer and optical simulation of the structure of the GaP (111) B surface

A Huber Infrared spectroscopy of mesoscopic Si systems

M Jackson Ultrafast optical studies of excitonic absorption in CdTe-CdZnTe superlattices

O Kühn Quasi-2D magneto-polarons in a tilted magnetic field

Y Kumzerov Ultrathin filaments (diameter 2–10 nm)

C Lettau Spectroscopy on GaAs-AlGaAs heterostructures laterally patterned by low-energy ion beam exposure

L Martin-Moreno Quantisation of the conductance in units of $e^2/2h$ in a ballistic quasi-1D channel, produced by strong electric and magnetic fields

H S Nguyen Magneto-optical phenomena in semiconductor supperlattices

J Nicholls Properties of split-gate devices in high magnetic fields

A Nogaret Calculated contribution of LO localised phonon modes to the valley current of a double barrier diode

A Orlov Mesoscopic effects in hopping conduction of n-GaAs samples with small dimensions

I S Osborne Observation of quantised resistance in a-Si memory structures

J J Palacios Effects of curvature on edge states

A J Peck Hydrostatic pressure studies of an asymmetrically doped resonant tunnelling diode

C Peters Raman and luminescence spectroscopy of electrically tunable electron systems on GaAs

M Portnoi Optical orientation and alignment in quantum wells

F Rodriguez Fermi edge singularities of a 1D electron gas

P Selbmann Magnetisation oscillations of a two-dimensional electron gas in an oriented magnetic field

E Skuras Magneto-transport measurements in spike-doped InSb and InAs grown by MBE

A Smart Optical spectroscopy of reactive ion etched microstructures in wide band-gap II-VI semiconductors

M Suhrke Quantum interference effects in the conductance of a quantum wire with electron-phonon scattering

T Swahn Magnetotransport in edge states (a self-consistent approach)

M Tewordt Fine structure in the current-voltage characteristics of a 200 nm diameter GaAs-AlGaAs resonant tunneling diode in high magnetic fields

M Trott Generalised Thomas-Fermi treatment of nanostructures

M Trott Full self-consistent, magnetic field dependent variational solutions

A Uhrig Linear and non-linear properties of quantum dots

N C van der Vaart Quantised current in a quantum dot turnstile

M van Hove Zero-dimensional staes in PdGe-contacted double-barrier diodes laterally constricted by H-plasma isolation

P D Wang Radiative recombination in GaAs-AlGaAs quantum dots and wires

S L Wang Spin dependence of inter-edge-channel scattering in Si MOSFETs in the quantum Hall regime

X Wu Electron-phonon interaction in double-barrier tunnelling

Participants

•Omar Abu-Zeid
Eindhoven University of Technology
Department E-EEA, Building EH8.34
PO Box 513
5600 MB Eindhoven, The Netherlands

•Dr Haroun Ahmed
Cambridge University
Microelectronics Research Centre
Cavendish Laboratory
Cambridge CB3 0HE, UK

•Flavio Aristone
CNRS/INSA & CNRS/SNCI
Department of Physique
Complexe Scientifique
156 Avenue de Ranquel
F-31077 Toulouse Cedex, France

•Pratheep Balasingham
Stanford University
PO Box 7728
Stanford, California 94309, USA

•Dr Gerald Bastard
Laboratoire de Physique
de la Matière Condensée
de l'Ecole Normale Supérieure
24 Rue Lhomond
F-75231 Paris Cedex 05, France
EMAIL: physol@frulm11.bitnet

•Vincent Bayot
Princeton University
Department of Electrical Engineering
Princeton, New Jersey 08544, USA

•Prof. Steven Beaumont
Department of Electronics
& Electrical Engineering
University of Glasgow
Glasgow G12 8QQ, UK
EMAIL: spb@nano.eng.glasgow.ac.uk

•John Roderic Beresford
Brown University
Division of Engineering
Box D
Providence, Rhode Island 02912, USA
EMAIL: rberes@brownvm.brown.edu

•Gunther Berthold
Walter Schottky Institut
Technische Universität München
Am Coulombwall
D-8046 Garching, Germany

•Ulrich Bockelmann
LPMC-ENS
24 Rue Lhomond
F-75005 Paris, France

•Colombo R Bolognesi
University of California
Department of Electrical
& Computer Engineering
Santa Barbara, California 93106, USA

•Fintan Bolton
Universität Regensburg
Institut Fur Theoretische Physik
D-8400 Regensburg, Germany

•Karl Brunner
Walter Schottky Institut
Technische Universität München
Am Coulombwall
D-8046 Garching, Germany

•Henrik Bruus
NORDITA
Blegdamsvej 17
DK-2100 Copenhagen 0, Denmark

•Philip D J Calcott
Royal Signals and Radar Establishment
K310 (S), St Andrews Road
Great Malvern, Worcs WR14 3PS, UK

•Maria Chamarro
Groupe de Physique des Solides
Universite Paris 7
Tour 23, 2 Place Jussieu
F-75251 Paris Cedex 05, France

•C Chapelier
Centre d'Etudes Nucleaires
de Grenoble
SPSMS/LCP
BP 85X-38041 Grenoble, France

•Sylvain Charbonneau
National Research Council
Institute for Microstructural Sciences
Ottawa, KIA OR6, Canada

•Doug Collins
California Institute of Technology
Caltech MS 128-95
Pasedena, California 91125, USA

•Prof. Harold G Craighead
National Nanofabrication Facility
Knight Laboratory, Cornell University
Ithaca, New York 14853, USA
EMAIL: craighead%nnfvax.dnet
@lonvax2.tn.cornell.edu

•David Cumming
Cambridge University
Microelectronics Research Centre
Cavendish Laboratory
Cambridge CB3 0HE, UK
EMAIL: drsc@uk.ac.cambridge.phoenix

•Ramon Cusco-Cornet
University of Barcelona
c/o Department of Electronics
& Electrical Engineering
University of Glasgow
Glasgow G12 8QQ, UK

•Maurizio Dabbicco
Universita Degli Studi Di Bari
Dipartimento di Fisica
Trav. 200 Re David
I-70124 Bari, Italy

•Dr John Davies
Department of Electronics
& Electrical Engineering
University of Glasgow
Glasgow G12 8QQ, UK
EMAIL:
jdavies@nano.eng.glasgow.ac.uk

•Martin Dawson
Sharp Laboratories
Neave House, Winsmore Lane
Abingdon, Oxon OX14 5UD, UK

•Wolfgang Demmerle
Walter Schottky Institut
Technische Universität München
Am Coulombwall
D-8046 Garching, Germany
EMAIL: e25@wsi.physik.
tu-muenchen.dbp.de

•Alexey Dmitriev
Moscow State University
Lenin Prospect, 69-2-281
117296 Moscow, Russia

•Prof. Laurence Eaves
Department of Physics
University of Nottingham
Nottingham NG7 2RD, UK

•Cris Eugster
Massachusetts Institute of Technology
MIT 13-3018
Cambridge, Massachusetts 02139, USA

•Ruth Eyles
Department of Physics
University of Nottingham
Nottingham NG7 2RD, UK

•Dr David Finlayson
Department of Physics and Astronomy
University of St Andrews
St Andrews KY16 9SS, UK
EMAIL: dmf@st-andrews.ac.uk

•Dr Alan Fowler
IBM T J Watson Research Centre
PO Box 218
Yorktown Heights, NY 10598, USA
EMAIL: fowler@watson.ibm.com

•Cecilia Noguez Garrido
National University of Mexico
Instituto de Fisica
Departamento de Edo Solido
Apartado Postal 20-364
CP01000, Mexico D.F.
EMAIL: rbarrera@unamum1.bitnet

•Dr Bart Geerligs
Department of Physics
University of California at Berkeley
Berkeley, California 94720, USA
EMAIL: geerligs@physics.berkeley.edu

•Michael Geller
University of California
Department of Physics
Santa Barbara, California 93106, USA

•Maria Gerling
University of Lund
Box 118, S-22100 Lund, Sweden

•Guido Goldoni
SISSA/ISAS
V Beirut 2-4
Trieste, Italy
EMAIL: goldoni@itssissa

•Harald Grossner
Walter Schottky Institut
Technische Universität München
Am Coulombwall
D-8046 Garching, Germany

•Oguz Gulseren
Bilkent University
Department of Physics
06533 Ankara, Turkey

•Guiennadii Gusev
Institute of Semiconductor Physics
Academy of Sciences, Siberian Branch
630090 Novosibirsk, Russia

•Anders Gustafsson
University of Lund
Department of Solid State Physics
Box 118, S-22100 Lund, Sweden

•Niaz Haider
Solid State Theory Group
Blackett Laboratory, Imperial College
Prince Consort Road
London SW7 2BZ, UK

•Dr Wolfgang Hansen
Sektion Physik
der Universität München
Geschwister-Scholl-Platz 1
D-8000 München 22, Germany

•Richard Heemskerk
University of Groningen
Applied Physics Laboratory
Nyenborgh 18
9747 AG Groningen, The Netherlands

•Prof. Detlef Heitmann
Max-Planck Institut für
Festkörperforschung
Heisenbergstrasse 1
D-7000 Stuttgart 80, Germany

•Tony Herbert
NMRC
Prospect Row
Cork, Ireland

•Franz Hirler
Walter Schottky Institut
Technische Universität München
Am Coulombwall
D-8046 Garching, Germany

•Richard Hornsey
Cambridge University
Microelectronics Research Centre
Cavendish Laboratory
Cambridge CB3 0HE, UK
EMAIL: rih12@cam.phx.ac.uk

•Xiao Yuan Hou
Fudan University
Surface Physics Laboratory
Shanghai, China

•Andreas Huber
Sektion Physik
der Universität München
Geschwister-Scholl-Platz 1
D-8000 München 22, Germany

•Michael Jackson
Ecole Polytechnique, ENSTA
Laboratory D'Optique Appliquee
Batterie de L'Yvette
F-91120 Palaiseau, France

•Oliver Kühn
Humboldt University
Department of Physics
Invalidenstrasse 110
O-1040 Berlin, Germany

•Youri Kumzerov
A F Ioffe Institute
Polytechnicheskya St. 26
194021 St Petersburg, Russia

•Christoph Lettau
Sektion Physik
der Universität München
Geschwister-Scholl-Platz 1
D-8000 München 22, Germany

•Dr Andrew Long
Department of Physics and Astronomy
University of Glasgow
Glasgow G12 8QQ, UK
EMAIL:
arlong@nano.eng.glasgow.ac.uk

•Josep Pages-Lozano
Universitat Autònoma de Barcelona
Department de Fisica
Grup D'Electomagnetisme
Edifici C, 08193 Bellaterra
Barcelona, Spain

•Dr Peter Main
Department of Physics
University of Nottingham
Nottingham NG7 2RD, UK

•Ron Marquardt
California Institute of Technology
Caltech, MS 128-95
Pasedena, California 91125, USA

•Luis Martin-Moreno
Cavendish Laboratory
Madingley Road
Cambridge CB3 0HE, UK
EMAIL: lm105@phx.cam.ac.uk

•Stephen Millidge
Royal Signals and Radar Establishment
PA ZZ1, St Andrews Road
Great Malvern, Worcs WR14 3PS, UK

•James Nicholls
Cambridge University
Cavendish Laboratory
Cambridge CB3 0HE, UK
EMAIL: jtnll@phx.cam.ac.uk

•Konstantin Nikolic
Solid State Theory Group
Blackett Laboratory, Imperial College
Prince Consort Road
London SW7 2BZ, UK

•Alain Nogaret
INSA
Department de Physique Complete
Scientifique de Rangueil
F-31077 Toulouse, France

•Alexey Orlov
Institute of Radio Engineering
& Electronics
Academy of Sciences
Moscow GSD-3 103907, Russia

•Ian Stewart Osborne
University of Dundee
Department of APEME
Perth Road
Dundee DD1 4HN, UK

•Monica Pacheco
Universidad de Santiago
de Chile
Casilla 307
Santiago, Chile
EMAIL: mpacheco@usachum1

•Juan Jose Palacios
Universidad Autonoma de Madrid
Department de Fisica
de la Materia Condensada
Cantoblanco 28049, Spain

•Andrew John Peck
Max-Planck Institute für
Festkörperforshung
Heisenbergstrasse 1
D-7000 Stuttgart 80, Germany

•Christian Peters
Sektion Physik
der Universität München
Geschwister-Scholl-Platz 1
D-8000 München 22, Germany

•Mikhail Portnoi
A F Ioffe Physico-Technical Institute
Academy of Sciences
194021 St Petersburg, Russia

•Steven Richardson
Howard University
School of Engineering
Materials Science Research Center
2300 6th Street, NW
Washington, DC 20059, USA

•Ferney Rodriguez
Universidad Autonoma de Madrid
Department de Fisica
de la Materia Condensada
Cantoblanco 28049, Spain

•Sergey Romanov
A F Ioffe Institute
Polytechnicheskya St. 26
194021 St Petersburg, Russia

•Prof. Hiroyuki Sakaki
Research Centre for Advanced
Science & Technology
University of Tokyo
4-6-1 Komabu, Meguro-ku
Tokyo 153 Japan

•Marcel F C Schemmann
Eindhoven University of Technology
Department E-EEA, Building EH8.34
PO Box 513
5600 MB Eindhoven, The Netherlands

•Peter Selbmann
Humboldt University
Department of Physics
Invalidenstrasse 110
O-1040 Berlin, Germany

•Helmut Silberbauer
Universität Regensburg
Institut für Theoretische Physik
Universitätsstrasse 31
D-8400 Regensburg, Germany

•Tomasz Skrabka
Technical University of Wroclaw
Institute of Electron Technology
Janiszewskiego 11/17
Wroclaw, Poland
EMAIL: skrabka@plwrtu11

•Eleftherios Skuras
University of Glasgow
Department of Physics and Astronomy
Glasgow G12 8QQ, UK

•Andrew Smart
University of Glasgow
Department of Electronics
& Electrical Engineering
Glasgow G12 8QQ, UK

•Dr Clivia Sotomayor Torres
Department of Electronics
& Electrical Engineering
University of Glasgow
Glasgow G12 8QQ, UK

•Dr Frank Stern
IBM T J Watson Research Centre
PO Box 218
Yorktown Heights, NY 10598, USA
EMAIL: stern@watson.ibm.com

•Prof. A Douglas Stone
Department of Physics
Yale University
PO Box 6666 Yale Station
New Haven, Connecticut 06511, USA
EMAIL: stone@venus.ycc.yale.com

•Michael Suhrke
Humboldt University
Department of Physics
Invalidenstrasse 110
O-1040 Berlin, Germany

•Thomas Swahn
Chalmers University of Technology
Institute of Theoretical Physics
S-41296 Goteborg, Sweden
EMAIL: swahn@fy.chalmers.se

•Matthias Tewordt
University of Cambridge
Cavendish Laboratory
Madingley Road
Cambridge CB3 0HE, UK

•Dr Gregory Timp
AT&T Bell Laboratories
Crawfords Corner Road
Holmdel, New Jersey 07733, USA
EMAIL: glt1@spin.att.com

•Michael Trott
Institut für Physik
TH Ilmenau, PSF 327
O-6300 Ilmenau, Germany

•Aiga Uhrig
Universität Kaiserlautern
Department of Physics
Erwin-Schrödinger Strasse 1
D-6750 Kaiserlautern, Germany
EMAIL: kphy01@earn.dklunI01

•Nijs Van Der Vaart
Delft University of Technology
Department of Applied Physics
PO Box 5046
2600 GA Delft, The Netherlands

•Marleen Van Hove
IMEC
Kapeldreef 75
B-3001 Leuven, Belgium

•Pei Dong Wang
University of Glasgow
Department of Electronics
& Electrical Engineering
Glasgow G12 8QQ, UK

•Shing-Lin Wang
University of Groningen
Department of Applied Physics
& Materials Science Centre
Nijenborgh 18
9747 AG Groningen, The Netherlands

•Tony Warwick
Lawrence Berkeley Laboratory
Building 2-414, 1 Cyclotron Road
Berkeley, California 94720, USA

•Dr Jon Weaver
Department of Electronics
& Electrical Engineering
University of Glasgow
Glasgow G12 8QQ, UK

•Xiaoguang Wu
Ohio University
Department of Physics
Athens, Ohio 45701, USA
EMAIL: wu-xiao@helios.phy.ohiou.edu

•Hai Ping Zhou
University of Glasgow
Department of Electronics
& Electrical Engineering
Glasgow G12 8QQ, UK

Index